THE PICKERING MASTERS

THE NARRATIVE OF THE BEAGLE VOYAGE, 1831–1836

CONTENTS OF THE EDITION

THE NARRATIVE OF THE BEAGLE VOYAGE, 1831–1836

EDITED BY
Katharine Anderson

Volume 1
Proceedings of the First Expedition, 1826–1830, Under the Command of Captain Phillip Parker King, Part I

Routledge
Taylor & Francis Group
LONDON AND NEW YORK

First published 2012 by Pickering & Chatto (Publishers) Limited

Published 2016 by Routledge
2 Park Square, Milton Park, Abingdon, Oxon OX14 4RN
605 Third Avenue, New York, NY 10017

Routledge is an imprint of the Taylor & Francis Group, an informa business

BRITISH LIBRARY CATALOGUING IN PUBLICATION DATA

The narrative of the Beagle voyage, 1831–1836. – (The Pickering masters)
1. Beagle Expedition (1831–1836) 2. Adventure and Beagle Expedition (1826–1830) 3. Scientific expeditions – South America – History – 19th century – Sources. 4. Fitzroy, Robert, 1805–1865. 5. King, Philip Parker, 1793–1856. I. Series II. Fitzroy, Robert, 1805–1865. III. King, Philip Parker, 1793–1856. IV. Anderson, Katharine, 1965–
910.4'5–dc22

ISBN 13: 978-1-138-76168-1 (hbk)
ISBN 13: 978-1-85196-844-2 (set)

Typeset by Pickering & Chatto (Publishers) Limited

CONTENTS

ACKNOWLEDGEMENTS

As all nineteenth-century scholars will know, this edition depends on the studies of Darwin's life and work, including most importantly the Darwin Correspondence Project (www.darwinproject.ac.uk) and the gathering of all things Darwinian in the website known familiarly as Darwin Online (www.Darwin-online.org.uk). I owe my interest in this project, however, not to Darwin, but to Fitzroy and the work of scientific voyages in this period. I am grateful to Mark Pollard and Paul Lee at Pickering and Chatto for the conception of the project and their hard work in carrying it to print. The librarians of the York University resource-sharing department have been, as usual, expert and efficient and this sort of work could not have been done without their support. My colleagues in the Science and Technology Studies department and the Victorian Studies Network at York University provide an unfailingly stimulating place to work. My undergraduate students are also an inspiration, whether they know it or not: their reactions to this famous voyage make it a fresh puzzle every year. I have received assistance and advice at critical points from many individuals, including Peter Anderson, Brendan Cook, Paul Delaney, Carla Hustak, Bernard Lightman, Matthew McKenzie, Ray McKinnon, Helen Rozwadowski, Anne Secord, James Secord, Alistair Sponsel, Neil Safier, Simon Schaffer, Amy Teitel and John van Whye. The range of subjects involved in this editorial work and the strong communities of interest in this famous voyage leaves me more than usually certain both that inaccuracies and incomplete references remain, and that I will hear about them. But, of course, none of the errors or omissions that remain are the fault of those who so cheerfully gave me the benefit of their expertise.

I would not have embarked on this project without the experience in 2006 of participating in a historical cruise with Sea Education Association in Wood's Hole, Massachusetts; spending a few days with a 360° view of the horizon renewed my interest in Fitzroy's perspective. The project also emerged from a second interest, the study of marine meteorological networks and the use of ships' logs as historical sources in the environmental sciences. For the support of my work in this area, I am indebted to a research development initiative grant from the Social Sciences and Humanities Research Council of Canada, 2007–9. Finally, I would have given up this project many times without three people who care very little for sea voyages or the nineteenth century, but give me abundant support and love anyway: William, Nicky and Hope Trischuk.

GENERAL INTRODUCTION

Published in 1839 in four hefty volumes of several hundred pages each, the *Narrative of the Surveying Voyages of HMS Adventure and Beagle* should not seem insubstantial. It covered what is undoubtedly the most famous scientific voyage in history, that of the *Beagle*'s circumnavigation. Yet in terms of literary history, the *Narrative* has a shadowy presence, seldom examined or read as a whole. Put out by the London publisher Henry Colburn, its bulk was overshadowed almost immediately by the independent life acquired by the third of the four volumes, the young Charles Darwin's *Journal and Remarks*, which was reissued within weeks as the *Journal of Researches into the Geology and Natural History of the Various Countries visited by HMS Beagle under the Command of Captain Fitzroy*. This single volume of the larger work was revised by Darwin for a second edition in 1845 and has been constantly in print since then.[1] The voyages of the *Beagle*, in short, are famous because of Charles Darwin, invited as a companion for Captain Fitzroy in 1831, and returning in 1836 stuffed with experiences from what he called in old age 'by far the most important event of my life'.[2] The original *Narrative*, spanning a full decade, two distinct voyages and three captains, sank and largely disappeared from view.[3]

The common explanation for this fate is a literary assessment. The original *Narrative* was dominated by the captains' account of the voyages and surrounded by lengthy appendices of measurements, vocabularies, technical navigation notes, memoranda and philosophical musings. In 1837, Darwin groaned that it was the heaviest sort of pudding. A modern reader might find this an apt description of the detail, digression and habit of allusion and extract that characterize the writing. Yet this judgement, positioned differently, is an important clue for the interpretation of the *Narrative*. Unlike Darwin's volume, the other three volumes did not present a single and compelling point of view, but the official record of a sustained enterprise. It is a work with many voices who speak through the captains according to the hierarchies of the British navy. For this reason, apart from its important connection with Darwin himself, the volumes compiled by Captains Phillip Parker King and Robert Fitzroy give us a revealing picture of scientific voyages.

The first volume of the original four covered the voyage commanded by King. The voyage lasted four years and three months from 1826 to 1830, and was focused on a close survey of the southern tip of South America. The text was prepared by Fitzroy after 1836, using the journals of King and other officers. It was the least polished of the volumes, a chronological summary that seems drawn straight from the journals. It concluded with a technical chapter for navigators to guide them the passage through the Strait Magellan. Its several elements were aimed at different audiences, ranging from the general reader interested in exploration, natural curiosities and encounters with 'savages' to the most sophisticated investigators of geo-magnetic phenomena.

The appendices (which have been neglected more than the voyage narrative itself) give us a view of the various audiences for such work. The first volume contained tables of latitude and longitude, meteorological observations, and copies of 'orders' (the official instructions for the voyage). There was also a lengthy discussion on magnetic variation in the southern hemisphere covering the observations made during both voyages with an instrument supplied by the noted Norwegian natural philosopher, Christopher Hansteen. This memoir on magnetic variation – a subject of intense interest to both those who used compasses for navigation and philosophers interested in magnetic and electrical phenomena – was written by Edward Sabine, a naval officer who had himself carried out similar researches in the Arctic in 1822–3. The memoir showed Sabine in pursuit of his ambitious scientific goals – a global network of magnetic and meteorological observation stations organized through the British Association for the Advancement of Science, as well as a centralized facility for standard instruments in Britain.[4] Then there was a catalogue of zoological specimens collected by King and other officers that was prepared with the assistance of William John. Broderick, a leading naturalist, the vice-president of the Zoological Society in London, and a catalogue of shells, prepared by another eminent naturalist, George Sowerby. (A botanical catalogue, in which King was especially interested, did not materialize.) Finally, the first volume included a paper King had delivered to the Royal Geographical Society in 1831 and a 'game book' submitted by Fitzroy listing the fish and fowl caught and cooked in the Strait in 1830. ('In common fairness', wrote the *Quarterly Review* in 1839, 'the name of the cook of the *Beagle* should have been blazoned—he must have been born with the genius of a *cordon bleu*.')[5] The volume, as King had instructed Fitzroy, sought to be comprehensive, a kind of reference volume.

The second volume gave an account of the voyage led by Captain Fitzroy from December 1831 to October 1836. On this circumnavigation, the *Beagle* spent forty-five months in South America but continued through the Pacific to Tahiti, New Zealand, Australia before returning by way of the Cape of Good Hope, after four years and ten months altogether. Literally and figuratively, the second volume covered much of the same ground as King's volume, and wove together the separate expeditions of different officers in the same manner, if more seamlessly. Like the first volume, it met the conventions of the narrative genre, with its mix of description, historical summary, natural history, anecdote and scientific memoir. But it was also a more ambitious narrative in a philosophical sense, as reviewers noticed.[6] Three chapters on 'South American aborigines' as well as concluding chapters on the origins of human society and Mosaic geology made this ambition abundantly clear. Its appendices, collected originally in the fourth volume of the set, Fitzroy himself called 'disorderly'.[7] They covered the standard materials (observations of latitude, longitude, meteorology and land surveys; orders; and technical navigation advice). But they also included documents related to political incidents in the Falkland islands, Tahiti and New Zealand, Fitzroy's views on cloud classifications, weather prediction and tidal variation, notes on the novel instruments he carried, and extracts from previous explorers.

The *Narrative* thus covers what were in the end two quite different voyages, not least because of the differences in the personalities and interests of their commanding officers. King was a seasoned officer, the son of the second governor of New South Wales. He had a survey of the Australian coasts under his belt, and on that voyage had seen much more violence, and faced his dangerous task with fewer resources. He had an estate near Sydney, a large young family and the life of a colonial landowner waiting for him. He returned to them and left Fitzroy to finish up the *Narrative* of the first voyage in London as soon as he had completed the all-important charts and navigation remarks. Fitzroy, by contrast, was a young aristocrat, the second son of a son of a duke, well-educated and intelligent. He had been at sea for several years, but had barely tasted his first command in the *Beagle* at the end of its first voyage. He had influential family connections and some wealth, but no extensive property of his own, and he therefore hoped for a long and illustrious naval career. Proud and often impulsive, he often acted too independently for the Admiralty's complete satisfaction. But he was a skilled captain and a formidable hydrographer who won the admiration of his scientific colleagues and inspired abiding loyalty in his men.

Read again at a distance of close to two centuries, the *Narrative* vividly captures the attitudes of early nineteenth-century Englishmen toward nature, society and history. It demonstrates how contemporaries saw an indissoluble connection between natural knowledge, political change and economic development. The primary task of a hydrographer was to produce accurate charts to support shipping and political expansion in a critical era of British imperialism. The work trained naval officers who then, in other ships on other duties, carried the British flag over the globe. But King, Fitzroy and their officers were also charged to act as professional observers of the natural and social world. During the decade from 1826 to 1836, the societies of South America and the Pacific were experiencing rapid political and economic changes, and the political and social order of Britain was shifting as well. King and Fitzroy were disposed to find significance in changes in the natural world, just as they were disposed to find evidence of progress or decline in human societies. Disciplined and curious, the officers of the *Adventure* and *Beagle* accumulated measurements, facts and specimens with dedicated energy and ambition. They were confident that such accumulation was a transformative force – for their own careers, for Britain's global authority and for the emerging shape of the colonial worlds they encountered.

This historical sensibility and approach to the work of scientific observation gives us a richer context for Darwin's work. Although all modern scholars of Darwin accept the significance of the voyage for his later work in natural history, the relatively inaccessible *Narrative* has made it too easy to pull Darwin out of the framework of a wider enterprise. Darwin consciously chose to write his own volume in tribute to the *Personal Narrative* of that other South American explorer Alexander von Humboldt. The other volumes of the *Narrative*, in contrast, present the scientific voyage as a collective endeavour. Before turning to King's and Fitzroy's separate voyages, we can first outline some key areas of common ground between the texts. Much of this context is less visible or simply absent from Darwin's well-known account of the *Beagle* and so presents us with an interesting counter-perspective to add to his own unusual experience as gentleman and layman on a discovery ship of the Royal Navy.

Surveying Voyages, South America and Scientific Reform

Two of the most important features of the *Narrative* – firstly, its picture of the labour of surveying and, secondly, the constant literary presence of other travellers of all kinds – help to explain its structure, style and reception. More generally, the context for both voyages includes the condition of the post-Napoleonic British navy, the political turmoil of South American republics, the anxiety of naval leadership, and the tradition of admiralty science.

Firstly, and most obviously, the *Narrative* preserves a chronological record of events of a working survey. The primary task of the officers on the voyages was to product charts of coastlines and take measurements of longitude. The purpose of both voyages was the production of charts and navigational advice to support increasing maritime traffic through the Strait of Magellan and on the east and west coasts of South America. The *Narrative* thus sat within a cluster of other publications: Admiralty charts, remark books and logs that went to the Admiralty archives, and published technical navigation literature for navigators.[8] The officers built up a detailed picture of the coastline through soundings of depth through a lead-line and the triangulation methods of field surveying. They worked from landmark to landmark to measure distances in a set of triangles, using either a directly measured baseline (for smaller areas like a river mouth or cove) or a distance based on astronomical calculation. They then ascended whatever high points the region afforded, and took lines of sight to other regional landmarks, using a sextant or the more sophisticated but cumbersome theodolite to get the angles. They used thermometers to verify heights by calculations of barometric pressure, and sketched profile views and took descriptions at every possible point in a field (or 'bearings') notebook.[9] It was, as Fitzroy stated, 'slow and monotonous' work.[10] It also involved much dangerous sailing – the cliffs in parts of Tierra del Fuego, especially what became know as the Beagle Channel, astonished all observers. In the splintered tip of the continent, the harbours for a sailing ship were small and poor. Even when safely berthed within one, contrary winds and the notorious South Atlantic storms – even in midsummer – often made it hard to leave.

Because the *Narrative* grew from the ships' logs, its repetitions mirrored the coming and going of the ships as well as the monotony of surveying. There were in fact several ships involved in the enterprise besides *Adventure* and *Beagle*. Each vessel carried small boats, which the captains sometimes sent off for several days to tackle measurements of bays or river mouths that were inaccessible to a larger vessel, or simply to cover more ground more quickly in the time at their disposal. Then there were the supplementary vessels. King had a decked boat called *Hope* along on the *Adventure* and later hired a supply ship that he named the *Adelaide*. Fitzroy hired two small schooners, *La Paz* and *La Liebre* in September 1832 and sent them off with his lieutenants to survey south of the River Plate. The following year, Fitzroy bought another much larger ship, the *Unicorn*, about half the size of the *Beagle*, and renamed her *Adventure*, to replicate the team of the first voyage. Unlike King, he had not waited for permission from the Admiralty. He was forced to sell off the unauthorized purchase at Valparaiso in 1834, but promptly borrowed, then hired the *Constitucion* a few months later. When the secondary survey vessels of the survey reported back, the log of each is drawn into the flow of events in the *Narrative*, sometimes reversing the chronology by several months.

The structure of the text reflects the complications of hydrographical work in other ways as well. King had tried to stay in the Strait and survey through the winter, and the hardships of that expedition, including the tragedy of Stokes's suicide, were a lesson for Fitzroy. He chose therefore to proceed in seasons, with the milder summer months of December to March or April devoted to the challenges of Tierra del Fuego, after which the ships would sail north and resume surveying work in milder latitudes or settle in port for chart-work. The ships also occasionally retraced their steps to confirm an observation – for instance, Fitzroy turned back for several weeks when he first got to Rio in 1832 and realized there was a large discrepancy between his longitude measurements and those of a recent French expedition. Both authors at times recognize the repetitive quality of their texts. At the end of 1834, Fitzroy, finally finished with over two full seasons of surveying in Tierra del Fuego, his charts posted off to London in November after several hard months of desk work, turned to the Chonos archipelago in southern Chile. As he wrote in a gloomy Christmas note, 'all looked dismal around us ... our immediate task was the survey of another Tierra del Fuego, a place swampy with rain, tormented by storms, without the interest even of population.'[11]

There are notable exceptions to this monotony when the *Narrative* gives the work of measurement in vivid detail. One such occasion is King's ascent of what he named Mount Tarn after the ship's surgeon. Struggling though stunted beeches and a jungle of thorny bushes, sleeping overnight on beds of branches that slowly sank into the boggy ground, a small group of officers and men reached the snowy peak in early evening the second day of climbing. There, as King records

> we immediately set up the instruments. I was obliged to avail myself of Mr. Tarn's assistance to hold the barometer, whilst two of my boat's crew held the legs of the theodolite-stand, for the wind was blowing very strongly, and the edge of a precipice was close to us, perpendicular for many hundred feet, and thence downwards so steep, that any body going over would fall at least a thousand feet. The theodolite-stand was unavoidably placed within a very few inches of the edge, and I took a round of angles, suffering, however, intense pain from the piercing coldness of the wind, which, heated as we were by the ascent, was much felt.[12]

But these pictures are exceptions. For the most part, both King and Fitzroy give their work the steady and repetitious literary form that it had in practice, in deference to the chronological conventions of the genre.

At the same time, the *Narrative* conveys a different sort of temporality, with its attention to incidents that punctuated their steady work, such as encounters with natives, political insult or negotiation, the disaster of earthquake or shipwreck. For the officers, these aspects of the voyages underlined a narrative of modern progress, and Britain's role as a civilizing force. Indeed, the survey always involved a profoundly historical sensibility, informed by the changes the travellers witnessed in the new republics of South America and by the transformation of indigenous societies and natural environments there and across the Pacific.[13] The day-to-day practice of hydrography constantly renewed this sense of history, as the officers combed out details from texts of past navigators and the archives of Spanish colonial cities to compare with their own views and measurements. The very names of the landmarks they measured were a record of the past, and they added to these with a conscious sense of the genealogies involved. The *Beagle*'s instructions, in fact, dealt explicitly with this practice of naming, urging Fitzroy to tie names to local terms or visible landmarks (not use them to score political favours), but above all to respect any existing designation. He sometimes complied with the former injunction, but always with the latter. Names recalled the story of the modern world from a thoroughly whiggish point of view: Port Famine pointed out Spanish failure and suffering while Donkin Cove commemorated the inventor of food preserving, a small triumph of modern science and manufacturing.

As this business of naming suggests, the mental geography of King and Fitzroy was at once global and relentlessly English. Every detail of the coast added shape to immensely large vista of continents, archipelagos and oceans. But at the same time, that view was evaluated according to its ability to support miniature versions of home. This chimes not so much with the spatial metaphor of centre and periphery in the history of imperialism as it does with metaphors of circulation, loops and transplantation.[14] And the voyage encountered many such pockets of familiar order: English merchant communities in Monte Video or Valparaiso, tidy missionary settlements in New Zealand, the agricultural promise of the pampas near Buenos Aires or the farmland of northern Chile, the 'post office' box of the Galapagos. The most significant and enduring of these miniature worlds were of course the ships themselves. On the first voyage the *Adventure* carried about fifty seamen and fifteen marines and eighteen officers; the *Beagle* carried twelve officers and about fifty others. On the second voyage the *Beagle* was more crowded, carrying seventy-three people, including officers, seamen, marines and 'supernumeraries' like the Fuegians and Darwin.[15] These were almost exclusively male worlds, except for Fuegia Basket and the occasional visitor, but worlds that otherwise contained many of the classes of British society, overlaid with the hierarchies of the royal navy.

Perhaps most obvious stylistic feature of the *Narrative* beyond its chronological form is its dense web of allusions to other documents. The fourth volume, Fitzroy's set of appendices, is the chief example of this sort of literary collage, in which Fitzroy bundled together a diverse collection of supporting material. But in addition to those appendices, the *Narrative* volumes of both captains constantly cite and extract from other works of exploration, whether English, Spanish, French and Dutch. This comparison to other voyages was not, it must be emphasized, just a sign of literary taste and the need to while away dull hours at sea – although this was undoubtedly one reason they came so readily to the authors of the *Narrative*. Instead, the constant references also point to a fundamental part of the hydrographer's work. King and Fitzroy both puzzled over interpretations of old charts and old narratives and sought to reconcile the descriptions of previous navigators with the view facing them in the present. Hydrography and other kinds of exploration, from the eighteenth century onwards, were very much literary exercises, and shipboard life involved many hours reading and rereading.[16] We have no information about the books on board the first voyage, but for the second voyage, we are lucky to have a good description of the ship's library, which contained more than 400 volumes.[17]

Besides references to works of naval exploration, the *Narrative* was suffused with evidence of another lively genre. Dozens of British contemporaries had taken the pulse of the new republics of South America; a handful had lived with Pacific islanders. London publishers vied to fill their lists with their productions.[18] This flood of print joined standard eighteenth-century historical and ethnographic authorities, like the Yorkshire Jesuit Falkner's *Description of Patagonia* and Molina's *Compendio* for Chile. To a certain extent, we can explain the reception of the *Narrative* in 1839 with its potted histories of Chiloe, the Falkland islands or Tahiti to market saturation (although, to be fair, Darwin's volume surmounted this difficulty). A final category of references was to natural history literature. More common in the first volume, covering King's voyage, rather than the second voyage, in which Fitzroy left much of these discussions to Darwin,[19] these references demonstrated the range of knowledge expected of a naval commander on such voyages.

Taken together, the use of authorities in the *Narrative* gives a glimpse of how a diverse range of information was accessed, compared, synthesized and re-circulated by early nineteenth century readers. It strikingly illuminates the early Victorian project of knowledge and its approaches to testimony and fact. This is a question we will return to below, in treating King's investigation of the Patagonians. Here we will emphasize only that is too simple to characterize this atmosphere of scissors-and-paste as a literary and intellectual shortcoming – as evidence of haste, or ill-digested knowledge. James Secord's recent analysis of scrapbooks as a vogue of the period is especially helpful here, showing how collection and compilation in England was a response to cultural change and the opening up of many fields of knowledge.[20] Like the scrapbook, the *Narrative* is a text that shows the dramatic profusion of early nineteenth-century print culture in action. It was rooted in eighteenth-century exploration narrative, and also in the growing set of technical navigation guides and charts sponsored by the Admiralty and the East India Company. Then it dipped in and out of the latest encyclopaedias and compendiums, popular history, political and social analysis. Finally, it was grafted to specialist scientific investigations, to the catalogues and journals of scientific societies.

In addition to these features of style and genre, these volumes of King and Fitzroy also share a critical historical context. This was the experience of defining Britain's role in South America and the Pacific during an extraordinary and volatile decade. Latin American countries had won independence from Spain and Portugal in a generation of conflict from 1804 to1825, often supported officially and unofficially by Britain. (After the end of the Napoleonic wars, this support included hundreds of recruits from the British navy to Chile and Brazil, which meant King and Fitzroy's encounters with 'foreign' navies in South America often meant meeting compatriots.)[21] At the end of the first voyage in 1826, the new republics and the independent monarchy of Brazil were far from stable, often on the brink of civil war. There were also ugly conflicts between the new states. From 1825 to 1828, Buenos Aires was at war with Brazil over the future what is now Uruguay. General Rosas, not yet dictator, but clearly the most formidable military and political figure in the Argentine provinces, was engaged in a brutal war against the native groups in the pampas, some of whom were reputedly encouraged by Chile, hungry for territory in the east. The slave trade had been declared illegal, but remained widely condoned in the busy ports of Brazil and elsewhere on the eastern coast. British forces gave increasingly attention in this period to policing the Atlantic to prevent the trade, which complicated their relationship to the Latin American governments, and was part of the ideological reason for maintaining a strong naval presence, with the commanding admiral usually based in Rio. But the navy was also there to promote and protect British economic interests more generally. Joining established merchant communities in the major ports (Rio, Montevideo, Valparaiso), British commercial interests in the continent grew rapidly as mining and railway development increased in the 1830s and 40s. In economic terms, however, the big boom of this decade was whaling, which was shifting in the 1820s from the South Atlantic to the rich whaling grounds of the Pacific. Everywhere the survey vessels went, there the whalers from Connecticut and Massachusetts were too. The presence of whalers in the *Narrative* makes obvious how their activities drove the commercial expansion of the whole Pacific region.

Although they were on a scientific voyage, as British naval vessels the *Beagle* and *Adventure* were constantly on the thin edge of diplomatic and military action. (The ships each carried some trained marines as well as ordinary seamen, and small deck guns for defence.) Two examples are familiar from Darwin's account of the episodes: the *Beagle*, heading to Buenos Aires in 1832, was refused entry to the port and fired upon by an Argentinian guardship. The explanation for this outrage (quarantine rules and fears of cholera) did not satisfy Fitzroy who maneouvred the *Beagle* out again past the guardship with a threateningly close display of her 'broadside' guns. Returning the short distance to Monte Video, Fitzroy had a British frigate sail back to demand an apology (and to collect the charts he had been seeking from the city archives). In the frigate's absence, the Monte Video authorities arrived to request the Beagle's marines in order to control a mutiny among the soldiers in the city citadel. Over fifty armed men, including Darwin, disembarked to lend a British presence to the authorities and prevent looting.

However exciting Darwin found these events,[22] they were not unexpected to the commanders of the expeditions. Fitzroy called the Buenos Aires guardship affair merely 'vexatious' and commented on the succeeding incident with complacency: 'In these unsettled states ... [revolt and mutiny] recur so frequently, that even on the spot they cause little sensation, and excepting by those personally concerned, are scarcely remembered afterwards.'[23] There were many other instances to add to those events of 1832. King, surveying Chiloe in September and October 1829, needed a new mast for the tender *Adelaide* at short notice, and accepted the island governor's offer of the new town flagpole so as to prevent delay finding another suitable timber. The gesture started a rumour that reached authorities on the mainland, and King had to reassure them that he had not removed the flagpole as the first step in British annexation of Chiloe. Similar rumours spread about the English and the island of Gorriti, near Monte Video, when the second *Adventure* spent some time there in October 1833 to take longitude measurements. More dramatically, Fitzroy became engaged in a violent crisis in the Falkland islands in 1833 and 1834. Residents at Port Louis were attacked in August 1833, and several murdered, including a remarkable ex-royal navy officer Brisbane, who had been earlier rescued from shipwreck by Captain Stokes and the *Beagle* in 1827. The *Beagle* arrived in the area early in 1834 to find British authority barely in place, Port Louis destroyed and the culprits hunted as outlaws. Fitzroy's expedition took on William Low, a refugee from the violence and the former owner of the *Unicorn*, as a pilot for the next year, and Fitzroy added much to his knowledge of Fuegians from conversations with Low.

The knife-edge of political instability deserves mention because it throws light on a distinctive thread of the *Narrative*. There was very real anxiety in being in command of these isolated areas, an anxiety that combined physical and mental dangers. In what is perhaps the best-known event of the first voyage (because it led to Fitzroy's appointment, and thus becomes part of the story of Darwin), the *Beagle*'s Captain Pringle Stokes committed suicide in August 1828 at the close of nearly six months' of solitary survey work on the western tip of the continent, just before rejoining the *Adventure*. King, recounting the loss, cited Stokes's 'excitable mind' and his tortured sense of 'hair-breadth escapes' in the hardships of the western archipelago. Stokes's lieutenants concealed his state of mind from the senior captain until the very end. [24] But Stokes's experience, though it ended in tragedy, was not unique or even unusual. Fitzroy, for instance, invited Darwin on the second voyage in hope of some company to assist similar difficulties of command. Nevertheless, he fell into a state of despair in late 1834 – significantly, not at sea like Stokes, but in port in Valparaiso, putting in endless hours with work on the charts, seeing only the gaps in his knowledge of the fractured coastlines of the south, and frustrated with the Admiralty. The point here is that these episodes were a recognized part of the voyages' hazards and Darwin's *post-facto* assessments of Fitzroy (seen below) as a troubled character have distorted the historical picture. King's sympathy for Stokes in the *Narrative* seems absolutely genuine; his *Narrative* shows that he was himself already preoccupied because the morale of his own crew was approaching crisis. Francis Beaufort at the Admiralty encouraged the young Fitzroy in 1831 to find someone like Darwin because mental pressure was a recognized hazard on a long voyage, not because he considered Fitzroy an unstable personality. Fitzroy's lieutenants were patient and loyal during the breakdown in 1834 for the same reason.

Apart from encouraging historians to re-evaluate Darwin's impressions, attention to the theme of anxiety is important because of its connection to scientific work. If the ship was a microcosm of British society, the captain's mind stood for the critical judgement and knowledge necessary to produce, and justify, the leadership of British civilization in the world. The English philosopher John Frederick William Herschel expressed this link between knowledge and leadership vividly in his *Preliminary Discourse on the Study of Natural Philosophy* (1830). The age of 'mere wonder', Herschel argued, has been replaced by systematic calculations and facts. This meant that people in a modern age must be 'guided and enlightened, rather than astonished and dazzled'. Herschel, the foremost natural philosophical of his day, used a recent feat of navigation by sextant and nautical almanac from west coast of Mexico around Cape Horn to Rio, 8,000 miles without touching land, as his example. 'How essentially', he remarked, 'the authority of a commanding officer over his crew may be strengthened by the occurrence of such incidents, indicative of a degree of knowledge and consequent power beyond their reach.'[25] The work of naval leadership and the collection of natural fact shared an unspoken but evident logical connection. Balanced judgement was based on complete knowledge of the world, an ability to compare and calculate among all its parts.[26] Knowledge was the essential quality of modern authority.

This grandiose principle had of course underwritten scientific voyages for several decades before Herschel gave it expression. But the timing of his remark in 1830 is significant, because it signals the position of these particular voyages in the movement to reform British science. The eighteenth-century voyages of discovery undertaken by many European nations as well as Britain had made a place for astronomy, botany, zoology, enthnology, mineralogy and meteorology as official pursuits.[27] The first voyage of James Cook in 1768 to 1771, for instance, transported astronomers to Tahiti for a particular astronomical event, the transit of Venus, carried on to chart the east coast of Australia and made Joseph Banks's lasting reputation as a naturalist. By King's day, natural history was business as usual, as his instructions in 1826 to collect anything 'new, rare or interesting' showed. He carried a botanist and several of his officers made collections that they deposited in the nation's museums, in accordance with the rule that collections were to be 'public'.[28]

The appointment of Francis Beaufort as Hydrographer in 1829 shifted this aspect of naval exploration into a higher gear. As an early historian of the office put it, Beaufort arrived to find 'a mere map-office' and left in 1855 a legacy of a great programme of surveys in Britain and around the world.[29] Beaufort was also part of a group of men who were impatient with the indulgent routine (as they saw it) of the London scientific community. He had been appointed to the council of the Royal Society in 1827, immediately joining a committee to restrict membership of that august body to active scientific contributors, and campaigning to make Herschel, Britain's most eminent natural philosopher, its president. As Hydrographer, Beaufort wanted to make the scientific work of the navy more systematic and authoritative as well. He helped found the Royal Geographical Society in 1830 and the British Association for the Advancement of Science in 1831. He promoted voyages to explore the arctic and terrestrial magnetism, and a voyage to take geodesic measurements with pendulums. He sponsored the global investigation of tides in cooperation with the best mathematicians of the day. He cultivated young scientific naval officers like Fitzroy. Last but surely not least, he probed his reform-minded friends in Cambridge for the name of a well-trained, promising young man with geological interests to suggest to Fitzroy, securing Darwin's invitation.[30] The voyages of King and Fitzroy, then, were part of a decisive new era of scientific work at the Admiralty, which was itself an expression of tightening relationship between the nation's government and men of science.

The First Voyage: Phillip Parker King

The first voyage originated in the aftermath of the Latin American wars of independence. As Spanish and Portuguese rule faded in the continent and trade grew, Britain saw a pressing need to update the inadequate charts, especially those of the Strait of Magellan. Although sealers and whales had gained great familiarity with the region in the last decade, none of their knowledge was published or generally available. Ships still preferred to brave the longer, less protected route around Cape Horn rather than risk the narrow passage through the islands discovered over three centuries earlier. King's instructions were therefore to concentrate on eastern and western Patagonia, the many islands of Tierra del Fuego and a detailed survey of the Strait.

King was thirty-four years old when he left Plymouth, an experienced surveyor. His father, Philip Gidley King, was a naval officer who travelled with Arthur Phillip to establish the penal colony in New South Wales in 1788, and then became liuetentant-governor of the isolated Norfolk Island. His son Phillip Parker (named for Admiral Phillip) was born there in 1791, but returned to England for his education as a young boy. His father tried unsuccessfully to get him placed at the Royal Naval College, a new school for the training of elite naval officers, especially for members of the aristocracy. King joined the navy in 1807, and became a lieutenant in 1814, and was appointed to a survey of the Australian coast in 1817. He was one of the first surveyors in a post-Napoleonic navy, which was shrinking rapidly, and counted himself lucky to have a commission at all. In *Mermaid* and *Bathurst* he spent five years charting the coast, building the first detailed picture of many areas, and returning to England in 1823 with charts that were used for the rest of the century. His account of work was published in 1827. By that date, ten years after his marriage, King had a large family of seven sons and an estate near Sydney. His appointment to the *Adventure* and *Beagle* expedition was his last, and he retired from the navy in 1832. A leading figure in Sydney, he continued to support surveying work and naval education in Australia until his death in 1856.[31]

Both of King's *Narratives* (the one published in 1827 as well as the volume here of 1839) show his wide-ranging curiosity. He had worked closely on his Australian survey with the botanist Allan Cunningham, who had became a lasting friend, and he seems to have embraced the natural history observations of his official instructions in 1826 cheerfully. He was a fellow of the Royal Society (1824), and presented some of his notable finds for the newly founded Zoological Society in London after his return to England. But his perspective was firmly limited, or possibly oriented to the more spectacular tropical finds of the antipodes. He commented on the sterility of the landscape without any sign of Fitzroy's interest geological speculation and his account of the steamer-duck focused as much on its culinary value as on its unusual anatomy and appearance.[32] The most engaging observations of his *Narrative* dealt with ethnography - not the Fuegians that so captured Darwin's and Fitzroy's imaginations, but the Patagonians, or Telhuelche. Here his encounters give us a fascinating glimpse of his experience. Investigating these encounters illuminates the nineteenth century practices of scientific observation.

The Patagonians and 'Vigia': Observation in the Nineteenth Century

Apart from warning about armed populations, and enjoining the captains to avoid any acts of hostility, the official instructions in 1826 and 1831 did not mention native populations directly, yet King clearly interpreted his Admiralty instructions to include ethnographic observations. His mandate to collect information of use to navigators – including climate, anchorage and availability of provisions – encompassed information about the geographical distribution of different tribes, about trading relationships, language and culture. This was a model consistent with his first Australian voyage, on which the populations were almost totally unknown to Europeans. King there carried an aboriginal from Sydney as a potential interpreter and contact. For intensely practical reasons, then – including the immediate one of ensuring their crews fresh supplies of meat as an anti-scorbutic – both expeditions to South America sought out almost every opportunity to meet with native populations.

The Telhuelche were a nomadic people living just north of Tierra del Fuego, on the southern pampas of the continent. They had been a subject of ethnographical fascination since the earliest days of European explorers, who described them as a race of giants. King's descriptions betray a close preoccupation with sorting out a particular Enlightenment sensation that pitted Spanish tales of the fabulous against sceptical observation. Magellan had first named the inhabitants, supposedly after their enormous feet, and John Byron, visiting the region some two hundred and fifty years later in the *Dolphin* in 1766, was accused by sceptics of recycling naïve Spanish fantasies. In fact, Byron's reports, when finally published, were far more cautious than early rumours had suggested.[33] King was tackling this history of sensation, and restoring the reputation of British evidence and British observers. To re-interpret these tall tales, King and Fitzroy took measurements, evaluated visual sensibilities and weighed testimony.

King's first encounter opens with the detailed measurements of some very large trees, including one enormous specimen, 21 feet in circumference, in a bay that Byron had visited some sixty years earlier. He then turns from timber to men, describing the sight from *Adventure*'s deck of three or four Patagonians standing with their horses on a hill. The men 'loomed very large' says King and we 'should certainly have thought them giants' if this was his first experience with the population. Instead, he is confident that this view was 'an optical deception', a mirage caused by the haze in the moist air, and the perspective of ship deck to headland.[34] Patagonians were like 'vigia': the reports of shoals or rocks that the hdydrographer must investigate and pin-point on the charts, or dismiss as error. Fitzroy, too repeatedly measured and discussed visual accuracy: He estimates the average height of natives, and compares the tallest he has seen to the earlier accounts of Falkner and Byron. He explains that the Tehuelhet are unusually proportioned, with long torsos but shorter legs, and wear bulky mantles about their upper bodies: both give them an appearance of height, an appearance that Fitz-Roy was able to correct with measurement. Similarly, he noted their small hands and feet – Magellan being deceived by the large footprints that come from the rough coverings of animal hide around the feet. The featureless landscape of the region – Fitzroy notes that 'the eye wanders over an apparently boundless extent of semi-desert' contributes as well to this idea of mirage – an approaching Tehuelhet on horseback will dominate the scenery – as King noted, 'looming large.'[35]

These examples show that the observers' concern is not simply the factual record (how tall are the Patagonians?) but the reliability of the historical record and modern eyewitnesses, by exploring the nature of (mis)perception. A directly parallel analysis of discussion of the height of the Patagonians could be carried out for the subject of cannibalism and matricide in the Fuegian tribes, which equally preoccupied Fitzroy (and Darwin as well). In both matters, the *Narrative* claims implicitly that the marine observer, trained in visual perception and misperceptions, and committed to testing and combining testimony from past exploration, had a special kind of expertise. The disciplined observer himself is the thread that moves among the bits of information, connecting different places and times and types of detail.

In a particularly striking passage, FitzRoy interrupts his description of the skin colour of the natives, to remind the reader that his eyes have gone to South America and back. 'The colour of these aborigines is extremely like that of the Devonshire breed of cattle. From the window of a room in which I am sitting, I see some oxen of that breed passing through the outskirts of a wood, and the partial glimpses caught of them remind me strongly of the South American red men.'[36] The traveller's expertise means even his partial glimpses enable him to connect distant phenomena and species. The official instructions from Francis Beaufort to Fitzroy in 1831 make this issue clear: The *Beagle* was to correct the 'the motley appearance of alternate error and accuracy' along the coast of South America, and replace the 'half-knowledge' of the past with an integrated view.[37] The careful description of Patagonians, then, communicates a sense of the unity of the *Adventure* and *Beagle*'s work. Ethnography and natural history were part of a wider set of observing practices.

A second, and completely different, reason for the vivid interest of the Patagonian parts of King's *Narrative* is their ability to turn the reader's attention to the observer. In the characteristic trope of travel and natural history writing, it is not the exotic quality of the people, their clothing and their behaviour alone that are fascinating; rather it is the witness's reaction to them. We can gain a dizzying sense of several points of view at once, a kind of dislocation that conveys the nineteenth-century experience of exploration for men like King and Fitzroy. The Patagonian tribes that King described were ruled by a woman called Maria, who had a far larger experience with Europeans than King had with Patagonians. 'A rather remarkable woman', King admits, the *Narrative* shows that she took charge of their encounters. She controlled the bargaining, invited herself to meals and staged a display of religious ritual for King's edification. (It ended with a reminder that 'her Christ', a small totem, liked tobacco.) In these scenes, it is King and his men who are the curiosities, and their ship, meals, tobacco and trinkets seem like specimens that Maria and her people are collecting.[38]

This example could be multiplied by many later encounters between King and Fitzroy's men further south and west, as the natives in Tierra del Fuego, assess the outsiders' hair, skin colour, gender, music and tools as carefully they are themselves studied. Recognizing these multiple perspectives should not romanticize the encounters. Decisive power lay with King and Fitzroy, however bewildered they seem at times. Indeed, the cycle of observer and observed seems reminiscent of Darwin's description of an incident in the Chiloe islands a few years later.

> A fox (*Canis fulvipes*), of a kind said to be peculiar to the island, and very rare in it ... was sitting on the rocks. He was so intently absorbed in watching the work of the officers, that I was able, by quietly walking up behind, to knock him on the head with my geological hammer. This fox, more curious or more scientific, but less wise, than the generality of his brethren, is now mounted in the museum of the Zoological Society.[39]

Just like Darwin, the captain held the hammer: the Telhuelche population was steadily and permanently eroded by European contact and the pressure of settlement from the north in the rest of the century.[40]

This reality lies behind another set of memorable events in the first voyage, events which led directly to the second. In 1829, Fitzroy set off in Beagle to survey alone in Tierra del Fuego for the last season of the expedition. He was new to his command, and not accustomed to dealing with the natives of the regions – they were, for him, as later for Darwin, the first complete 'savages' of his experience. In a storm at the end of January 1830, one of her valuable whaleboats, indispensable for close surveys (as well as a rough kind of insurance in case of shipwreck), was washed away. A few days later, natives stole a second boat from a small surveying party under the command of the master, Mr Murray. Fitzroy, angry and worried, searched for the boat for weeks, taking several hostages as he pursued signs of the boat's remains at encampments through the islands. Some of his hostages escaped but when he eventually abandoned the search, he still held three individuals, a young girl and two young men, whom he named Fuegia Basket, York Minster and Boat Memory. Two months later, in May, he took a fourth individual from a different group of the Yamana people, whom he named Jemmy Button. Although Fitzroy describes the events with restraint in the *Narrative*, the whole episode shows him at his worst – impatient, vengeful, and rather stupidly persistent. His subsequent decision to turn the events into a social experiment – to educate the Fuegians in England, then return them to their native land as future interpreters and support for a Protestant mission – put a philanthropic and scientific face on the humiliating theft of his equipment. It is hard to imagine the same set of events unfolding under the more experienced King. The *Beagle* rejoined the *Adventure* at their rendezvous in Rio to sail home to England in August 1830, but at no point does the *Narrative* record King's reaction to the hostages.

The Second Voyage: Robert Fitzroy

The second voyage of 1831 to 1836 arose from the decisions of Fitzroy to keep the natives on board. Both he and King had confidently anticipated that a further survey of the region would be commissioned quickly, as so much remained to be done, especially on the outer edge of the western islands and north towards Chonos. The recent work of Fitzroy and his Lieutenant Skyring had also raised the tantalizing possibility of an unknown short-cut to the west coast (which eventually proved false). The Admiralty at home in 1830 proved less enthusiastic and more parsimonious. After a few months in England, Fitzroy's plan to return the natives had unravelled. Considering it a matter of personal honour, he hired a ship himself, before his uncle[41] intervened and the Admiralty gave in. This time, however, it was to be a single ship, the *Beagle*, and its survey aims were supplemented by a plan hatched by Beaufort to carry out a complete chain of longitude measurement by chronometer, point to point around the globe. This particular ambition became one of Fitzroy's triumphs: he claimed to have measured by chronometer, longitude to longitude, across the globe, with a resulting discrepancy of only thirty-three seconds over the twenty-four hours.[42]

The *Beagle* voyage is deservedly the more famous of the two voyages recorded by the *Narrative*, and not only because of Darwin. It was slightly longer in duration, and visited a far greater variety of regions – Tahiti, Cocos-Keeling, the Galapagos, New Zealand, Australia and South Africa as well as South America. It recorded many remarkable episodes – the rebellion and imposition of English rule in the Falklands, the failed settlement in Tierra del Fuego, an attempt to penetrate up the Santa Cruz river to the Andes, the dramatic earthquake of February 1835, the shipwreck and rescue of the *Challenger* a few months later, and political negotiations with the queen of Tahiti. It was also a far more philosophical voyage – again, not simply because of Darwin, but because of Fitzroy's own range of interests. Partly led by Beaufort, and partly by his own inclinations, he was keenly interested in geology, fossils, the questions surrounding coral reefs, phrenology and meteorology. The *Beagle* voyage of 1831 to 1836 was the first Admiralty ship to implement Beaufort's system of notation for wind force and weather conditions, and Fitzroy was a quick convert, already using the barometer to predict the weather and speculating about electrical effects in the atmosphere in his first voyage.[43] The *Beagle* carried other novel technologies in addition to the Beaufort scale: twenty-two chronometers, some government issue and some belonging to the captain, a new form of lightning conductor, a new stove design for the galley and plenty of tinned goods. The philosophical turn is evident in the structure of the volume. The *Narrative* ends, not with the return to Plymouth, but with two unusual chapters that set out Fitzroy's views on human migration and mosaic geology. In all these ways, it was a remarkable voyage and a distinctive *Narrative*.

But the second voyage is principally more famous because we know so much more about the characters of the men on board through the window of the relationship of two men, Robert Fitzroy and Charles Darwin. The evolution of their relationship, Whig country gentleman and Tory aristocrat naval officer, lies behind the text. Fitzroy wrote up the *Narrative* between 1837 and 1839, when he had left a Tory world and come back to a Whig ascendancy. Darwin's standing and confidence as a naturalist rocketed after they returned, while Fitzroy's own career stumbled against the political obstacles and his religious beliefs hardened. He was unmistakably on the defensive. For historians, then, part of the value of the *Narrative* is the intricate connections it shows between politics, the religion and natural philosophy. In his chapter on the deluge, Fitzroy referred pointedly to geology as 'a young branch of the tree of knowledge, which has yet to undergo the trial of experience.'[44] The remark could not strike a more different tone from his earnest resolution in 1829 to benefit in any future voyage from someone qualified to carry out scientific researches.

Fitzroy was born in 1805, the second son of Lord Charles Fitzroy. His grandfather was Augustus Henry, the Duke of Grafton. He was named after his maternal grandfather, the Marquess of Londonderry and his uncle, Robert Stewart, the Viscount Castlereigh, foreign secretary and leader of the House of Commons from 1812 to his death by suicide in 1822. His mother died when he was five years old, and his father died in December 1829 (about the time his son was setting off for his first lengthy independent survey in Tierra del Fuego). He was educated at Harrow and the Royal Naval College, and he graduated with perfect marks in his leaving examination. He went to sea aged fifteen. Before his appointment as commander of the *Beagle* he had been a midshipman on four different ships in South America and the Mediterranean; when he was promoted to lieutenant in 1824 (again with top marks in the examination), he joined the *Thetis* and sailed in her to Rio, where he joined the flagship of the South American command, the *Ganges,* under Admiral Robert Otway. It was Otway who gave him the *Beagle*, over-ruling King's appointment of Lieut. Skyring. Fitzroy's career was rapid, but not unusually so, given his aristocratic connections and his obvious ability. In 1831, as the *Beagle* left Plymouth, he was twenty-six years old. He had used some of his own wealth to re-fit the *Beagle* to create more space for survey work, and he paid for an instrument assistant and a personal servant, a steward, as well as arranging for the artist and draughtsman, Augustus Earle (later succeeded by Conrad Martens) and a companion, Charles Darwin.[45]

Towards the end of the voyage, Darwin wrote that the three high points of his experience were seeing the Andes, the Galapagos and the Fuegians. It was a significant trio, marking his passion for grand geological theory and his future work on human evolution. Fitzroy never singled out such a list of highlights, but, to judge by the events he described most energetically in the *Narrative*, the list would have been equally revealing. The expedition up the Santa Cruz in 1834 stands out – one of the few times in the voyage that Fitzroy got to do what Darwin had been enjoying for the past eighteen months, an inland exploration in terrain made for geological speculation on a grand scale. Like Darwin, he was stunned and fascinated by the earthquake they witnessed in 1835. But perhaps the episode that most reveals his character was his intervention in the rescue of the shipwrecked *Challenger*.

As narrated by Fitzroy, the wreck of the *Challenger* demonstrated what it meant to be a naval captain and a hydrographer. In May 1835, Captain Michael Seymour sailed the frigate *Challenger* onto a shoal off Tucapel, an isolated region of Chile. Seymour managed to land most of his men and some stores ashore, but was unable to salvage the ship or risk his men on a long march to the nearest settlement. His predicament was complicated by the coming of the winter rains and simmering conflict between rival groups of natives, some supporting the republic and some hostile. The nearest Chilean authority, a fort ninety miles away, was handicapped by earthquake damage and the tense political situation. Fitzroy heard of the event a few weeks later, and spared no effort to come to Seymour's assistance, showing his deep sense of responsibility and his energy. He rode overland to assess Seymour's situation, and then helped direct difficult near-shore operations to collect the men and equipment. Shipwreck, near mutiny and threatening bands of Indians made for the kind of adventure designed to appeal to a wide audience at home. Fitzroy met this expectation and more by mixing epic poetry and the story of sixteenth-century Araucanian resistance to Spanish rule with a description of his wild ride to Seymour's camp. Yet the event was also intensely significant on a personal level. Fitzroy had known Seymour since his time at the Royal Naval College. After many pages of detail and historical digression, Fitzroy is almost wordless: 'Old friends, meeting under such circumstances, can say but little.'[46]

The shipwreck episode is especially characteristic of the *Narrative,* however, because its mixture of anecdote, crisis and history smoothed down to naturalistic explanation. The episode made clear why the close study of natural phenomena was so crucial: shipwreck was what hydrography worked to prevent. Fresh from the shock and fascination of the natural disaster that brought the city of Concepcion to ruins, Fitzroy interpreted this professional disaster as a direct consequence of that geological event. The natural disaster had changed the local currents so unpredictably that the *Challenger* had drifted far from her apparent route. At Seymour's subsequent court-martial, faced with Fitzroy's explanation, the Admiralty authorities agreed. The centralized management of empire, symbolized by the quarter-deck in Portsmouth where Seymour formally accounted for an event over 7,000 miles away, had its counter-part in the universal knowledge represented by Fitzroy's geological generalization.

The frequent descriptions of native societies throughout Fitzroy's volume similarly met the expectations of his audience while giving Fitzroy scope for his own preoccupations. He was deeply interested in ethnography and all questions about change – progress and degeneration – in human societies under the pressure of history and the environment. Fitzroy believed in the unity of humankind – that humans came from one stock, and were not separate races. Like many contemporaries, his views displayed profound religious and social beliefs. The monogenesis theory, or single origin of human beings, fit the creation myth of Christianity. God had created man, and his descendants had dispersed and, some advancing and some declining, gradually filled every step in the ladder of civilization. Monogenesis, as the theory of single origin was known, therefore also supported attitudes towards social development and progress that were so much a part of the British global world view in the nineteenth century. On an individual level, as with the Fuegian hostages, human beings could be transformed from savage to civilized. They could also degenerate, or in that evangelical term, backslide, just as Jemmy Button did when he was returned to Tierra del Fuego, or as shipwrecked sailors could when forced to live like animals in desolate coasts of the world. Change was an obvious fact of bio-geography. 'There is abundant proof', Fitzroy noted, 'that animals have changed their habits, shape, coat, colour, or size, in consequence of migration, or transportation to different climates.'[47]

On a wider level, Fitzroy believed human societies were similarly in flux, progressing or degenerating. He repeatedly drew attention to the parallel between the 'savages' he saw, and the ancient Britons in the eyes of the Romans, influenced by the work of the historian and political philosopher, James Mackintosh. Above all, he knew that the native societies he saw in South America and the Pacific were acutely affected by contact with Europeans – by disease, by wars of extermination, by changes to their traditional food resources. In more detail than many other observers, including King, he mapped the distinctions between different tribes and regions of Tierra del Fuego. He knew that they stole the *Beagle* whale-boat because the sealers from North America were rapidly eliminating near-shore seal populations which they could reach in their own canoes. He estimated the changes to the population of the Patagonians and speculated that inter-breeding had reduced their height over several generations since the origins of the 'giant' rumours. He was acutely interested in evidence of past, more complex native societies, or what he would have called more 'civilized' forms. He was also fascinated by linguistic evidence for common roots among geographically or temporally distant populations and in other evidence of human migration. He believed that phrenology could be a guide to understanding human variety and could explain the human capacity for transformation as individuals and as societies. Such views were summarized in his penultimate chapter of 'early migrations of the human race'; they are also scattered throughout the *Narrative* as he described the native communities in the lands the *Beagle* touched.

Fitzroy's views on humankind were obviously inseparable from his religious beliefs. For the reader of the *Narrative*, his religious views are significant in two different ways. They are the background to an important strand in his volume, the story of the Fuegian hostages, the missionary Richard Matthews and the failed settlement at Woollya in Tierra del Fuego. But they are also a critical part of his relationship to Darwin and his reputation after 1839.

First, the missionaries. The *Narrative* is genuinely valuable as a record of nineteenth-century British missionary work and its place in the history of the Pacific region. When he arrived back in Plymouth in 1830 at the end of the first voyage, Fitzroy enlisted the assistance of a prominent evangelical, the Rev. Joseph Wigram, to carry out his plans for the Fuegian hostages. The education and return of the Fuegians to their home quickly changed from a private project of aristocratic philanthropy into mission work, as quickly as it had earlier changed from hostage-taking to philanthropy. Richard Matthews, a young and completely inexperienced man from the Church of England's missionary organization, the Church Missionary Society, sailed to Tierra del Fuego on the second voyage. The project to build the nucleus of a settlement at Woollya was a complete failure, described by both Darwin and Fitzroy in some detail. Matthews ended up settled with his brother, also a missionary, in New Zealand. The aftermath of the failure for Fitzroy, however, is less well known. Fitzroy was deeply disappointed, even shaken by the evidence of the impermanence of the changes that education had effected. It coloured his interest in missionary work in Tahiti, in New Zealand and in South Africa. Like Darwin, he found the evidence of piety in Tahiti moving and impressive. In New Zealand, among cannibalistic and warring Maori, he saw the missionaries in a favourable light as well. He defended their work against the attacks of the Russian navigator Capt. Kotzebue and his former artist, Augustus Earle, who had published a denunciation of their venality in 1827 that stirred controversy about Pacific missions at home.[48] Fitzroy's and Darwin's views coincided on these questions in 1836 as they sailed home, as they jointly published a letter on the value of missionary work that was later largely incorporated into the *Narrative*.[49] After his return, while he was writing the *Narrative*, Fitzroy gave thoughtful evidence to parliamentary inquiries into mission work and politics in the Pacific region.[50] By then, missionary work was fixed as part of his hopes of gradual social progress in human society.

The odd final words of his volume show the depth of this preoccupation with civilization and its changes. The closing refers abruptly to Patagonian 'courts of justice' and games of hockey, as an addenda to a defence of the geological evidence for the deluge and the literal truth of the Bible. Had Fitzroy simply lost track of his notes, cramming a last remark into the proofs of the *Narrative*? Perhaps, but these remarks on courts and hockey were consistent with the theme of the final two chapters on the deluge and human migration in a way that shows their relationship to the ethnographic and historical discussions elsewhere in the volume. It is a call to interpret evidence carefully, based on a Christian and intellectual humility:

> Having no pretension to more knowledge than any observant seaman may acquire in the course of a few years active employment afloat, it would be as vain as presumptuous in me, were I to offer any conjecture about the central mass of the earth. Perhaps, at a future day, when the nature of aerolites; the agency of electricity; and of electric communication through the superficial, if not through the interior regions of the globe, are better known, other opinions, respecting this wonderful world which we inhabit, may be formed by philosophers.[51]

This call for humility is easily dismissed because Fitzroy was himself proud to a fault. But the opinion was genuine, a revealing mixture of the practical seaman, the British conservative and the speculative philosopher in the 1830s.

Fitzroy's later career was far more chequered than that of King's. He did not immediately receive another commission in the navy, and (as he had done in the hiatus between the first and second voyage) sought a political career instead. Supported by his uncle Lord Londonderry, he was elected as a Tory to the House of Commons in 1841 for Durham. In 1842 he became a conservator for the Mersey River, reporting on its navigation and suggesting improvements. In 1843, he was appointed as the second governor of New Zealand, but was quickly relieved of the post in 1845 after his support for the Maori population worsened the tensions between settlers, merchants and missionaries and the aboriginal population. Three years after this wounding experience, which damaged him politically and financially, Fitzroy finally received another naval commission, as Superintendent of Woolwich Dockyard, and the job of trying out a new steam-powered vessel named *Arrogant*. He resigned suddenly two years later, temporarily incapacitated by another bleak depressive illness. As the Crimean war began in 1853, he hoped for another appointment, and eyed Beaufort's post as Hydrographer, but nothing materialized. Finally, he was appointed as head of a new body in the Marine Department of the Board of Trade, a department for meteorological statistics that became Britain's first Meteorological Office. There, surrounded by what must have been a combination of details and philosophizing, his former initiative and energy returned. He pushed the department beyond statistical collection to telegraphic reports and then weather predictions, writing several pamphlets on meteorology for mariners from 1857 and a lengthy text on the subject in 1863. These efforts won mixed reviews from the public and from the scientific community; some prominent members of the Royal Society, where he was now a fellow (1851), found the practice of weather prediction dismayingly unscientific. His opposition to Darwin's theory of evolution, published in 1859, also put him on the defensive with the same group of scientific leaders that opposed his forecasting innovations. Disheartened by a mixture of private and professional troubles, he committed suicide in 1865, aged sixty. The manner of his death has often coloured his reputation as a hydrographer and a meteorologist. Yet he was a remarkable man. Every aspect of his career after 1836 from New Zealand politics to experimental naval architecture to weather science can be traced in the wide-ranging, endlessly curious discussions in his 1839 *Narrative*.

The *Narrative* and its Reception

Because Darwin and Fitzroy later differed so greatly on religious matters, and on evolutionary theory, the congruity in their religious and scientific attitudes during the voyage is often underplayed. It points, again, to their changing relationship after the end of the voyage – Darwin apparently disliked Fitzroy's pious new wife, Mary Henrietta O'Brien, whom Fitzroy married in 1836, and certainly rejected his chapter on Deluge geology, which explicitly repudiated what must have been many earlier shipboard discussions of Lyell.[52] More importantly for our analysis here, this relationship has created problems of interpretation for the historian of Fitzroy and the *Narrative*. Darwin's assessment of Fitzroy's character later in life so caricatured the divide between them – Darwin the patient scientific thinker and Fitzroy the impassioned Biblical literalist – that an evaluation of their shared attitudes during earlier times is difficult. In Darwin's words to a colleague after Fitzroy's death in 1865, 'I never knew in my life so mixed a character. Always much to love & I once loved him sincerely; but so bad a temper & so given to take offence, that I gradually quite lost my love & wished only to keep out of contact with him.'[53] In 1860 he had written to another scientific colleague that 'his mind is often on Verge of insanity.'[54] The very force of Darwin's opinion makes the *Narrative* more valuable as an opportunity to compare and re-evaluate.

Sailing home in 1836, however, Darwin and Fitzroy still saw eye to eye on many matters, scientific and other, such as missionary work and emigration to Australia (not an attractive idea for a gentleman, they agreed). They made the decision to publish their accounts of the voyage together, with Darwin's description of natural history researches, organized by geographical region, joining the chronological narratives of King and Fitzroy. They hoped to complete the work by November 1837, although Darwin negotiated from the start the right to publish his volume separately once the first edition had appeared. Darwin later regretted this decision to make his work part of the official narrative, in part because Fitzroy was so slow to complete his volume and in part because he and his family feared it would get lost amid the other volumes. He was also, once back in Cambridge and London, more alive to the difference between his career as a naturalist and Fitzroy's as a naval officer than he had been at sea. In November 1837, his own volume finished, Darwin sent it to Fitzroy to read. He received in return a stiff lecture on his superficial acknowledgment of the assistance given to him by officers of the *Beagle*. Darwin was in the wrong, and apologized; in truth, as Janet Browne's recent biography of Darwin shows very well, he had by then lost sight of his *Beagle* life.[55] In April 1838, he told a family member that he had visited Fitzroy and read over a few pages of King's volume. It was, he thought, dreadful. 'I was absolutely forced against all love of truth to tell the Captain I supposed it was very good, but in honest reality, no pudding for little school boys was ever so heavy. – It abounds with Natural History of a very trashy Nature.' He trusted that 'Captain's own volume will be better.'[56] By February the following year, Darwin and his scientific friends were still waiting for Fitzroy to finish writing. In the presidential address to the Geological Society that month, the Cambridge philosopher William Whewell publicly expressed 'disappointment and regret' that Darwin's account was delayed, due to the 'customs of the Service' by which the Captain controlled its publication.[57]

When the work did appear, under the publisher Henry Colburn, later in 1839, it did not, according at least to Darwin's letters, return a profit.[58] It was favourably reviewed in the major periodicals, which printed large extracts and showed greatest interest in the Fuegian episodes. Reviewers directed some criticism at Fitzroy's Deluge chapter. As one noted severely, 'On this subject the gallant Captain has got quite beyond his depth.'[59] But all reviews paid tribute to the comprehensive achievement of Fitzroy both in the survey itself and in writing up the *Narrative*. As William Hamilton, president of the Royal Geographical Society, put it, 'Henceforth, the names of Humboldt and FitzRoy must be inseparably connected, as the chief authorities for our knowledge of the Geography of South America.'[60] The work was a reference point, a signal of the achievements of British scientific voyages in the early nineteenth century.

It remains a complicated work, combining the many perspectives and tasks of its captains. It was clearly directed to many different audiences at once, which made it difficult for readers at the time to consider as a whole, and even more so for readers in the present day. But it is these complications which make the work so revealing as a record of early nineteenth-century scientific enquiry and empire. The work of precision measurement was coupled to a wide range of other kinds of observations and tasks which severally illuminate each other. Travelling through the spheres of the former Spanish empire, the exotic worlds of little-known native societies, and the new zones of Pacific commercial and missionary enterprise, these scientific voyages showed the dominant assumptions of their culture about time, place and knowledge. Fitzroy and King saw themselves as part of the dispersed threads of the British navy and part of a long heritage of European global exploration. Their work was part of a systematic and sustained endeavour. But they also shared a powerful sense of novelty and change – a speculative spirit that showed their consciousness of a rapidly changing world – and they sought out the patterns of natural knowledge and social practices that could make sense of it.

The Pickering-Chatto edition of the *Narrative of the Surveying Voyages of HMS Adventure and Beagle*

This edition consists of four volumes, two for the narrative of the first voyage and its appendices, and two for the narrative of the second voyage and its appendices. It thus presents the material from volumes 1, 2 and 4 of the original set of 1839, and excludes the well-known text of Darwin's *Journal*. The engravings of Earle and Martens, too, have become widely available in other publications.[61] But one key element of the original text has not been reproduced here: the large maps which accompanied the original edition. The maps (four altogether) were put together by the London cartographer, John Arrowsmith, who spent two years on their preparation, combining the information from Fitzroy and King's surveys with the best maps of the interior of the South American continent available.[62] These visual elements were powerful parts of reading the work for contemporaries, but have not been reproduced here.

The annotations in this edition seek to identify authorities cited by King and Fitzroy, to identify the individuals they mention, and to contextualize the chief activities and incidents of the voyages. References to members of the ships' company in general have not been annotated in the text; instead, a list with biographical details is given separately. The extensive cross-referencing carried out by Fitzroy to King's volume and to the appendices to his own volume have been preserved. Beyond the cross-referencing, the appendices of King and Fitzroy (contained in the concluding pages of volume 1 and the whole of volume 4 of the 1839 edition) have not been annotated in detail. In any references to Darwin's writings in the annotations, I have followed the most accessible print versions available: Janet Browne and Michael Neve's *Voyage of the Beagle* (Penguin, 1989), which omits some parts of Darwin's original, and Keynes's *Diary of the Beagle* (Cambridge, 1988). The presentation of the original edition of the *Narrative* in *Darwin Online* is complete and easy to use. The original *Beagle* and *Adventure* logs have not been exploited for the annotations here, but they are now available online as apart of the CORRAL project to recover meteorological data from ships' logs.[63] Finally, it should be noted that references to Darwin's work in this edition have taken advantage of editorial privilege: I pointed to connections that interested me, and have not attempted to be comprehensive, as a detailed study of the inter-relationships of Fitzroy's and Darwin's points of view would be a vast project of its own.

Key biographical sources have been the *Oxford Dictionary of National Biography*, the *Dictionary of New Zealand Biography*, the *Australian Dictionary of Biography*: G. Anderson, *Biographical Dictionary of Christian Missions* Grand Rapids, MI: Eerdhams, 1999); S. Bizarro, *Historical Dictionary of Chile*, 3rd edn (Lanham, Md: Scarecrow Press, 2005); R. Desmond *Dictionary of British and Irish Botanists*, 2nd edn (London: Taylor and Frances/Museum of Natural History, 1994); W. Edmundson, *A History of the British Presence in Chile* (New York: Palgrave MacMillan, 2009); P. Figuero, *Diccionario biográphico de Chile* (Santiago, 1897); P. O'Brien and R. Tessier, *Tahitiens: Reportoire biographique de la Polynesie francaise* 2nd edn (Paris: Musee de l'homme, 1975) W. O'Byrne, *Naval Biographical Dictionary* (London, 1849); and I. Wright and L. M. Nekhom, *Historical Dictionary of Argentina* (Metuchen, NJ: Scarecrow Press, 1978) as well as the biographies compiled by the Darwin Correspondence Project.

Notes

1. See the editors' introduction to C. Darwin, *Voyage of the Beagle*, ed. J. Browne and M. Neve (London and New York: Penguin Books, 1989).
2. Darwin's autobiographical 'Recollections', in J. Secord (ed.), *Evolutionary Writings* (Oxford: Oxford University Press, 2008), p. 387.
3. The exceptions are a facsimile reprint of the 1839 edition by AMS Press in 1966 and the online presentation of the edition in *Darwin Online*. A work of extracts from Fitzroy's volume was published by David Stanbury in 1977. Also well, several editions of Darwin's writings have extracted from the second volume collated with Darwin's diary and published *Journal of Researches*; see R. D. Keynes, *Selections from the Original Pictorial Records and Written Accounts of the Voyage of the H.M.S. Beagle* (Cambridge: Cambridge University Press, 1979).
4. Both goals were met in the following decade, with the British Association for the Advancement of Science taking over Kew Observatory in 1842 and Sabine's so-called 'magnetic crusade' of the 1840s.
5. Anon., '*Narrative of the Surveying Voyages of HMS Adventure and Beagle*', *Quarterly Review*, 65 (1839), p. 211.
6. The *Athenaeum* thought the journals of the 'surveyors' had 'encroach[ed] on the domain of the naturalist', 607 (15 June 1839), pp. 446–9, on p. 446.
7. See Volume 4, p. 155.
8. See King, *Sailing Directions* (London: Hydrographic Office, 1832) and Fitzroy, *South American Pilot* (London: Hydrographic Office, 1848).
9. A good description of their practices can be found in Fitzroy and Raper, 'Hints to Travellers' as well as in his appendix, 'Notes on Surveying a Wild Coast'. The Admiralty later issued a manual outlining how to conduct observations: *Admiralty Manual of Scientific Inquiry* (London: John Murray,1849).
10. See Volume 3, p. 81.
11. See Volume 3, p. 306.
12. See Volume 1, pp. 44–5.

13. On history and memory as a tool for the interpretation of travel narratives, see Schmitt, *Darwin* (Cambridge: Cambridge University Press, 2009).
14. For a recent view of these debates in history of science, see the focus section, Sivasundarum et al., 'Global Histories of Science', *Isis* (2010).
15. The history of the ship itself before during and after the second voyage has been studied intensively; see Darling, 'H.M.S. Beagle' (1984).
16. Safier, writing on La Condamine's expedition of 1735–44, has an illuminating analysis of literary exploration, *Measuring the New World* (Chicago, IL: University of Chicago Press, 2008), pp. 57–92; the wider availability of texts and translations made this aspect of voyages even more marked a century later. Darwin acquired his habits of systematic reading and note-taking on the *Beagle* as well: see J. Browne, *Charles Darwin: Voyaging* (Princeton, NJ: Princeton University Press, 1995), pp. 192–4.
17. 'Books on the Beagle' at www.darwinproject.ac.uk/books-on-the-beagle.
18. See Lamb, *Exploration and Exchange* (Chicago, IL: University of Chicago Press, 2000) and Hume and Youngs, *Travel Writing* (Cambridge: Cambridge University Press, 2002) for an introduction to this genre. Cohen, 'Traveling Genres' (2004) discusses the representation of travel in the nineteenth-century novel and its relationship to non-fictional material.
19. Darwin wrote himself or coordinated the analysis of the zoological specimens by leading naturalists which were published in parts over several years. See C. Darwin (ed.), *Zoology of HMS Beagle* (London, Smith Elder, 1837–43).
20. J. Secord, 'Scrapbook Science: Composite Portraits in Late Georgian England', in B. Lightman and A. Shteir (eds), *Figuring It Out: Science, Gender and Visual Culture* (Lebanon, NH: Dartmouth College Press, 2006), pp. 164–91; see also, Lamb, 'Minute Particulars' (2004).
21. Vale, *Independence* (London: British Academic Press, 1996).
22. See his *Diary*, 5 July 1832.
23. See Volume 3, p. 66.
24. See Volume 1, p. 131.
25. J. F. W. Herschel, *Preliminary Discourse* (Chicago IL: University of Chicago Press, 1830), pp. 27–9. He referred specifically to the voyage of Capt. Basil Hall from west coast of Mexico to Rio de Janeiro without touching land.
26. On measurement and empire, see Schaffer, 'When the Stars Threw Down Their Spears' (2010) and Bourget, Licoppe and Sibum, *Instruments, Travel and Science* (London and New York: Routledge, 2002).
27. Sorrenson, 'The Ship as a Scientific Instrument' (1996).
28. See Volume 1, p. 12.
29. L. S. Dawson, *Memoirs of Hydrography* (Part 2) (Eastbourne: H. W. Keay, 1969), p. 3.
30. A. Friendly, *Beaufort of the Admiralty* (New York: Random House, 1977), pp. 276–300; G. S. Ritchie, *Admiralty Chart* (New York: Elsevier, 1967); Browne, *Voyaging*, pp. 144–61; Reidy, *Tides of History* (Chicago, IL: University of Chicago Press, 2008).
31. M. Hordern, *King of the Australian Coast* (South Carlton: Melbourne University Press, 2002); *ODNB*.
32. See Volume 1, p. 41.
33. This explanation based on *pata* as the Spanish for foot is now rejected by scholars; instead, the name is now traced to a character in a sixteenth century Spanish romance named Patagones. On the myth, see P. G. Adams, *Travelers and Travel Liars* (Berkeley, CA: University of California Press, 1962), and C. McEwan et al., *Patagonia* (Princeton, NJ: Princeton University Press, 1996).

34. See Volume 1, p. 78.
35. See Volume 3, p. 115.
36. See Volume 3, p. 113.
37. See Volume 3, p. 30.
38. On the shifting perspectives of the European encounter, see S. Banner, *Possessing the Pacific* (Cambridge, MA: Harvard, 2007); G. Beer, 'Travelling the Other Way', in N. Jardine, J. A. Secord and E. C. Spary, *Cultures of Natural History* (Cambridge: Cambridge University Press, 1996). For King's description of his meeting with Maria, see pp. 78–83.
39. Darwin, *Voyage of the Beagle*, ed. Browne and Neve, p. 221. On foxes as a subject of biogeographical interest for Fitzroy and Darwin, see Bourne, 'Fitzroy's foxes' (1992).
40. A. Chapman, *European Encounters* (New York: Cambridge University Press, 2010); McEwan et al., *Patagonia*.
41. Fitzroy does not specify the uncle. He had two powerful uncles, the Duke of Grafton on his father's side of the family and the Lord Londonderry on his mother's; H. E. L. Mellersh, *Fitzroy of the Beagle* (London, R. Hart-Davies, 1969) p. 60, has suggested it was the more politically engaged of the two, Londonderry.
42. A. von Humboldt, 'On the Longitude of Valparaiso and Callao', *Journal of the Royal Geographical Society*, 9 (1839), pp. 502–6; for a later discussion about Fitzroy's treatment of the different instruments and the data, see, A, Auwers, 'Chain of Meridian Distances', *Monthly Notices of the Royal Astronomical Society*, 44 (1884), pp. 303–46.
43. On the history of the Beaufort scale, see S. Huler, *Defining the Wind* (New York: Three Rivers Press, 2004); for Fitzroy's meteorological work generally, see K. Anderson, *Predicting the Weather* (Chicago, IL: University of Chicago Press, 2005).
44. See Volume 4, p. 132.
45. On Fitzroy's life, see the *Oxford Dictionary of National Biography*; Mellersh, *Fitzroy of the Beagle* and a recent biography, J. and M. Gribbin, *FitzRoy* (London: Hoddart, 2003). On the redesign of the *Beagle*, see L. Darling, *HMS Beagle* (1984).
46. See Volume 3, p. 367.
47. See Volume 4, p. 142.
48. A. Earle, *Narrative of a Nine Months' Residence in New Zealand* (London: Longman, Rees, Orme, Brown, Green and Longman, 1832); Stock, *Church Missionary Society* (Unknown Journal, ca. 1888); Brown, *Providence and Empire* (London: Longman, 2008).
49. Fitzroy and Darwin 'A Letter Containing Remarks on the Moral State' (1836).
50. See Parliamentary Papers, *Report ... on the Aborigines* (1836); *Report ... of an Enquiry into the Present State of the islands of New Zealand* (1837–8).
51. See Volume 4, p. 149.
52. Fitzroy had given Darwin the first volume of Lyell's work; on their post-voyage meetings, see the biography by Browne, *Charles Darwin: Voyaging*.
53. Darwin to J. D. Hooker, 4 May 1865, Darwin Correspondence Project Database. http://www.darwinproject.ac.uk/entry-4827 (letter no. 4827; accessed 25 July 2011)
54. Darwin to J.S. Henslow, 15 July 1860, Darwin Correspondence Project Database. http://www.darwinproject.ac.uk/entry-2869 (letter no. 2869; accessed 25 July 2011)
55. See Browne, *Charles Darwin: Voyaging*, pp. 370–1.
56. Darwin to S.E. Darwin, 1 April 1838. Darwin Correspondence Project Database. http://www.darwinproject.ac.uk/entry-407 (letter no. 407; accessed 25 July 2011)
57. Darwin to William Whewell, 16 February, 1839. Darwin Correspondence Project Database. http://www.darwinproject.ac.uk/entry-496 (letter no. 496; accessed 25 July 2011)

58. Darwin to Susan Darwin 22 February 1842. Darwin told his sister Colburn billed him for presentation copies, and that only 1,337 copies had been sold. Darwin Correspondence Project Database. http://www.darwinproject.ac.uk/entry-621 (letter no. 621; accessed 25 July 2011). On the publisher and the travel narratives, see Sutherland, 'Henry Colburn' (1986).
59. *Quarterly Review*, 65 (1839) pp. 195–234, p. 212; other major reviews were *Athenaeum*, 605 (1 June 1839), pp. 403–5, 607 (15 June 1839), pp. 446–9 and *Edinburgh Review*, 69 (1839) pp. 467–93.
60. J. Hamilton, 'President's Anniversary Address to the Royal Geographical Society', *Journal of the Royal Geographical Society*, 9 (1839), pp. xlvv–lxxxvi, lxxxix.
61. e.g., Keynes, *The Beagle Record*, Taylor, *Voyage of the Beagle* (London: Conway, 2008).
62. Hamilton, 'President's Anniversary Address', p. lxxx.
63. CORRAL – UK Colonial Register and Royal Navy Logbooks and ACRE historical data recovery, imaging, digitization and visualizations, part of the British Atmospheric Data Centre (http://www.corral.org.uk).

BIBLIOGRAPHY AND FURTHER READING

Adams, P. G., *Travelers and Travel Liars* (Berkeley, CA: University of California Press, 1962).

A Diary of the Wreck of HMS Challenger on the Western Coast of South America in May 1835 (London: Lees, Orme, Brown, Green, Longman: 1836).

Allen, D. E., *The Naturalist in Britain: a Social History* (London: A. Lane, 1976).

Anderson, K., *Predicting the Weather: Victorians and the Science of Meteorology* (Chicago, IL: University of Chicago Press, 2005).

Armstrong, P., *Darwin's Desolate Islands: A Naturalist in the Falklands* (Chippenham: Picton, 1992).

Auwers, A., 'Some Remarks on a Chain of Meridian Distances Measured Around the Earth by H. M. S. *Beagle* in the Years 1831 to 1836', *Monthly Notices of the Royal Astronomical Society* 44 (1884), pp. 303–47.

Banner, S., *Possessing the Pacific: Land, Settlers and Indigenous People from Australia to Alaska* (Cambridge, MA: Harvard, 2007).

Ballantyne, T. (ed.), *Science, Empire and the European Exploration of the Pacific* (Burlington, VT and Aldershot, Hants, England: Ashgate, 2004).

Beer, G., 'Travelling the Other Way', in N. Jardine, J. A. Secord and E. C. Spray (eds), *The Cultures of Natural History* (Cambridge: Cambridge University Press, 1996).

Bollaert, W., *Antiquarian, ethnological and other researches in New Granada, Equador, Peru and Chile* (London: Trubner, 1860).

Bourguet, M. -N. and Licoppe, C. and Sibum, O., eds, *Instruments, Travel and Science: Itineraries of Precision from the Seventeenth to the Twentieth Century* (London, New York: Routledge, 2002).

Bourne, W. R. P., 'Fitzroy's Foxes and Darwin's Finches', *Archives of Natural History* 19 (1992): pp. 29–37.

Brown, S. *Providence and Empire Religion, Politics and Society in the United Kingdom, 1815–1914* (London: Longman, 2008).

Browne, J. *Charles Darwin: Voyaging* (Princeton, NJ: Princeton University Press, 1995).

Bud, R., and D. J. Warner (eds), *Instruments of Science: an Historical Encyclopedia* (London: The Science Museum, 1998).

Burkhardt, F., and S. Smith (eds), *The Correspondence of Charles Darwin, volume 1: 1821–1836* (Cambridge: Cambridge University Press, 1985), at http://www. darwinproject. ac. uk

— (eds), *The Correspondence of Charles Darwin, volume 2: 1837– 1843* (Cambridge: Cambridge University Press, 1986), at http://www.darwinproject.ac.uk

Caldcleugh, A., 'An Account of the great earthquake experienced in Chile on the 20th of February, 1835, with a map', *Philosophical Transactions of the Royal Society of London*, 126 (1836), pp. 21–6.

Chapman, A., *European Encounters with the Yamana People of Cape Horn, before and after Darwin* (New York: Cambridge University Press, 2010).

Coan, Titus, *Adventures in Patagonia: a Missionary's Exploring Trip* (New York: Dodd and Mead, 1880).

Cohen, M. 'Traveling Genres' *New Literary History*, 34 (2003), pp. 481–99.

Cooter, R., *The Cultural Meaning of Popular Science: Phrenology and the Organization of Consent in Nineteenth-Century Britain* (Cambridge: Cambridge University Press, 1984).

'Courts martial,' *Nautical Magazine*, 4 (1835), pp. 695–6.

Darling, L., 'H M S *Beagle* 1820–1870', *Sea History*, 31(1984), pp. 27–42.

Darwin, C., *Voyage of the Beagle: Charles Darwin's Journal of Researches*, ed. J. Browne and M. Neve (London and New York: Penguin Books, 1989).

Darwin, C., ed. *Zoology of the Voyage of HMS Beagle* (London: Smith Elder, 1838–1843).

Dawson, L. S., *Memoirs of Hydrography: Including Brief Biographies of the Principal Officers who have served in H. M. Naval Surveying Service between the Years 1750 and 1885* (Eastbourne: H. W. Keay, 1885).

Day, A., *The Admiralty Hydrographic Service 1795–1919* (London: Stationary Office, 1967).

Douglas, B. 'Climate to Crania: the Racialization of Human Difference' in Douglas, ed. Douglas, B. and C. Ballard, *Foreign Bodies: Oceania and the Science of Race 1750–1940* (Australian National University, 2008) at http://epress. anu. edu. au/foreign_bodies_ citation. html.

Driver, F. and Martins, L. 'Shipwreck and Salvage in the tropics: The Case of HMS Thetis 1830–1854', *Journal of Historical Geography* 32 (2006), pp. 539–562

Ellis, J., *Polynesian Researches During a Residence of Nearly Six Years in the South Sea Island*, 2 vols, (London Fisher, 1829).

Fitzroy, R. *Remarks on New Zealand* (London: W. and H. White, 1846).

—, 'Sketch of the Surveying Voyages of His Majesty's Ships *Adventure* and *Beagle*, 1825–1836', *Journal of the Royal Geographical Society of London*, 6 (1836), pp. 311–42.

—, *The Weather Book: A Manual of Practical Meteorology* (London: Longman, Green, Longman, Roberts, & Green, 1863).

Fitzroy, R. and Darwin, C. 'A Letter, Containing Remarks on the Moral State of Tahiti, New Zealand, &c', *South African Christian Recorder*, 2 (1836), pp. 221–38, at www.darwin-online.org.uk.

Fitzroy, R. and King, P. P., *South American Pilot: Sailing Directions for South America,* 2nd edn (London: Hydrographic Office 1848).

Friendly, A., *Beaufort of the Admiralty* (New York: Random House, 1977).

Gillispie, C. C., *Genesis and Geology* (Cambridge: Harvard University Press, 1951

Gribbon, J. and M., *Fitzroy: the Remarkable Story of Darwin's Captain and the Invention of the Weather Forecast* (London: Hoddart, 2003).

Herbert, S., 'Doing and Knowing: Charles Darwin and Other Travellers', in P. N. W. Jackson

(ed.), *Four Centuries of Geological Travel: The Search for Knowledge on Foot, Bicycle, Sledge and Camel* (London: Geological Society, 2007), pp. 311–23.

Herschel, J. F. W. *Preliminary Discourse on the Study of Natural Philosophy* (1830; Chicago, IL: University of Chicago Press, 1987).

Hordern, M., *King of the Australian Coast: the work of Phillip Parker King in the Mermaid and Bathhurst 1817–22* (South Carlton: Melbourne University Press, 2002).

Huler, S., *Defining the Wind: the Beaufort Scale, and how a Nineteenth-Century Admiral turned Science into Poetry* (New York: Three Rivers Press, 2004).

Hulme, P. and Youngs, T., *The Cambridge Companion to Travel Writing* (Cambridge: Cambridge University Press, 2002).

Humboldt, A., 'On the Longitude of Valparaiso and Callao', *Journal of the Royal Geographical Society of London*, 9 (1839), pp. 502–6

Kartunnen, F., 'Interpreters Snatched from the Shore: The Successful and the Others', in E. Gray and N. Fiering, *The Language Encounter in the Americas: 1492–1800* (New York: Berghahn, 2000), pp. 215–29.

Keynes, R. D., *The Beagle Record: Selections from the Original Pictorial Records and Written*

Accounts of the Voyage of the H. M. S. Beagle (Cambridge: Cambridge University Press, 1979).

—, *Charles Darwin's Beagle Diary* (Cambridge: Cambridge University Press, 1988).

King, P. P., *Narrative of a Survey of the Intertropical and Western Coasts of Australia performed between the years 1818 and 1822*, 2 vols (London: John Murray, 1827).

King, P. P., *Sailing directions for the coasts of Eastern and Western Patagonia, from Port St.Elena on the East side of Cape Tres Montes on the West Side; including the Strait of Magalhaens and the sea coast of Terra del Fuego* (London: Hydrographic Office, 1832).

King, P. P., 'Some Observations upon the Geography of the Southern Extremity of South America, Tierra del Fuego and the Strait of Magalhens', *Journal of the Royal Geographic Society of London*, 1 (1832), pp. 155–75.

Lamb, J. *Exploration and Exchange: British and American Narratives of the Pacific 1680–1900* (Chicago, IL: University of Chicago Press, 2000).

Lamb, J., 'Minute Particulars', in Ballantyne, T. (ed.), *Science, Empire and the European Exploration of the Pacific* (Burlington, VT: Ashgate, 2004), pp. 345–58.

Lawrence, C., 'Disciplining Disease: Scurvy, the Navy and Imperial Expansion, 1750–1825', in D. Miller and P. Reill (eds), *Visions of Empire: Voyages, Botany and Representations of Nature* (Cambridge: Cambridge University Press, 1996).

Lyell, C., *Principles of Geology*, ed. with an introduction by J. Secord (Harmondsworth: Penguin, 1997).

McEwan, C., L. A. Borrero and A. Prieto (eds), *Patagonia: Natural History, Prehistory and Ethnography at the Uttermost End of the Earth* (Princeton, NJ: Princeton University Press, 1997).

Mellersh, H. E. L., *Fitzroy of the Beagle* (London: R. Hart-Davies, 1968).

Prichard, J. C., *The Natural History of Man: Comprising Inquiries in to the Modifying Influence of Physical and Moral Agencies*, 2 vols, 4th edn (London: Bailliere 1855).

Raper, H. and Fitzroy, R., 'Hints to Travellers', *Journal of the Royal Geographical Society* 24 (1854), pp. 328–58.

Reidy, M., *Tides of History: Ocean Science and Her Majesty's Navy* (Chicago, IL: University of Chicago, 2008).

Report from the Select Committee on the Aborigines (British settlements) together with Minutes of Evidence House of Commons Papers 1836 (58).

Report from the Select Committee of the House of Lords Appointed to Inquire into the Present state of the Islands of New Zealand 1837–8 (680).

Ritchie, G. S. *The Admiralty Chart British Naval Hydrography in the Nineteenth Century* (New York: Elsevier, 1967).

Roussin, A. R., *Le Pilote du Brésil, ou description des côtes de l'Amérique méridionale* (Paris: Imprimerie Royale, 1826–7).

Rudwick, M. J. S., *Bursting the Limits of Time: The Reconstruction of Geohistory in the Age of Revolution* (Chicago, IL: Chicago University Press, 2005).

Safier, N. *Measuring the New World: Enlightenment Science and South America* (Chicago, IL: University of Chicago Press, 2008).

Sands, J. O., 'The Sailor's Perspective: British Naval Topographic Artists', in D. Howse (ed.), *Background to Discovery: Pacific Exploration from Dampier to Cook* (Berkeley, CA: University of California Press, 1990).

Schaffer, S., 'When the Stars Threw Down Their Spears: Histories of Astronomy and Empire. ' Tarner Lectures, Trinity College, University of Cambridge, 2010, at http://www. trin. cam. ac. uk/index. php?pageid=243

Schmitt, C., *Darwin and the Memory of the Human: Evolution, Savages, and South America* (Cambridge: Cambridge University Press, 2009).

Secord, J. (ed.), *Evolutionary Writings* (Oxford: Oxford University Press, 2008).

Secord, J., 'Scrapbook Science: Composite Portraits in Late Georgian England', in B. Lightman and A. Shteir (eds), *Figuring It Out: Science, Gender and Visual Culture* (Lebanon, NH: Dartmouth College Press, 2006), pp. 164–91.

Sivasundaram, S., et al., 'Global Histories of Science', *Isis*, 101 (2010), pp. 95–158.

Smith, B., *European Vision and the South Pacific, 1768–1850: A Study in the History of Art and Ideas* (Oxford: Clarendon Press, 1960).

Sorrenson, R., 'The Ship as a Scientific Instrument', *Osiris* 11(1996), pp. 221–36.

Stanbury, D., *A Narrative of the Voyage of HMS Beagle being Passages from the 'Narrative' written by Captain FitzRoy selected and edited by David Stanbury* (London: The Folio Society, 1977).

Starbuck, A., *History of the American Whale Fishery From its Earliest Inception to the Year 1876* (Washington: Government Printing Office, 1878).

Stock, E., *The History of the Church Missionary Society: its Environment, its Men and its Work*, 3 vols (London: Church Missionary Society, 1899–1916).

Stoddart, D. R., 'Darwin, Lyell, and the Geölogical Significance of Coral Reefs', *British Journal for the History of Science*, 9 (1976), pp. 199–218.

Sulloway, F. J., 'Darwin's Early Intellectual Development: An Overview of the *Beagle* Voyage (1831–1836)', in D. Kohn (ed.), *The Darwinian heritage* (Princeton, NJ: Princeton University Press, 1985), pp. 121–54.

Sutherland, J., 'Henry Colburn, Publisher', *Publishing History*, 19 (1986), pp. 59–84.

Taylor, J., *Voyage of the Beagle: Darwin's Extraordinary Adventure Aboard FitzRoy's Famous Survey Ship* (London: Conway, 2008).

Taussig, M., 'Tierra del Fuego – Land of Fire, Land of Mimicry', in McEwan, C., L. A. Borrero and A. Prieto (eds), *Patagonia: Natural History, Prehistory and Ethnography at the Uttermost End of the Earth* (Princeton, NJ: Princeton University Press, 1997), pp. 153–72.

Turner, G., *Nineteenth Century Scietific Instruments* (Berkeley, CA: University of California Press 1983).

Vale, B., *Independence or Death! British Sailors and Brazilian Independence 1822–25* (London: British Academic Press, 1996).

Wyhe, J., van (ed.), *The Complete Work of Charles Darwin Online* (2002), at

http://www.darwin-online. org. uk/ [last updated 10 June 2011].

Whye, J., van, *Phrenology and the Origins of Victorian Scientific Naturalism* (Burlington, VT and Aldershot: Ashgate, 2004).

Young, V. (ed.), *The Journal of Syms Covington* (1995), at http://www.asap.unimelb.edu.au/bsparcs/covingto/chap_1.htm [accessed 9 June 2007].

LIST OF SHIPS' COMPANIES AND PASSENGERS

King and Fitzroy refer often by surname to members of the ships' officers. Their identities and subsequent histories, where known, are summarized below rather than in the endnotes.

The First Voyage 1826–30

HMS *Adventure*

Ainsworth, B.	Master; drowned in February 1827.
Anderson, James	1797–1842. Botanical collector, born in Scotland, died in Sydney. He had collected in South America and Africa before joining the *Adventure*. Became assistant superintendent, then superintendent to the Botanic Garden at Sydney, 1835–42.
Brand, J. F.	Mate. Moved to *Beagle* to replace Skyring after Stokes's death. Returned to England after December 1828.
Cooke, John	First lieutenant, appointed 1824. Invalided home in June 1827. Did not serve again after this voyage. Died in 1846.
Graves, Thomas	Mate and Assistant Surveyor. Entered navy 1816. Joined *Adventure* in 1827. Later worked on survey of Mediterranean (1832 to 1836). Became a captain 1846.
Harrison, George	Mate.
Hodgskin, G.	Clerk.
King, Philip Gidley.	Capt. King's son. See below.
Mellersh, Arthur	Volunteer First Class. See below.
Millar, Alexander	Master's Assistant. Moved to the *Beagle* as Master after Stokes' death. Returned to *Adventure* in 1828. Died in Tierra del Fuego February 1830.
Mitchell, Thomas	Second lieutenant. Entered navy in 1811. Served in anti-piracy mission in Caribbean in the early 1820s. Made lieutenant in 1824. Joined *Adventure* in June 1829. Later served on surveys with Edward Belcher in Africa and Mediterranean. Became commander in 1842.

Park, J.	Assistant Surgeon.
Rowlett, George	Purser. See below.
Russell, J.	Volunteer Second class.
Sholl, Robert H.	Midshipman. Became Lieutenant on the *Beagle* after Hawes left in September 1826. Died at Port San Julian in in January 1828.
Tarn, John	Surgeon. Appointed a naval surgeon in 1818; later served on *Donegal*.
Wickham, J. C.	Mate. See below, second *Beagle* voyage.
Williams, E.	Second master; he became acting master upon Ainsworth's death, then returned to his former position.
Wilson, William Watts	1804–45. Entered navy in 1821. Midshipman. Became Mate after December 1828 and lieutenant in 1841. Did much of the sketching for the charts on the first voyage. Died in service in Africa.

HMS *Beagle*

Atrill, J.	Purser. Left *Beagle* by end of 1827.
'Boat Memory'	Fuegian (Yamana/Alikhoolip) estimated aged twenty. Died in Plymouth of smallpox, 1830.
Bennet, James	Coxswain. See below, second *Beagle* voyage.
Bowen, Evan	Surgeon. Left *Beagle* July 1828 .
Braily, D.	Joined *Beagle* by end of the 1827 as volunteer second class.
Bynoe, B.	Assistant surgeon. See below, second *Beagle* voyage.
Flinn, S. S.	Master. Became Master of the *Adventure* after Ainsworth's death in February 1827. Moved to the *Adventure* after Stokes's death.
Hawes, Edward	Lieutenant. Left *Beagle in* September 1826. Served subsequently in Africa, Mediterranean and China. Ended career as superintendent of packet service at Portpatrick, 1844–51.
Jones, W.	Volunteer second class. Left *Beagle* by end of 1827.
Kempe, J.	Lieutenant after Dec 1827.
Kirke, J.	Mate.
Lunie, R. F.	Volunteer first class.
MacDouall, J.	Clerk. Left *Beagle* by end of 1827.
May, Jonathan	Carpenter. See below, second *Beagle* voyage.
Megget, J.	Clerk.
Mogg, W.	Acting purser. Joined *Beagle* to replace Atrill, remained acting purser for the rest of the first *Beagle* voyage.
Murray, James	Master. Joined with *Beagle* in 1828 with Fitzroy.

Skyring, William George	Lieutenant and assistant surveyor. Acting commander after Stokes's death, then returned to his former position when Fitzroy was appointed.
Stokes, John Lorkes	Midshipman. See below, second *Beagle* voyage.
Stokes, Pringle	Appointed commander and surveyor in 1826. Committed suicide in August 1828.
Wilson, John	Surgeon. Appointed after Stokes's death.

The Second Voyage 1831–6

HMS *Beagle*

Bennett, James	Coxswain on first and second voyages.
Bynoe, Benjamin	1804–65. Born in Barbados. Qualified as a surgeon in London 1825. Appointed to *Beagle* in 1825 as assistant surgeon. Became acting surgeon of the *Beagle* in 1827 after Bowen fell ill. Remained acting surgeon until 1828. Collected geological and zoological specimens. On second voyage, became acting surgeon after the departure of McCormack* to England in April 1832. Promoted to surgeon on Fitzroy's recommendation at the end of the second voyage; joined *Beagle*'s Australian surveying under Wickham and Stokes on which he was the principal natural history collector. He later served as surgeon to convict ships, and assisted in the epidemics following the Irish famine. He retired in 1863.
Chaffers, Edward Main	Master.
Covington, Syms	1816?–61. Ship's boy; became Darwin's servant in 1833; stayed on with Darwin as assistant and servant until 1839, then emigrated to Australia.
Darwin, Charles	1809–82. Gentleman naturalist, companion to Fitzroy.
Derbishire, Alexander	Mate. Left voyage at Rio de Janeiro in 1832.
Dring, John Edward	Acting purser (after death of Rowlett).
Earle, Augustus	1793–1838. Artist and draughtsman, left *Beagle* at Monte Video due to ill-health in 1832. Lived several months in Monte Video before returning to England. Widely travelled, and a stepbrother to the renowned hydrographer William Smyth.

'Fuegia Basket'	Fuegian (Yamana/Alikhoolip), named Yokcushlu, estimated aged nine in 1830. Married 'York Minster'.
Forsyth, Charles C.	Midshipman. Entered navy 1826; took his lieutenant's examination at end of the voyage in 1836; promoted 1843. Served as mate on later *Beagle* surveys of Australia under Wickham and Stokes. In 1844, stationed at Cape of Good Hope.
Fuller, Harry	Fitzroy's steward or personal servant.
Hamond, Robert Nicholas	1809–83. Lieutenant. Joined *Beagle* temporarily in November 1832 from *Druid* August 1832 to May 1833, then returned to England.
Hellyer, Edward H.	Clerk. Drowned in 1833 in Falklands.
'Jemmy Button'	Fuegian (Yamana/Tekeenica) named Orundellico, estimated age fourteen in 1830. *c.* 1815–64. He had contact with an English mission (South American Missionary Society) to the area, 1855 to 1859; and was accused of involvement in an attack on this group in late 1859.
Johnson, Charles R.	d. 1882. Midshipman, later mate. Joined navy, 1826. Joined *Beagle* in 1832. Became lieutenant in 1840. Commander of a steamer, *Comet*, in 1846; then of *Driver* in Pacific 1848 to 1852; made captain in 1856. Retired in 1864.
Kent, William	Assistant surgeon. Joined *Beagle* in September 1833
King, Philip Gidley	1817–1904. Oldest son of Capt. P. P. King. Went on both voyages. Accompanied his father in *Adventure*, 1826–30, then midshipman, 1831–6. Left *Beagle* to rejoin his family in Australia in 1836. Became an agriculturalist and estate manager; he and Wickham married sisters.
MacCormick, Robert	1800–90. Surgeon; left *Beagle* and returned to England in April 1832. Had been on voyage to Spitsbergen in 1827 with William Parry and later joined another Arctic expedition, with James Clark Ross, in 1839 to 1843. MacCormick had had yellow fever in 1825 and was invalided home from other voyages besides the *Beagle*; he was an unpopular shipmate.

Martens, Conrad	1801–78. Joined the *Beagle* as artist and draughtsman in place of Earle in 1832 in Monte Video, left *Beagle* at Valpairaso in 1834, settled in Australia.
Matthews, Richard	1811–1893. Son of a master weaver in Oxfordshire. Church Missionary Society missionary, settled in New Zealand in 1835.
May, Jonathan	Carpenter on both *Beagle* voyages.
Mellersh, Arthur	1812–94. Midshipman, became Mate.
Musters, Charles	Volunteer first class, son of John Musters of Colwich Hall, Nottinghamshire; died of fever in 1832 near Rio de Janeiro after a snipe shooting expedition up the Macao River.
Rowlett, George	Purser. Served on both voyages, in *Adventure* in 1826–30. Died in June 1834, buried at sea.
Sorrell, Thomas	Acting Boatswain.
Stebbing, George James	Fitzroy's assistant, in charge of chronometers, son of Portsmouth instrument-maker.
Stewart, Peter Benson	d. 1864. Mate on *Beagle*. Lieutenant, 1837, serving Australia and East Indies.
Stokes, John Lort	1812–85. On first and second voyage. Mate and assistant surveyor. No relation to Captain Stokes. Made Lieutenant in 1837, and joined *Beagle* again to survey in Australia under Wickham, succeeded Wickham as commander in 1841–3. Published an account of his Australian and New Zealand voyages in 1846. Made captain 1847, led survey of New Zealand in *Acheron*, 1847–51, and ended his career as admiral in 1877.
Sulivan, Bartholomew James	1810–90. Second Lieutenant and surveyor, served on both voyages. Son of a naval officer. Attended Royal Naval College, in 1823; served in *Thetis* 1824–28 where he met Fitzroy. Joined *Beagle* when Fitzroy was appointed, and left in 1829 to take lieutenant's examination in England. Second lieutenant in voyage of 1831–6. Surveyed in Falklands in 1838–9, made commander in 1841 and continued South American service in *Philomel* in 1842–6. Carried out distinguished surveying work in the Crimean war and became naval officer in charge of the marine department of the Board of Trade in 1856, retiring in 1865. Knighted, 1869.

Usborne, Alexander Burns	d. 1885. Master's assistant, then master. Surveyed west coast of Peru in *Constitucion* 1836 and returned to England separately from *Beagle*. Master on *Beagle* in Australia, 1838–9; surveyed British coasts, 1847–65. Became a commander in 1863, captain, 1867.
Wickham, John Clements	1798–1864. Entered the navy in 1812. Appointed mate on *Adventure* in 1826. Became lieutenant after Cooke left the *Adventure* in June 1827. First lieutenant on the second *Beagle* voyage. Appointed commander of the next *Beagle* voyage, surveying Australian coast 1837–41 and left due to ill-health. Emigrated to Australia, 1842 and married the sister of his *Beagle* shipmate P. G. King's wife. Became magistrate and government official at Moreton Bay, but retired to Biarritz, France in 1860.
'York Minster'	Fuegian (Yamana/Alikhoolip), named El'leparu, age estimated at twenty-six in 1830. Married 'Fuegia Basket.'

CHRONOLOGY OF THE FIRST VOYAGE

The following outline is based on the dates in *Narrative*. When looking for particular dates, it is helpful to know that the practice of King and Fitzroy was to give the year and month only for the first mention of each, and thereafter only the date of the month, i.e. for a reference to events on 1 January 1834, the text would indicate' 1 January 1834', but for the rest of the month, simply 'the 15th' until the first mention of an event in February, which would be 'the 3rd February' and thereafter 'the 6th', etc.

22 May 1826	*Adventure* and *Beagle* depart from Plymouth.
November 1826	Ships leave Monte Video to commence survey in Strait of Magellan.
December 1826–April 1827	Survey work in strait, with the three ships (*Beagle, Adventure, Hope*) in different parts of the Strait.
March 1827	Beagle rescues Capt. Brisbane and crew from shipwreck after over three months marooned.
April–May 1827	First meetings with the Telhuelche chief Maria and her people in Gregory Bay .
May–December 1827	Leave Strait, return to Rio, refitting and waiting for permission from London for purchase of *Adelaide*.
January 1828	All ships back to Strait, again surveying in different regions.
March–August 1828	Although *Adelaide* and *Adventure* are in some contact, in March the *Beagle* heads to western part and is not seen again until August.
June–July 1828	Winter; scurvy begins to appear among all crews. *Adventure* awaits *Beagle* at Port Famine.
August 1828	*Beagle* finally returns from western survey.
1–12 August 1828	Stokes shoots himself, and dies slowly. Lieutenant Skyring becomes acting commander. *Adventure, Beagle* and *Adelaide* return to Rio.
October 1828	Fitzroy is appointed to *Beagle*.
January 1829	Dangerous storm ('pampero') at Maldonado on 30 January.
March 1829	All ships leave again for the Strait.
April–May 1829	*Adventure* joins *Chanticleer* for magnetic observations.

June–July 1829	*Adventure* heads for Valparaiso and Santiago. Beagle continues surveys in Otway and Skyring Water, hoping to find western passage.
August–September 1829	Ships are in Chiloe region (San Carlos), surveying and refitting.
November 1829	*Beagle* leaves Chiloe for the southern coast of Tierra del Fuego alone until August 1830.
December–April 1830	*Adelaide* surveys in Gulf of Penas and Strait.
January–February 1830	*Adventure* in Valparaiso then heads to Strait.
May 1830	*Adelaide* and *Adventure* meet at Port Famine and sail together to Monte Video and Rio.
November 1829–May 1830	*Beagle* surveying in Tierra del Fuego. Loss of one boat, theft of another. Hostages seized. Beagle Channel discovered.
June 1830	*Beagle* leaves Strait for Rio.
14 October 1830	*Beagle* and *Adventure* arrive back in Plymouth and Woolwich.

CHRONOLOGY OF THE SECOND VOYAGE

The following chronological outline of the second voyage is based on the *Narrative*. See above, *Chronology of First Voyage,* for dating in the text. This outline can be compared to the itineraries of Darwin's travels by Nora Barlow (ed.), *Darwin and Henslow: The Growth of an Idea* (London: Bentham-Moxon Trust, John Murray 1967; also available in transcription, darwin-online.org.uk) and by the chronological register of Darwin's field notebooks by Kees Rookmaaker. (Chronological register. In Gordon Chancellor and John van Wyhe (eds) with the assistance of Kees Rookmaaker. 2009. *Charles Darwin's notebooks from the voyage of the Beagle* (Cambridge: University Press), pp. 570–82.) Darwin, however, spent much time off the *Beagle*, spending an estimated 34 per cent of his nights on land, so his itinerary is not a guide to the *Beagle*'s voyage.

27 December 1831	*Beagle* leaves England.
January 1832	Visit to Cape Verde Islands and St Paul Rocks.
February–July 1832	At Bahia (San Salvador) and Rio de Janeiro. Survey work checking existing charts. Fitzroy returns to Bahia to verify measurement of longitude. The young volunteer Musters dies of fever (malaria) in May. *Beagle* holds a regatta with rest of British fleet in Rio. The unpopular surgeon McCormick leaves *Beagle* to return to England.
July–August 1832	*Beagle* heads to Monte Video. Her visit to Buenos Aires is prevented on 2 August; the mutiny in Monte Video follows; *Beagle* marines asked to assist the local authorities.
August–October 1832	Surveying work on the eastern coast of Argentinian pampas. First visit to Bahia Blanca (Blanco Bay), south of the Rio Plata; first meeting with Harris (pilot and merchant). Darwin, Fitzroy and Harris make shore excursion and are taken for spies. Fossil find of *Megatherium* and others at Punta Alta in September. Fitzroy decides to rent vessels from Harris: *La Paz* and *Liebre*. The artist Earle leaves *Beagle* due to illness and is replaced by Conrad Martens.

October–November 1832	*Beagle* returns to Buenos Aires and Monte Video, leaving *La Paz* and *Liebre* to continue working. In November Lieut. Robert Hamond joins *Beagle.*
December 1832–February 1833	Left Monte Video, briefly meeting with *La Paz* and *Liebre*, then *Beagle* continues south for first survey season in Tierra del Fuego. Darwin's first sight of inhabitants of region in mid-December.
January 1833	The *Beagle* weathers a severe gale, which has caused many wrecks in the area, including La Magellan in Falkland Islands. Settlement at Woollya Sound established, but Matthews returns of board a few weeks later. The *Beagle* takes leave of the three re-settled Fuegians in February.
February–April 1833	First visit to the Falkland Islands. Fitzroy decides to purchase *Unicorn*, renamed *Adventure*, from the merchant captain Mr Low. Edward Hellyer is drowned in March. Fitzroy meets Brisbane, the explorer and whaler.
April–July 1833	At Maldonado and Monte Video, restocking *Beagle* and re-fitting the schooner *Adventure*. Fitzroy accepts brief loan of a small schooner to keep the survey work in progress while re-fitting. Hamond decides to return to England.
July–October 1833	The Beagle heads to River Negro region, engaged in soundings in bays and dangerous banks there. *La Paz* and *Liebre* contract ends, and the *Beagle* takes leave of these vessels in August. Second visit to Bahia Blanco. In August and September Darwin travelled overland to Buenos Aires, with a second long excursion near Montevideo in November. The *Beagle* continues close survey of shoals near Cape Corrientes.
August 1833	Rebellion in Port Louis, Falkland Island, leaves town destroyed; *Challenger* arrives and takes possession of disputed islands again for Britain in January 1834.
October 1833–November 1834	The *Beagle* returns to Maldonado and Monte Video for stores, chart-work, and fitting out the *Adventure*. They collect Darwin; stir up political tension about British 'settlement' while taking measurements on island of Gorriti.
December 1833	*Beagle* and *Adventure* leave for a second season in Tierra del Fuego.

January 1834	*Beagle* strikes a rock and is damaged. Fitzroy, Darwin and others make an unsuccessful search for fresh water near Port San Julian. *Adventure* makes for the Falkland Islands at the end of the month while the *Beagle* continues work in the Strait. They meet Maria in Gregory Bay.
March 1834	*Beagle* visits Woollya, and the officers learn of York Minster and Fuegia Basket's treachery towards Jemmy Button. They exchange gifts and leave him for the last time.
March–April 1834	The *Beagle* makes a second visit to survey in Falkland Islands; they learn of the massacre at Port Louis and Brisbane's death. Mr Low, escaping from the violence, joins the *Adventure* in February.
April–May 1834	Expedition up Santa Cruz River, hoping to reach Andes.
May–June 1834	Survey through Strait of Magellan. The *Beagle* rejoins *Adventure* and heads for Chiloe. In June Rowlett dies at sea. Low is commissioned to make a short independent survey with local help in a *Beagle* whaleboat.
July–November 1834	*Beagle* heads to Valparaiso and stays there several months. Darwin made overland excursions; Martens left the *Beagle*. Fitzroy became ill working on his 'arrears' of charts. He receives a reprimand regarding the purchase of *Adventure,* and sells her. Darwin is laid up with an unknown illness for several weeks in October–November as well.
November–February 1835	Survey work in Chiloe, the Chonos archipelago and coast of Chile. The *Beagle* picks up some deserters from a whaling vessel who had been stranded several months. Darwin makes trips to the interior of the island.
February 1835	*Beagle* is near Valdivia. Earthquake centred northward in Concepcion occurs 20 February.
Mar-April 1835	Except for a brief visit to Valparaiso for anchors, the *Beagle* remains in and near Concepcion until mid-April, touring the damaged city and harbour.
April 1835	Survey work on coast of Chile near river Maule. The *Beagle* meets Darwin briefly at Valparaiso while collecting mail. Fitzroy learns of his promotion to post-captain.
May 1835	Anchored near Coquimbo. Arranges to borrow *Constitucion* in order to complete survey work on coast of Chile.

June–July 1835	*Beagle* sails to Valparaiso, *Constitucion* carries on under Liuet. Sulivan. Fitzroy now learns of wreck of the *Challenger* nearly a month earlier (19 May) and throws himself into rescue arrangements in association with HMS *Blonde*. Fitzroy is occupied with this event until July 10; having sent *Beagle* off to continue the survey and collect Darwin, he rejoins it on 9 August.
August 1835	Fitzroy purchases *Constitucion* and decides to leave it in the charge of Usborne to complete the coastal work in Peru and Chile and sail back east to England separately from the *Beagle*.
September 1835	*Beagle* leaves for the Galapagos.
September–October 1835	*Beagle* in Galapagos Islands.
November–December 1835	*Beagle* in Tahiti; Fitzroy carries out diplomatic work, meeting with the Queen and her council.
December 1835–February 1836	*Beagle* briefly visits New Zealand, Australia and Van Dieman's Land (Tasmania). The young midshipman King leaves the *Beagle* to remain with his father, Capt King, in Australia.
April 1832	*Beagle* visits Cocos-Keeling Islands.
April–June 1836	*Beagle* continues west across Pacific with brief stops in Mauritius and Cape of Good Hope.
July 1836	*Beagle* visits St Helena, Ascension islands, then to Bahia, Brazil
August–September 1836	*Beagle* is on the homeward leg from Brazil via Cape Verde and Azores.
2 October 1836	*Beagle* arrives in Falmouth, England.

GUIDE TO AUTHORITIES

This list gives the works mentioned frequently by King and Fitzroy. * indicates that the works were part of the *Beagle* library (1831–6). In the case of a voyage, the parentheses give the name of the ship, if King and Fitzroy used the name of the ship as a shorthand.

Agüeros	Agüeros, P. G. de, *Descripción historial de la provincial y archipiélago de Chiloé en el reyno de Chile y obispado de la Concepción* (Madrid: B. Cano, 1791).
*Beechey (*Blossom*)	Beechey, F., *Narrative of a Voyage to the Pacific and Bering's Strait, to Co-operate with the Polar Expeditions Performed in His Majesty's Ship Blossom, in the Years 1825, 26, 27, 28*, 2 vols (London: Henry Colburn and Richard Bentley, 1831).
*Blainville	Blainville, H.M.D. de, 'Rapport sur les resultants scientifiques du voyage de M. Alcide d'Orbigny ... par MM. De Blainville, Brongniart, Savary, Cordier,' *Nouvelles annales du Muséum National d'Histoirs Naturelle*, 3 (1834), pp. 84–115.
*Bory de Saint-Vincent	Bory de Saint-Vincent, J. B. G. M. (ed.), *Dictionnaire classique d'histoire naturelle*, 17 vols (Paris: Rey, Gravier; Baudoin Frères, 1822–31).
*Bougainville	(*La Bouduese, L'Etoile*) Bougainville, L.A. de, Count, *A Voyage Round the World: Performed by order of His Most Christian Majesty, in the years 1766, 1767, 1768, and 1769* translated from the French by John Reinhold Forster (London: Printed for J. Nourse and T. Davies, 1772).
Burney	Burney, J., *A Chronological history of Voyages and Discoveries in the South Sea or Pacific Ocean*, 5 vols (London: Hansard, 1803–17).
Byron (*Wager*)	Byron, J., *Byron's narrative of the loss of the Wager: with an Account of the Great Distresses Suffered by himself and his Companions on the Coast of Patagonia from the Year 1740 till their Arrival in England 1746* (1786; London: H. Leggatt, 1832).

*Caldcleugh	Caldcleugh, A., *Travels in South America, during the Years 1819–20–21: containing an Account of the Present State of Brazil, Buenos Ayres, and Chile*, 2 vols (London: John Murray, 1825).
Candolle	Candolle, A. P. de, 'Géographie botanique', in G. Cuvier (ed.), *Dictionnaire des Sciences Naturelles*, 18 (1832), pp. 359–422.
*Colnett	Colnett, J., *A Voyage to the South Atlantic and Round Cape Horn* ... London: W. Bennett, 1798).
Cook (*Endeavour, Resolution)*	See Hawkesworth (Cook's first voyage in *Endeavour)*; Forster or Douglas (Cook's second voyage, in *Resolution*) and Wallis (a compendium of several voyages covering the first of Cook's).
Cordova (*Santa Maria*)	Cordova, A., *A Voyage of Discovery to the Strait of Magellan with an Account of the Manners and Customs of the Inhabitants and of the Natural Productions of Patagonia*, translated from the Spanish (London: Richard Phillips, 1820).
*Cuvier	Cuvier, G., *The animal kingdom arranged in conformity with its organization ... with additional descriptions of all the species hitherto named, and of many not before noticed. By Edward Griffith and others. Supplementary volume on the fossils* (London: Whittaker, Treacher and Co., 1830).
D'Azara	Azara, F. d', *Essais sur l'histoire naturelle des quadrupèdes de la Province du Paraguay écrits depuis 1783, jusqu'en 1796, avec une appendice sur quelques reptiles; et formant suite nécessaire aux Oeuvres de Buffon. Traduits sur le manuscrit inédit de l'Auteur, par M. L. E. Moreau-Saint-Méry* (Paris: Imprimerie de C. Pougens, Librairie de Madame Huzard, 1801).
Douglas (Cook, *Resolution*)	Douglas, J. (ed.), *A Voyage towards the South Pole and Round the World...by Captain James Cook*, 2 vols (London, Strahan and Cadell, 1777).
*Duperrey (*La Coquille*)	Duperrey, L. I., *Voyage autour du monde ... sur la Corvette de sa Majesté, La Coquille, pendant les années 1822, 1823, 1824 et 1825. Zoologie, par MM. Lesson et Garnot*, 3 vols (Paris: Arthus Bertrand,1826–30).
*Ellis	Ellis, W., *Polynesian Researches, during a Residence of Nearly Six Years in the South Sea Islands, including Descriptions of the Natural History and Scenery of the Islands, with Remarks on the History, Mythology, Traditions, Government, Arts, Manners, and Customs of the Inhabitants*, 2 vols (London: Fisher, Son, & Jackson, 1829).
*Encyc. Brit.	*Encyclopaedia Britannica*, 20 vols (6th edn).

*Falkner	Falkner, T., *A Description of Patagonia and the Adjoining parts of South America* (Hereford: Printed by C. Pugh, 1774).
Febres	Febres, A., *Arte de la lengua general del reyno de Chile, con un dialogo chileno-hispano muy curioso* (Lima: En la calle de la Encarnaçion, 1765).
Forster (Cook, *Resolution*)	Forster, G., *A Voyage Round the World...commanded y Capt. James Cook during the Years 1772, 3, 4, 5* (London: White, 1777).
Foster	See Webster (*Chanticleer*).
*Frézier	Frézier, A.-F., *A Voyage to the South-Sea, and Along the Coasts of Chili and Peru, in the Years 1712, 1713, and 1714. Particularly describing the Genius and Constitution of the Inhabitants, as well Indians as Spaniards: their Customs and Manners; their Natural History, Mines, Commodities, Traffick with Europe, &c.* (London: Printed for J. Bowyer, 1717).
Funes	Funes, G., *Ensayo de la historia civil del Paraguay, Buenos Aires y Tucuman* (Buenos Ayres: M. J. Gandarillas, 1816–17).
*Gay	Gay, C., 'Aperçu sur les recherches d'histoire naturelle faites dans l'Amérique du Sud, et principalement dans le Chili, pendant les années 1830 et 1831', *Annales des Sciences Naturelles*, 28 (1833), pp. 369–93.
Gay	Gay, C., 'Tremblement de terre de Chili', *Comptes Rendus hebdomadaires de l'Academie des Sciences*, 6 (1838), p. 833.
*Hall	Hall, B., *Extracts from a Journal written on the coast of Chili, Peru, and Mexico, in the years 1820, 1821, 1822* (Edinburgh: Printed for Archibald Constable; London : Hurst, Robinson & Co., 1824).
*Hawkesworth (*Dolphin, Swallow, Endeavour*)	Hawkesworth, J. (ed.), *An Account of the Voyages Undertaken ... for Making Discoveries in the Southern Hemisphere ... Performed by Commodore Byron, Captain Wallis, Captain Carteret, and Captain Cook in the Dolphin, the Swallow, and the Endeavour drawn from the Journals which were kept by the Several Commanders and the Papers of Joseph Banks*, 3 vols (London: Strahan and Cadell, 1773).
Horsburgh	Horsburgh, J., *India Directory, or, Directions for Sailing to and from the East Indies, China, New Holland, Cape of Good Hope, Brazil, and the Interjacent ports: compiled chiefly from original journals at the East India House, and from observations and remarks made during twenty-one years experience navigating in those seas*, 3rd edn (London: Parbury, Allen and Co., 1826).

*Humboldt	Humboldt, F. W. H. A. von, *Personal narrative of travels to the equinoctial regions of the New Continent, during the years 1799–1804. By Alexander de Humboldt, and Aimé Bonpland,... translated by Helen Maria Williams*, 7 vols (London: Longman, Hurst, Rees, Orme and Brown, 1814–29).
*Kotzebue	Kotzebue, O. von, *Voyage of Discovery in the South Sea, and to Behring's Straits, in Search of a North-East Passage: Undertaken in the Years 1815, 16, 17, and 18*, 3 vols (London: Printed for Sir Richard Phillips and Co., 1821).
*Krusenstern	Krusenstern, A. J. von (Krusenstern, I. F.), *Atlas de l'Océan Pacifique dressé par M. de Krusenstern* (St Petersburg: Publié par ordre de Sa Majesté impériale, 1824–7).
Lamarck	Lamarck, J. B. A. de M., Chevalier de, *Philosophie zoologique, ou exposition des considérations relatives à l'histoire naturelle des animaux, à la diversité de leur organisation et des facultés qu'ils en obtiennent*, 2 vols (Paris: Dentu, 1809).
*Lamarck	Lamarck, J.B.A. de M., Chevalier de, *Histoire naturelle des animaux sans vertèbres, présentant les caractères généraux et particuliers de ces animaux, leur distribution, leurs classes, leurs familles, leurs genres, et la citation des principales espèces qui s'yrapportent; précédée d'une introduction offrant la détermination des caractères essentiels de l'animal, sa distinction du végétal et des autres corps naturels, enfin, l'exposition des principes fondamentaux de la zoologie*, 7 vols (Paris: Verdière, 1815–22).
*La Pérouse (*Boussole, Astrolade*)	La Pérouse, J.F. de G. de, *A voyage round the world performed in the years 1785, 1786, 1787, and 1788, by the Boussole and Astrolabe, under the commande of J.F.G. de la Pérouse*, 2 vols, atlas (London: Printed by A. Hamilton, for G. G. and J. Robinson; J. Edwards; and T. Payne, 1798–9).
*MacIntosh	MacIntosh, J., *The History of England* (London: Longman, Rees, Orme, Brown & Green; John Taylor, 1830).
*Mawe	Mawe, J., *Travels in the Gold and Diamond Districts of Brazil; describing the Methods of Working the Mines, the Natural Productions, Agriculture, and Commerce, and the Customs and Manners of the Inhabitants: to which is Added a Brief Account of the Process of Amalgamation Practised in Peru and Chili* (London: Printed for Longman, Hurst, Rees, Orme, Brown, and Green, 1825).

*Miers	Miers, J., *Travels in Chile and La Plata, Including Accounts Respecting the Geography, Geology, Statistics, Government, Finances, Agriculture, Manners and Customs, and the Mining Operations in Chile* (London: Printed for Baldwin, Cradock, and Joy, 1826).
*Molina	Molina, J.I., *Compendio de la historia geografica natural y civil del Reyno de Chile* (Madrid: Don Antonio de Sancha, 1794–95), vol. 1.
*Narborough	Narborough, J., *An Account of Several Late Voyages and Discoveries to the South and North* (London: Smith, Walford, 1694).
D'Orbigny	D' Orbigny, A.C.V.D. , *Voyage dans l'Amérique méridionale: le Brésil, la République orientale de l'Uruguay, la République argentine, la Patagonie, la République du Chili, la République de Bolivia, la République du Pérou-executé pendant les années 1826, 1827, 1828, 1829, 1830, 1831, 1832, et 1833* (Paris: Bertrand; Strasbourg: Levrault, [1834]–47).
Parish	Parish, W., *Buenos Ayres, and the Provinces of the Rio de la Plata: their Present State, Trade, and Debt. With Some Account from Original Documents of the Progress of Geographical Discovery in those Parts of South America during the Last Sixty Years* (London: John Murray, 1839).
Pennant	Pennant, T., *Synopsis of Quadrupeds* (Chester: By J. Monk, 1771).
*Quoy (*Astrolabe*)	Quoy, J. R. C. and J. P. Gaimard, 'Observations zoologiques faites à bord de l'*Astrolabe*, en Mai 1826, dans le Détroit de Gilbraltar', *Annales des sciences naturelles*, 10 (1827), pp. 232–3.
Southey	Southey, R., *History of Brazil*, 3 vols (London: Longman, 1810–19).
Stevenson	Stevenson, W. B., *A Historical and Descriptive Narrative of Twenty Years' Residence in South America* (London: Hurst, Robinson, 1825).
*Turnbull	Turnbull, J., *A voyage round the world, in the years 1800, 1801, 1802, 1803, and 1804*, 2nd edn (London: A. Maxwell, 1813).

*Ulloa	Juan, G., and A. de Ulloa, *A Voyage to South America: Describing at Large the Spanish Cities, Towns, Provinces, &c. on that Extensive Continent. Undertaken, by Command of the King of Spain, by Don George Juan, and Don Antonio de Ulloa, captains of the Spanish navy. Translated ...with notes and observations; and an account of the Brazils by John Adams, Esq*, 2 vols, 4th edn (London: J. Stockdale, 1806).
*Webster (*Chanticleer*)	Webster, W. H. B., *Narrative of a Voyage to the Southern Atlantic Ocean, in the Years 1828, 29, 30, performed in H.M. Sloop Chanticleer* ... 2 vols (London: R. Bentley, 1834).
*Weddell	Weddell, J., *A Voyage towards the South Pole Performed in the Years 1822–24* (London: Longman, Hurst, Rees, Orme, Brown, and Green, 1825).
Wallis &Carteret (*Dolphin, Swallow*)	Wilson, R. (ed.), *Voyages of Discoveries Round the World; Successively Undertaken by the Hon. Commodore Byron, in 1764; Captains Wallis and Carteret, in 1766; and Captain Cook in the Years 1768 to 1789 Inclusive. The Whole Carefully Selected from the Journals of the Respective commanders*, 3 vols (London: James Cundee, 1806).
Zool. or Zool. Soc.	*Transactions of the Zoological Society of London*. 1833–

VOYAGES

OF THE

ADVENTURE AND BEAGLE.

VOLUME I.

P. P. King. T. Landseer.

PATAGONIAN.

NARRATIVE

OF THE

SURVEYING VOYAGES

OF HIS MAJESTY'S SHIPS

ADVENTURE AND BEAGLE,

BETWEEN

THE YEARS 1826 AND 1836,

DESCRIBING THEIR

EXAMINATION OF THE SOUTHERN SHORES

OF

SOUTH AMERICA,

AND

THE BEAGLE'S CIRCUMNAVIGATION OF THE GLOBE.

IN THREE VOLUMES.

VOL. I.

LONDON:
HENRY COLBURN, GREAT MARLBOROUGH STREET.

1839. /

LONDON:
Printed by J.L. Cox and Sons, 75, Great Queen Street, Lincoln's-Inn Fields. /

VOLUME I.

PROCEEDINGS

OF

THE FIRST EXPEDITION,

1826 – 1830,

UNDER THE COMMAND OF

CAPTAIN P. PARKER KING,

R.N., F.R.S. / /

TO

THE RIGHT HONOURABLE

THE EARL OF MINTO, G.C.B.,

FIRST LORD COMMISSIONER

OF THE

ADMIRALTY.

MY LORD:

I have the honour of dedicating to your lordship, as Head of the Naval Service, this narrative of the Surveying Voyages of the Adventure and Beagle, between the years 1826 and 1836.

Originated by the Board of Admiralty, over which Viscount Melville presided, these voyages have been carried on, since 1830, under his lordship's successors in office.[1]

Captain King has authorized me to lay the results of the Expedition which he commanded, from 1826 to 1830, before your lordship, united to those of the Beagle's subsequent voyages.

I have the honour to be,

MY LORD,

Your lordship's obedient servant,

ROBERT FITZ-ROY. / /

PREFACE.

In this Work, the result of nine years' voyaging, partly on coasts little known, an attempt has been made to combine giving general information with the paramount object – that of fulfilling a duty to the Admiralty, for the benefit of Seamen.

Details, purely technical, have been avoided in the narrative more than I could have wished; but some are added in the Appendix to each volume: and in a nautical memoir, drawn up for the Admiralty, those which are here omitted will be found.

There are a few words used frequently in the following pages, which may not at first sight be familiar to every reader, therefore I need hardly apologize for saying that, although the great Portuguese navigator's name was Magalhaens – it is generally pronounced as if written Magellan:– that the natives of Tierra del Fuego are commonly called Fuegians; – and that Chilóe is thus accented for reasons given in page 384 of the second volume.

In the absence of Captain King, who has entrusted to me the care of publishing his share of this work, I may have overlooked errors which he would have detected. Being hurried, and unwell, while attending to the printing of his volume, I was not able to do it justice.[2] /

It may be a subject of regret, that no paper on the Botany of Tierra del Fuego is appended to the first volume. Captain King took great pains in forming and preserving a botanical collection, aided by a person embarked solely for that purpose. He placed this collection in the British Museum, and was led to expect that a first-rate botanist would have examined and described it; but he has been disappointed.[3]

In conclusion, I beg to remind the reader, that the work is unavoidably of a rambling and very mixed character; that some parts may be wholly uninteresting to most readers, though, perhaps, not devoid of interest to all; and that its publication arises solely from a sense of duty.

ROBERT FITZ-ROY.

London, March 1839. /

INTRODUCTION

In 1825, the Lords Commissioners of the Admiralty directed two ships to be prepared for a Survey of the Southern Coasts of South America; and in May, of the following year, the ADVENTURE and the BEAGLE were lying in Plymouth Sound, ready to carry the orders of their Lordships into execution.

These vessels were well provided with every necessary, and every comfort, which the liberality and kindness of the Admiralty, Navy Board, and officers of the Dock-yards, could cause to be furnished.

On board the Adventure, a roomy ship, of 330 tons burthen, without guns,** lightly though strongly rigged, and very strongly built, were[4] –

PHILLIP PARKER KING, Commander and Surveyor, Senior Officer of the Expedition.

J. COOKE . . . . Lieutenant.
B. AINSWORTH . . . . Master.
J. TARN . . . . Surgeon. /
G. ROWLETT . . . . Purser.
R. H. SHOLL . . . . Mate.
J. C. WICKHAM . . . . Mate.
J. F. BRAND . . . . Mate.
T. GRAVES . . . . Mate and Assistant Surveyor.
G. HARRISON . . . . Mate.
E. WILLIAMS . . . . Second Master.
J. PARK . . . . Assistant Surgeon.
W. W. WILSON . . . . Midshipman.
A. MILLAR . . . . Master's Assistant.
A. MELLERSH . . . . Volunteer 1st Class.
J. RUSSELL . . . . Volunteer 2d Class.
G. HODGSKIN . . . . Clerk.
J. ANDERSON . . . . Botanical Collector.

Gunner – Boatswain – and Carpenter.

Serjeant and fourteen Marines; and about forty Seamen and Boys.

* Excepting one for signals.

In the Beagle, a well-built little vessel, of 235 tons, rigged as a barque, and carrying six guns, were –

Pringle Stokes.	Commander and Surveyor.
E. Hawes	Lieutenant.
W. G. Skyring	Lieut. and Assist. Surveyor.
S. S. Flinn	Master.
E. Bowen	Surgeon.
J. Atrill	Purser.
J. Kirke	Mate.
B. Bynoe.	Assistant Surgeon.
J. L. Stokes	Midshipman.
R. F. Lunie.	Volunteer 1st Class.
W. Jones.	Volunteer 2d Class.
J. Macdouall	Clerk.
	Carpenter.

Serjeant and nine Marines; and about forty Seamen and Boys. /

In the course of the voyage, several changes occurred among the officers, which it may be well to mention here.

In September, 1826, Lieutenant Hawes invalided: and was succeeded by Mr. R.H. Sholl, the senior mate in the Expedition.

In February, 1827, Mr. Ainsworth was unfortunately drowned;[5] and, in his place, Mr. Williams acted, until superseded by Mr. S.S. Flinn, of the Beagle.

Lieutenant Cooke invalided in June, 1827; and was succeeded by Mr. J.C. Wickham.

In the same month Mr. Graves received information of his promotion to the rank of Lieutenant.

Between May and December, 1827, Mr. Bowen and Mr. Atrill invalided; besides Messrs. Lunie, Jones, and Macdouall: Mr. W. Mogg joined the Beagle, as acting Purser; and Mr. D. Braily, as volunteer of the second class.

Mr. Bynoe acted as Surgeon of the Beagle, after Mr. Bowen left, until December, 1828.

In August, 1828, Captain Stokes's lamented vacancy was temporarily filled by Lieutenant Skyring; whose place was taken by Mr. Brand.

Mr. Flinn was then removed to the Adventure; and Mr. A. Millar put into his place. /

In December, 1828, the Commander-in-chief of the Station (Sir Robert Waller Otway) superseded the temporary arrangements of Captain King, and appointed a commander, lieutenant, master, and surgeon to the Beagle. Mr. Brand then invalided, and the lists of officers stood thus –

Adventure (1828-30).

PHILLIP PARKER KING, Commander and Surveyor, Senior Officer of the Expedition.

T. GRAVES	Lieut. and Assist. Surveyor.
J. C. WICKHAM	Lieutenant.
S. S. FLINN	Master.
J. TARN	Surgeon.
G. ROWLETT	Purser.
G. HARRISON	Mate.
W. W. WILSON	Mate.
E. WILLIAMS	Second Master.
J. PARK	Assistant Surgeon.
A. MELLERSH	Midshipman.
A. MILLAR	Master's Assistant.
J. RUSSELL	Volunteer 2d Class.
G. HODGSKIN	Clerk.
J. ANDERSON	Botanical Collector.

Gunner – Boatswain – and Carpenter.

Serjeant and fourteen Marines: and about fifty* Seamen and Boys.

Beagle (1828-30).

ROBERT FITZ-ROY	Commander and Surveyor.
W. G. SKYRING	Lieut. and Assist. Surveyor.
J. KEMPE	Lieutenant.
M. MURRAY	Master. /
J. WILSON	Surgeon.
W. MOGG	(Acting) Purser.
J. KIRKE	Mate.
B. BYNOE	Assistant Surgeon.
J. L. STOKES	Midshipman.
J. MAY	Carpenter.
D. BRAILY	Volunteer 2d Class.
J. MEGGET	Clerk.

Serjeant and nine Marines: and about forty Seamen and Boys.

In June, 1829, Lieutenant Mitchell joined the Adventure; and in February, 1830, Mr. A. Millar died very suddenly:– and very much regretted.[6]

The following Instructions were given to the Senior Officer of the Expedition.

'By the Commissioners for executing the Office of Lord High Admiral of the United Kingdom of Great Britain and Ireland, &c.

'Whereas we think fit that an accurate Survey should be made of the Southern Coasts of the Peninsula of South America, from the southern entrance of the River Plata, round to Chilóe; and of Tierra del Fuego; and whereas we have been

* Twelve additional seamen having been ordered, by the Admiralty, for the Adelaide schooner.

induced to repose confidence in you, from your conduct of the Surveys in New Holland; we have placed you in the command of His Majesty's Surveying Vessel the Adventure; and we have directed Captain Stokes, of His Majesty's Surveying Vessel the Beagle, to follow your orders.

'Both these vessels are provided with all the / means which are necessary for the complete execution of the object above-mentioned, and for the health and comfort of their Ships' Companies. You are also furnished with all the information, we at present possess, of the ports which you are to survey; and nine Government Chronometers[7] have been embarked in the Adventure, and three in the Beagle, for the better determination of the Longitudes.

'You are therefore hereby required and directed, as soon as both vessels shall be in all respects ready, to put to sea with them; and on your way to your ulterior destination, you are to make, or call at, the following places, successively; namely; Madeira: Teneriffe: the northern point of St. Antonio, and the anchorage at St. Jago; both in the Cape Verd Islands: the Island of Trinidad, in the Southern Atlantic: and Rio de Janeiro: for the purpose of ascertaining the differences of the longitudes of those several places.

'At Rio de Janeiro, you will receive any supplies you may require; and make with the Commander-in-chief, on that Station,[8] such arrangements as may tend to facilitate your receiving further supplies, in the course of your Expedition.

'After which, you are to proceed to the entrance of the River Plata, to ascertain the longitudes of the Cape Santa Maria, and Monte Video: you are then to proceed to survey the Coasts, Islands, and Straits; from Cape St. Antonio, at the south side / of the River Plata, to Chilóe; on the west coast of America; in such manner and order, as the state of the season, the information you may have received, or other circumstances, may induce you to adopt.

'You are to continue on this service until it shall be completed; taking every opportunity to communicate to our Secretary, and the Commander-in-Chief, your proceedings: and also, whenever you may be able to form any judgment of it, where the Commander-in-Chief, or our Secretary, may be able to communicate with you.

'In addition to any arrangements made with the Admiral, for recruiting your stores, and provisions; you are, of course, at liberty to take all other means, which may be within your reach, for that essential purpose.

'You are to avail yourself of every opportunity of collecting and preserving Specimens of such objects of Natural History as may be new, rare, or interesting; and you are to instruct Captain Stokes, and all the other Officers, to use their best diligence in increasing the Collections in each ship: the whole of which must be understood to belong to the Public.[9]

'In the event of any irreparable accident happening to either of the two vessels, you are to cause the officers and crew of the disabled vessel to be / removed into the other, and with her, singly, to proceed in prosecution of the service, or return to England, according as circumstances shall appear to require; understanding that the officers and crews of both vessels are hereby authorized, and required, to continue to perform their duties, according to their respective ranks and stations, on board either vessel to which they may be so removed. Should, unfortunately, your own vessel be the one disabled, you are in that case to take the command of the Beagle: and, in the event of any fatal accident happening to yourself; Captain Stokes is hereby authorized to take the command of the Expedition; either on board the Adventure, or Beagle; as he may prefer; placing the officer of the Expedition who may then be next in seniority to him, in command of the second vessel: also, in the event of your inability, by sickness or otherwise, at any period of this service, to continue to carry the Instructions into execution, you are to transfer them to Captain Stokes, or to the surviving officer then next in command to you, who is hereby required to execute them, in the best manner he can, for the attainment of the object in view.

'When you shall have completed the service, or shall, from any cause, be induced to give it up; you will return to Spithead with all convenient expedition; and report your arrival, and proceedings, to our Secretary, for our information. /

'Whilst on the South American Station, you are to consider yourself under the command of the Admiral of that Station; to whom we have expressed our desire that he should not interfere with these orders, except under peculiar necessity.

'Given under our hands the 16th of May 1826.

(Signed) MELVILLE.
G. COCKBURN.[10]

To Phillip P. King, Esq., Commander
of His Majesty's Surveying Vessel
Adventure, at Plymouth.
By command of their Lordships.

(Signed) J. W. CROKER.'[11]

On the 22d of May, 1826, the Adventure and Beagle sailed from Plymouth; and, in their way to Rio de Janeiro, called successively at Madeira, Teneriffe, and St. Jago.

Unfavourable weather prevented a boat being sent ashore at the northern part of San Antonio; but observations were made in Terrafal Bay, on the south-west side of the island: and, after crossing the Equator, the Trade-wind hung so much to the southward, that Trinidad could not be approached without a sacrifice of time, which, it was considered, might be prejudicial to more important objects of the Expedition.

Both ships anchored at Rio de Janeiro on the / 10th of August, and remained there until the 2d of October, when they sailed to the River Plata.

In Maldonado,* their anchors were dropped on the 13th of the same month; and, till the 12th of November, each vessel was employed on the north side of the river, between Cape St. Mary and Monte Video.

* On the north side of the river Plata. /

C. Martens. T. Landseer

MONTE VIDEO.

CHAPTER I.

Departure from Monte Video – Port Santa Elena – Geological remarks – Cape Fairweather – Non-existence of Chalk – Natural History – Approach to Cape Virgins, and the Strait of Magalhaens (or Magellan).

WE sailed from Monte Video on the 19th of November 1826; and, in company with the Beagle, quitted the river Plata.[1]

According to my Instructions, the Survey was to commence at Cape San Antonio, the southern limit of the entrance of the Plata; but, for the following urgent reasons, I decided to begin with the southern coasts of Patagonia, and Tierra del Fuego, including the Straits of Magalhaens.* In the first place, they presented a field of great interest and novelty; and secondly, the climate of the higher southern latitudes being so severe and tempestuous, it appeared important to encounter its rigours while the ships were in good condition – while the crews were healthy – and while the charms of a new and difficult enterprize had full force.

Our course was therefore southerly, and in latitude 45° south, a few leagues northward of Port Santa Elena, we first saw the coast of Patagonia. I intended to visit that port; and, on the 28th, anchored, and landed there.

Seamen should remember that a knowledge of the tide is of especial consequence in and near Port Santa Elena. During a calm we were carried by it towards reefs which line the shore, and were obliged to anchor until a breeze sprung up.

The coast along which we had passed, from Point Lobos to the north-east point of Port Santa Elena, appeared to be dry and bare of vegetation. There were no trees; the land seemed to be one long extent of undulating plain, beyond which were high, flat-topped hills of a rocky, precipitous character. The shore was fronted by rocky reefs extending two or three miles from high-water mark, which, as the tide fell, were left dry, and in many places were covered with seals.

As soon as we had secured the ships, Captain Stokes accompanied me on shore to select a place for our observations. We found the spot which the Spanish astronomers of Malaspina's Voyage (in 1798) used for their observatory, the most convenient for our purpose.[2] It is near a very steep shingle (stony) beach

* Commonly called Magellan. See p. 11. /

at the back of a conspicuous red-coloured, rocky projection which terminates a small bay, on the western side, at the head of the port. The remains of a wreck, which proved to be that of an American whaler, the Decatur of New York,[3] were found upon the extremity of the same point; she had been driven on shore from her anchors during a gale.

The sight of the wreck, and the steepness of the shingle beach just described, evidently caused by the frequent action of a heavy sea, did not produce a favourable opinion of the safety of the port: but as it was not the season for easterly gales, to which only the anchorage is exposed, and as appearances indicated a westerly wind, we did not anticipate danger.

While we were returning on board, the wind blew so strongly that we had much difficulty in reaching the ships, and the boats were no sooner hoisted up, and every thing / made snug, than it blew a hard gale from the S.W. The water however, from the wind being off the land, was perfectly smooth, and the ships rode securely through the night: but the following morning the gale increased, and veered to the southward, which threw a heavy sea into the port, placing us, to say the least, in a very uneasy situation. Happily it ceased at sunset. In consequence of the unfavourable state of the weather, no attempt was made to land in order to observe an eclipse of the sun; to make which observation was one reason for visiting this port.

The day after the gale, while I was employed in making some astronomical observations, a party roamed about in quest of game: but with little success, as they killed only a few wild ducks. The fire which they made for cooking communicated to the dry stubbly grass, and in a few minutes the whole country was in a blaze. The flames continued to spread during our stay, and, in a few days, more than fifteen miles along the coast, and seven or eight miles into the interior were overrun by the fire. The smoke very much impeded our observations, for at times it quite obscured the sun.

The geological structure of this part of the country, and a considerable portion of the coast to the north and south, consists of a fine-grained porphyritic clay slate. The summits of the hills near the coast are generally of a rounded form, and are paved, as it were, with small, rounded, siliceous pebbles, imbedded in the soil, and in no instance lying loose or in heaps; but those of the interior are flat-topped, and uniform in height, for many miles in extent. The valleys and lower elevations, notwithstanding the poverty and parched state of the soil, were partially covered with grass and shrubby plants, which afford sustenance to numerous herds of guanacoes.[4] Many of these animals were observed feeding near the beach when we were working into the bay, but they took the alarm, so that upon landing we only saw them at a considerable distance. In none of our excursions could we find any water that had not a brackish taste. Several wells have been dug in the valleys, both near the sea and at a considerable distance from it, by the / crews of sealing vessels; but, except in the rainy season, they all contain saltish water. This observation is applicable to nearly the whole extent of the porphyritic country. Oystershells, three or four inches in diameter, were found, scattered over the hills, to the height of three or four hundred feet above the sea. Sir John Narborough, in 1652, found oyster-shells at Port San Julian; but, from a great many which have been lately collected there, we know that they are of a species different from that found at Port Santa Elena. Both are fossils.

No recent specimen of the genus *Ostrea* was found by us on any part of the Patagonian coast. Narborough, in noticing those at Port San Julian, says, 'They are the biggest oystershells that I ever saw, some six, some seven inches broad, yet not one oyster to be found in the harbour: whence I conclude they were here when the world was formed.'[5]

The short period of our visit did not enable us to add much to natural history. Of quadrupeds we saw guanacoes, foxes, cavies, and the armadillo; but no traces of the puma (*Felis concolor*), or South American lion, although it is to be met with in the interior.

I mentioned that a herd of guanacoes was feeding near the shore when we arrived. Every exertion was made to obtain some of the animals; but, either from their shyness, or our ignorance of the mode of entrapping them, we tried in vain, until the arrival of a small sealing-vessel, which had hastened to our assistance, upon seeing the fires we had accidentally made, but which her crew thought were intended for signals of distress. They shot two, and sent some of the meat on board the Adventure. The next day, Mr. Tarn succeeded in shooting one, a female, which, when skinned and cleaned, weighed 168 lbs. Narborough mentions having killed one at Port San Julian, that weighed, 'cleaned in his quarters, 268 lbs.' The watchful and wary character of this animal is very remarkable. Whenever a herd is feeding, one is posted, like a sentinel, on a height; and, at the approach of danger, gives instant alarm by a loud neigh, when away they all go, at a hand-gallop, to the next eminence, where they quietly resume their feeding, / until again warned of the approach of danger by their vigilant 'look-out.'

Another peculiarity of the guanaco is, the habit of resorting to particular spots for natural purposes. This is mentioned in the 'Dictionnaire d'Histoire Naturelle,' in the 'Encyclopédie Méthodique,' as well as other works.[6]

In one place we found the bones of thirty-one guanacoes collected within a space of thirty yards, perhaps the result of an encampment of Indians, as evident traces of them were observed; among which were a human jaw-bone, and a piece of agate ingeniously chipped into the shape of a spear-head.

The fox, which we did not take, appeared to be small, and similar to a new species afterwards found by us in the Strait of Magalhaens.

The cavia* (or, as it is called by Narborough, Byron, and Wood,[7] the hare, an animal from which it differs both in appearance and habits, as well as flavour), makes a good dish; and so does the armadillo, which our people called the shell-pig.† This little animal is found abundantly about the low land, and lives in burrows underground; several were taken by the seamen, and, when cooked in their shells, were savoury and wholesome.

Teal were abundant upon the marshy grounds. A few partridges, doves, and snipes, a rail, and some hawks were shot. The few sea-birds that were observed consisted of two species of gulls, a grebe and a penguin (*Aptenodytes Magellanica*).

We found two species of snakes and several kinds of lizards. Fish were scarce, as were also insects; of the last, our collections / consisted only of a few species of *Coleoptera*, two or three *Lepidoptera*, and two *Hymenoptera*.

Among the sea-shells, the most abundant was the *Patella deaurata*, Lamk.; this, with three other species of Patella, one *Chiton*, three species of *Mytilus*, three of *Mureœ*, one of *Crepidula*, and a *Venus*, were all that we collected.

About the country, near the sea-shore, there is a small tree, whose stem and roots are highly esteemed for fuel by the crews of sealing-vessels which frequent this coast. They call it 'piccolo.' The leaf was described to me as having a prickle upon it, and the flower as of a yellow colour. A species of berberis also is found, which when ripe may afford a very palatable fruit.

Our short visit gave us no flattering opinion of the fertility of the country near this port. Of the interior we were ignorant; but, from the absence of Indians and the scarcity of fresh water, it is probably very bare of pasturage. Falkner, the Jesuit missionary,[9] says these parts were used by the Tehuelhet tribes for burying-places: we saw, however, no graves, nor any traces of bodies, excepting the jaw-bone above-mentioned; but subsequently, at Sea Bear Bay, we found many places on the summits of the hills which had evidently been used for such a purpose, although then containing no remains of bodies. This corresponds with Falkner's account, that after a period of twelve months the sepulchres are formally visited by the tribe, when the bones of their relatives and friends are collected and carried to certain places, where the skeletons are arranged in order, and tricked out with all the finery and ornaments they can collect.[10]

---

* *Dasyprocta patachonica:* it is the *Patagonian cavy* of Dr. Shaw, and Pennant's Quadr., tab. 39, and the *liévre pampa* of D' Azara. M. Desmarest thinks that if the teeth were examined it would form a new genus, for which he proposes the name of *Doliohotis* (Ency. Meth. Mamm. p. 359). At present he has, from its external character, placed it amongst the genus *Dasyprocta* (agouti). The only one that was taken was not preserved, which prevented me from ascertaining the fact.[8]

† *Dasypus minutus*, Desm. *Tatou pichiy*, or *tatou septrême* of D'Azara, &c. &c. It has seven bands.

The ships sailed from Port Santa Elena on the 5th December, and proceeded to the southward, coasting the shore as far as Cape Two Bays.

Our object being to proceed with all expedition to the Strait of Magalhaens, the examination of this part of the coast was reserved for a future opportunity. On the 13th, we had reached within fifty miles of Cape Virgins, the headland at the entrance of the strait, but it was directly in the wind's eye / of us. The wind veering to S.S.W., we made about a west course. At day-light the land was in sight, terminating in a point to the S.W., so exactly like the description of Cape Virgins and the view of it in Anson's voyage,[11] that without considering our place on the chart, or calculating the previous twenty-four hours' run, it was taken for the Cape itself, and, no one suspecting a mistake, thought of verifying the ship's position. The point, however, proved to be Cape Fairweather. It was not a little singular, that the same mistake should have been made on board the Beagle, where the error was not discovered for three days.*

From the appearance of the weather I was anxious to approach the land in order to anchor, as there seemed to be every likelihood of a gale; and we were not deceived, for at three o'clock, being within seven miles of the Cape, a strong wind sprung up from the S.W., and the anchor was dropped. Towards evening it blew so hard, that both ships dragged their anchors for a considerable distance.

On the charts of this part of the coast the shore is described to be formed of 'chalk hills, like the coast of Kent.' To geologists, therefore, especially, as they were not disposed to believe that such was the fact, this was a question of some interest.[13] From our anchorage the appearance of the land favoured our belief of the existence of chalk. The outline was very level and steep; precipitous cliffs of whitish colour, stratified horizontally, with their upper part occasionally worn into hollows, strongly resembled the chalk cliffs of the English coasts.

The gale prevented our landing for three days, when (19th) a few minutes sufficed to discover that the cliffs were composed / of soft clay, varying in colour and consistence, and disposed in strata running horizontally for many miles without interruption, excepting where water-courses had worn them away. Some of the strata were very fine clay, unmixed with any other substance, whilst others were plentifully strewed with round siliceous gravel,† without any vestige of

* A similar error was made by one of the ships of the fleet under Loyasa in the year 1525. The Nodales also, in their description of the coast, mention the similarity of appearance in the two capes, Virgins and Fairweather. 'Y venido de mar en fuera à buscar la tierra facilmente podian bacer de Rio de Gallegos el Caboe de Virgenes,' (and in making the land Cape Virgins may easily be mistaken for the river Gallegos). – Viage de los Nodales, p. 53.[12]

† Some of the specimens of the clay strata consist, according to Dr. Fitton,[14] who has kindly examined my collection, of a white marl not unlike certain varieties of the lower chalk; and of a clay having many of the properties of fuller's earth. The pebbles on the beach consist of quartz, red jasper, hornstone, and flinty slate, but do not contain any stone resembling chalk flint.

organic remains. The sea beach, from high-water mark to the base of the cliffs, is formed by shingle, with scattered masses of indurated clay of a green colour.* Between the high and low tide marks there is a smooth beach of the same green clay as the masses above-mentioned, which appears to have been hardened by the action of the surf to the consistence of stone. Generally this beach extends for about one hundred yards farther into the sea, and is succeeded by a soft green mud, over which the water gradually deepens. The outer edge of the clay forms a ledge, extending parallel with the coast, upon the whole length of which the sea breaks, and over it a boat can with difficulty pass at low water.

The very few shells we found were dead. Strewed about the beach were numbers of fish, some of which had been thrown on shore by the last tide, and were scarcely stiff. They principally belonged to the genus *Ophidium*; the largest that we saw measured four feet seven inches in length, and weighed twenty-four pounds. Many caught alongside the ship were, in truth, coarse and insipid; yet our people, who fed heartily upon them, called them ling, and thought them palatable. The hook, however, furnished us with a very wholesome and well-flavoured species of cod (*Gadus*). Attached to the first we found two parasitical animals; one was a *Cymothoa*, the other a species of *Lernœa*, which had so / securely attached itself under the skin, as not to be removed without cutting off a piece of the flesh with it. An undescribed species of *Murœna* was also taken.

Whilst we were on shore, the Beagle moved eight or nine miles nearer to the Cape, where Captain Stokes landed to fix positions of remarkable land. One peaked hill, from the circumstance of his seeing a large animal near it, he called Tiger Mount. Mr. Bowen shot a guanaco; and being at a distance in shore, unable to procure assistance, he skinned and quartered it with his pocket-knife, and carried it upon his shoulders to the boat.

Next morning the ships weighed, and proceeded towards Cape Virgins.

When abreast of Cape Fairweather, the opening of the river Gallegos was very distinctly seen; but the examination of it was deferred to a future opportunity. Passing onward, the water shoaled to four fathoms, until we had passed extensive banks, which front the river.

Our approach to the entrance of the Strait, although attended with anxiety, caused sensations of interest and pleasure not easily to be described. Though dangers were experienced by some navigators who had passed it, the comparative facility with which others had effected the passage showed that, at times, the difficulties were easily surmounted, and we were willing to suppose that in the former case there might have been some little exaggeration.

* Dr. Fitton considers these masses of clay to bear a resemblance to the upper green sand of England.

The most complete, and, probably, the only good account of the navigation of the Strait of Magalhaens is contained in the narrative of Don Antonio de Cordova, who commanded the Spanish frigate Santa Maria de la Cabeza, on a voyage expressly for the purpose of exploring the strait. It was published under the title of 'Ultimo Viage al Estrecho de Magallanes.' That voyage was, however, concluded with only the examination of the eastern part, and a subsequent expedition was made, under the command of the same officer, the account of which was appended to the Cabeza's voyage; so that Cordova's expedition still retained the appellation of 'Ultimo / Viage, &c.' It is written in a plain and simple style, gives a most correct account of every thing seen, and should therefore be in the possession of every person who attempts the navigation of the strait.[15]

Cordova's account of the climate is very uninviting. Speaking of the rigours of the summer months (January, February, and March), he says, 'Seldom was the sky clear, and short were the intervals in which we experienced the sun's warmth: no day passed by without some rain having fallen, and the most usual state of the weather was that of constant rain.'*

The accounts of Wallis and Carteret are still more gloomy. The former concludes that part of his narrative with the following dismal and disheartening description: 'Thus we quitted a dreary and inhospitable region, where we were in almost continual danger of shipwreck for near four months, having entered the strait on the 17th of December, and quitted it on the 11th of April 1767: a region where, in the midst of summer, the weather was cold, gloomy, and tempestuous, where the prospects had more the appearance of a chaos than of nature; and where for the most part the valleys were without herbage and the hills without wood.'[16]

These records of Cordova and Wallis made me feel not a little apprehensive for the health of the crew, which could not be expected to escape uninjured through the rigours of such a climate. Nor were the narratives of Byron or Bougainville calculated to lessen my anxiety.[17] In an account, however, of a voyage to the strait by M.A. Duclos Guyot, [18] the following paragraph tended considerably to relieve my mind upon the subject:– 'At length, on Saturday the 23d of March, we sailed out of that famous Strait, so much dreaded, after having experienced that there, as well as in other places, it was very fine, and very warm, and that for three-fourths of the time the sea was perfectly calm.'

In every view of the case, our proximity to the principal scene of action occasioned sensations of a peculiar nature, in which, however, those that were most agreeable and hopeful / preponderated. The officers and crews of both ships were healthy, and elated with the prospect before them; our vessels were in every respect strong and sea-worthy; and we were possessed of every comfort

* Ultimo Viage al Estrecho de Magallanes, part ii. p. 298.

and resource necessary for encountering much greater difficulties than we had any reason to anticipate.

There has existed much difference of opinion as to the correct mode of spelling the name of the celebrated navigator who discovered this Strait. The French and English usually write it Magellan, and the Spaniards Magallanes; but by the Portuguese (and he was a native of Portugal) it is universally written Magalhaens. Admiral Burney and Mr. Dalrymple[19] spell it Magalhanes, which mode I have elsewhere adopted, but I have since convinced myself of the propriety of following the Portuguese orthography for a name which to this day, is very common both in Portugal and Brazil. /

CHAPTER II.

Enter the Straits of Magalhaens (or Magellan), and anchor off Cape Possession – First Narrow – Gregory Bay – Patagonian Indians – Second Narrow – Elizabeth Island – Freshwater Bay – Fuegian Indians – Arrival at Port Famine.

A CONTRARY tide and light winds detained us at anchor near Cape Virgins until four o'clock in the afternoon, when, with the turn of the tide, a light air carried us past Dungeness Point, aptly named by Wallis from its resemblance to that in the English Channel. A great number of seals were huddled together upon the bank, above the wash of the tide, whilst others were sporting about in the surf. Cape Possession was in sight, and with the wind and tide in our favour we proceeded until ten o'clock, when the anchor was dropped. At daylight we found ourselves six miles to the eastward of the cape. The anchor was then weighed, and was again dropped at three miles from the cape until the afternoon, when we made another attempt; but lost ground, and anchored a third time. Before night a fourth attempt was made, but the tide prevented our making any advance, and we again anchored.

Mount Aymond* and 'his four sons,' or (according to the old quaint nomenclature) the Asses' Ears, had been in sight all day, as well as a small hummock of land on the S.W. horizon, which afterwards proved to be the peaked hillock upon Cape Orange, at the south side of the entrance to the First Narrow.

At this anchorage the tide fell thirty feet, but the strength of the current, compared with the rate at which we afterwards found it to run, was inconsiderable. Here we first experienced / the peculiar tides of which former navigators have written. During the first half of the flood† or westward tide, the depth decreased, and then, after a short interval, increased until three hours after the stream of tide had begun to run to the eastward.

The following morning (21st) we gained a little ground. Our glasses were directed to the shore in search of inhabitants, for it was hereabouts that Byron,

* A hill on the north shore of Possession Bay, having near it, to the westward, four rocky summits, which, from a particular point of view, bear a strong resemblance to the cropped ears of a horse or ass. These are described less briefly in the Sailing Directions.[1]

† Flowing into the strait from the east towards the west.

and Wallis, and some of the Spanish navigators held communication with the Patagonian Indians; but we saw none. Masses of large sea-weed,* drifting with the tide, floated past the ship. A description of this remarkable plant, although it has often been given before, may not be irrelevant here.[2] It is rooted upon rocks or stones at the bottom of the sea, and rises to the surface, even from great depths. We have found it firmly fixed to the ground more than twenty fathoms under water, yet trailing along the surface for forty or fifty feet. When firmly rooted it shows the set of the tide or current. It has also the advantage of indicating rocky ground: for wherever there are rocks under water, their situation is, as it were, buoyed by a mass of seaweed† on the surface of the sea, of larger extent than that of the danger below. In many instances perhaps it causes unnecessary alarm, since it often grows in deep water; but it should not be entered without its vicinity having been sounded, especially if seen in masses, with the extremities of the stems trailing along the surface. If there be no tide, or if the wind and tide are the same way, the plant lies smoothly upon the water, but if the wind be against the tide, the leaves curl up and are visible at a distance, giving a rough, rippling appearance to the surface of the water.

During the last two days the dredge had furnished us with a few specimens of *Infundibulum* of Sowerby (*Patella trochi-formis*, Lin.),[3] and some dead shells (*Mureœ Magellanicus*) were brought up by the sounding-lead.[4]

We made another attempt next morning, but again lost / ground, and the anchor was dropped for the eighth time. The threatening appearances of the clouds, and a considerable fall of the barometer indicating bad weather, Captain Stokes agreed with me in thinking it advisable to await the springtides to pass the First Narrow: the ships were therefore made snug for the expected gale, which soon came on, and we remained several days wind-bound, with top-masts struck, in a rapid tide-way, whose stream sometimes ran seven knots. On the 28th, with some appearance of improving weather, we made an attempt to pass through the Narrow. The wind blowing strong, directly against us, and strengthening as we advanced, caused a hollow sea, that repeatedly broke over us. The tide set us through the Narrow very rapidly, but the gale was so violent that we could not show more sail than was absolutely necessary to keep the ship under command. Wearing every ten minutes,[5] as we approached either shore, lost us a great deal of ground, and as the anchorage we left was at a considerable distance from the entrance of the Narrows, the tide was not sufficient to carry us through. At slack water the wind fell, and as the weather became fine, I was induced to search for anchorage near the south shore. The sight of kelp, however, fringing the coast, warned me off, and we were obliged to return to an anchorage in Pos-

* Fucus giganteus.

† Usually called by seamen 'kelp.'

session Bay. The Beagle had already anchored in a very favourable berth; but the tide was too strong to permit us to reach the place she occupied, and our anchor was dropped a mile astern of her, in nineteen fathoms. The tide was then running five, and soon afterwards six miles an hour. Had the western tide set with equal strength, we should have succeeded in passing the Narrow. Our failure, however, answered the good purpose of making us more acquainted with the extent of a bank that lines the northern side of Possession Bay, and with the time of the turn of tide in the Narrow; which on this day (new moon) took place within a few minutes of noon.

As we passed Cape Orange, some Indians were observed lighting a fire under the lee of the hill to attract our notice; but we were too busily engaged to pay much attention to / their movements. Guanacoes also were seen feeding near the beach, which was the first intimation we had of the existence of that animal southward of the Strait of Magalhaens.

When day broke (29th) it was discovered that the ship had drifted considerably during the night. The anchor was weighed, and with a favourable tide we reached an anchorage a mile in advance of the Beagle. We had shoaled rather suddenly to eight fathoms, upon which the anchor was immediately dropped, and on veering cable the depth was eleven fathoms. We had anchored on the edge of a bank, which soon afterwards, by the tide falling, was left dry within one hundred yards of the ship. Finding ourselves so near a shoal, preparations were made to prevent the ship from touching it. An anchor was dropped under foot, and others were got ready to lay out, for the depth alongside had decreased from eleven to seven fathoms, and was still falling. Fortunately we had brought up to leeward of the bank, and suffered no inconvenience; the flood made, and as soon as possible the ship was shifted to another position, about half a mile to the S.E., in a situation very favourable for our next attempt to pass the Narrow. This night the tide fell thirty-six feet, and the stream ran six knots.

The ensuing morning we made another attempt to get through the Narrow, and, from having anchored so close to its entrance, by which the full benefit of the strength, as well as the whole duration of the tide was obtained, we succeeded in clearing it in two hours, although the distance was more than twenty miles, and the wind directly against us, the sea, as before, breaking repeatedly over the ship.

After emerging from the Narrow we had to pass through a heavy 'race' before we 'reached' out of the influence of the stream that runs between the First and Second Narrow,[6] but the tide lasted long enough to carry us to a quiet anchorage. In the evening we weighed again, and reached Gregory Bay, where the Beagle joined us the next morning.

Since entering the Strait, we had not had any communication / with the Beagle on account of the weather, and the strength of the tide; this opportunity was therefore taken to supply her with water, of which she had only enough left for two days.

The greater part of this day was spent on shore, examining the country and making observations. Large smokes* were noticed to the westward. The shore was strewed with traces of men and horses, and other animals. Foxes and ostriches were seen; and bones of guanacoes were lying about the ground.

The country in the vicinity of this anchorage seemed open, low, and covered with good pasturage. It extends five or six miles, with a gradual ascent, to the base of a range of flat topped land, whose summit is about fifteen hundred feet above the level of the sea. Not a tree was seen; a few bushes† alone interrupted the uniformity of the view. The grass appeared to have been cropped by horses or guanacoes, and was much interspersed with cranberry plants, bearing a ripe and juicy, though very insipid fruit.

Next day the wind was too strong and adverse to permit us to proceed. In the early part of the morning an American sealing vessel, returning from the Madre de Dios Archipelago on her way to the Falkland Islands, anchored near us. Mr. Cutler,[7] her master, came on board the Adventure, passed the day and night with us, and gave me much useful information respecting the nature of the navigation, and anchorages in the Strait. He told me there was an Englishman in his vessel who was a pilot for the strait, and willing to join the ship. I gladly accepted the offer of his services.

In the evening an Indian was observed on horseback riding to and fro upon the beach, but the weather prevented my sending a boat until the next morning, when Lieutenant Cooke went on shore to communicate with him and other Indians who appeared, soon after dawn, upon the beach. On landing, he was received by them without the least distrust. They were eight or ten in number, consisting of an old man and his wife, three young men, and the rest children, all mounted on / good horses. The woman, who appeared to be about fifty years of age, was seated astride upon a pile of skins, hung round with joints of fresh guanaco meat and dried horse-flesh. They were all wrapped in mantles, made chiefly of the skins of guanacoes, sewed together with the sinews of the same animal. These mantles were large enough to cover the whole body. Some were made of skins of the 'zorillo,' or skunk, an animal like a pole-cat, but ten times more offensive; and others, of skins of the puma.

The tallest of the Indians,[8] excepting the old man, who did not dismount, was rather less than six feet in height. All were robust in appearance, and with respect to the head, length of body, and breadth of shoulders, of gigantic size; therefore, when on horseback, or seated in a boat, they appeared to be tall, as

* Columns of smoke rising from large fires.

† Berberis.

well as large men. In proportion to the parts abovementioned, their extremities were very small and short, so that when standing they seemed but of a moderate size, and their want of proportion was concealed by the mantle, which enveloped the body entirely, the head and feet being the only parts exposed.

When Mr. Cooke landed, he presented some medals* to the oldest man, and the woman; and suspended them round their necks. A friendly feeling being established, the natives dismounted, and even permitted our men to ride their horses, without evincing the least displeasure, at the free advantage taken of their good-nature. Mr. Cooke rode to the heights, whence he had a distinct view of the Second Narrow, and Elizabeth Island, whither, he explained to the Indians who accompanied him, we were going.

Mr. Cooke returned to the ship with three natives, whom he had induced to go with us to Elizabeth Island; the others were to meet them, and provide us with guanaco meat, to which arrangement the elders of the family had, after / much persuasion, assented. At first they objected to their companions embarking with us, unless we left hostages for their safety; but as this was refused, they did not press the point, and the three young men embarked. They went on board singing; in high glee.

While the ship was getting under way, I went ashore to a larger number of Indians who were waiting on the beach. When my boat landed they were mounted, and collected in one place. I was surprised to hear the woman accost me in Spanish, of which, however, she knew but a few words. Having presented medals to each of the party, they dismounted (excepting the elders), and in a few minutes became quite familiar. By this time Captain Stokes had landed, with several of his officers, who increased our party to nearly double the number of theirs: notwithstanding which they evinced neither fear nor uneasiness. The woman, whose name was Maria,[9] wished to be very communicative; she told me that the man was her husband, and that she had five children. One of the young men, whom we afterwards found to be a son of Maria, who was a principal person of the tribe, was mounted upon a very fine horse, well groomed, and equipped with a bridle and saddle that would have done credit to a respectable horseman of Buenos Ayres or Monte Video. The young man wore heavy brass spurs, like those of the Guachos of Buenos Ayres.[10] The juvenile and feminine appearance of this youth made us think he was Maria's daughter, nor was it until a subsequent visit that our mistake was discovered. The absence of whiskers and beard gives all the younger men a very effeminate look, and many cannot be distinguished, in appearance, from the women, but by the mode in which they wrap their mantles around them, and by their hair, which is turned up and confined

* Previous to the expedition quitting England, I had provided myself with medals, to give away to the Indians with whom we might communicate, bearing on one side the figure of Britannia, and on the reverse 'George IV.' 'Adventure and Beagle,' and '1826.'

by a fillet of worsted yarn. The women cross their mantle over the breast like a shawl, and fasten it together with two iron pins or skewers, round which are twisted strings of beads and other ornaments. They also wear their hair divided, and gathered into long tresses or tails, which hang one before each ear; and those who have short hair, wear false tails made of horse-hair. Under / their mantle the women wear a sort of petticoat, and the men a triangular piece of hide instead of breeches. Both sexes sit astride, but the women upon a heap of skins and mantles, when riding. The saddles and stirrups used by the men are similar to those of Buenos Ayres. The bits, also, are generally of steel; but those who cannot procure steel bits have a sort of snaffle, of wood, which must, of course, be frequently renewed. Both sexes wear boots, made of the skins of horses' hind legs, of which the parts about the hock joints serve for the heels. For spurs, they use pieces of wood, pointed with iron, projecting backwards two or three inches on each side of the heel, connected behind by a broad strap of hide, and fastened under the foot and over the instep by another strap.

The only weapons which we observed with these people were the 'bolas,' or balls, precisely similar to those used by the Pampas Indians;[11] but they are fitter for hunting than for offence or defence. Some are furnished with three balls, but in general there are only two. These balls are made of small bags or purses of hide, moistened, filled with iron pyrites, or some other heavy substance, and then dried. They are about the size of a hen's egg, and attached to the extremities of a thong, three or four yards in length. To use them, one ball is held in the hand, and the other swung several times around the head until both are thrown at the object, which they rarely miss. They wind round it violently, and if it be an animal, throw it down. The bolas, with three balls, similarly connected together, are thrown in the same manner.

As more time could not be spared we went on board, reminding the natives, on leaving them, of their promise to bring us some guanaco meat. Aided by the tide, the ships worked to windward through the Second Narrow, and reached an anchorage out of the strength of tide, but in an exposed situation. The wind having been very strong and against the tide, the ship had much motion, which made our Patagonian passengers very sick, and heartily sorry for trusting themselves afloat. One of them, with tears in his eyes, begged to be landed, but was soon convinced of the difficulty of compliance, / and satisfied with our promise of sending him ashore on the morrow.

After we anchored, the wind increased to a gale, in which the ship pitched so violently as to injure our windlass. Its construction was bad originally, and the violent jerks received in Possession Bay had done it much damage. While veering cable, the support at one end gave way, and the axle of the barrel was forced out of the socket, by which some of the pawls were injured. Fortunately, dangerous consequences were prevented, and a temporary repair was soon applied.

The Beagle, by her better sailing, had reached a more advanced situation, close to the N.E. end of Elizabeth Island, but had anchored disadvantageously in deep water, and in the strength of the tide. Next morning we made an attempt to pass round Elizabeth Island, but found the breeze so strong that we were forced to return, and were fortunate enough to find good anchorage northward of the island, out of the tide.

The Patagonians, during the day, showed much uneasiness at being kept on board so much longer than they expected; but as they seemed to understand the cause of their detention, and as their sickness ceased when we reached smooth water, they gradually recovered their good-humour, and became very communicative. As well as we could understand their pronunciation, their names were 'Coigh,' 'Coichi,' and 'Aighen.' The country behind Cape Negro they called 'Chilpéyo;' the land of Tierra del fuego, 'Oschērri;' Elizabeth Island, 'Tŭrrĕtterr;' the island of Santa Magdalena, 'Shrēe-ket-tup;' and Cape Negro, 'Oērkrĕckur.' The Indians of Tierra del Fuego, with whom they are not on friendly terms, are designated by them 'Săpāllĭŏs.' This name was applied to them in a contemptuous tone.

Aighen's features were remarkably different from those of his companions. Instead of a flat nose, his was aquiline and prominent, and his countenance was full of expression. He proved to be good-tempered, and easily pleased; and whenever a shade of melancholy began to appear, our assurance of / landing him on the morrow restored his good-humour, which was shown by singing and laughing.

The dimensions of Coichi's head[12] were as follows:–

From the top of the fore part of the head to the eyes			4 inches.
Do	do	to the tip of the nose	6
Do	do	to the mouth	7
Do	do	to the chin	9
Width of the head across the temples			7 ½
Breadth of the shoulders			18 ½

The head was long and flat, at the top; the forehead broad and high, but covered with hair to within an inch and a half of the eyebrow, which had scarcely any hair. The eyes were small, the nose was short, the mouth wide, and the lips thick. Neck short, and shoulders very broad. The arms were short, and wanting in muscle, as were also the thighs and legs. The body was long and large, and the breast broad and expanded. His height was nearly six feet.

The next day we rounded Elizabeth Island, and reached Cape Negro, where we landed the Indians, after making them several useful presents, and sending some trifles by Aighen to Maria, who, with her tribe, had lighted large fires about the country behind Peckett's Harbour, to invite us to land. Our passengers frequently pointed to them, telling us that they were made by Maria, who had brought plenty of guanaco meat for us.

Our anxiety to reach Port Famine prevented delay, and, as soon as the boat returned, we proceeded along the coast towards Freshwater Bay, which we reached early enough in the afternoon to admit of a short visit to the shore.

From Cape Negro the country assumed a very different character. Instead of a low coast and open treeless shore, we saw steep hills, covered with lofty trees, and thick underwood. The distant mountains of Tierra del Fuego, covered with snow, were visible to the southward, some at a distance of sixty or seventy miles.

We had now passed all the difficulties of the entrance, and had reached a quiet and secure anchorage. /

The following day was calm, and so warm, that we thought if Wallis and Cordova were correct in describing the weather they met with, Duclos Guyot was equally entitled to credit; and we began to hope we had anticipated worse weather than we should experience. But this was an unusually fine day, and many weeks elapsed, afterwards, without its equal. The temperature of the air, in the shade on the beach, was 67 1/2°, on the sand 87 1/2°; and that of the water 55°. Other observations were made, as well as a plan of the bay, of which there is a description in the Sailing Directions.

Here we first noticed the character of the vegetation in the Strait, as so different from that of Cape Gregory and other parts of the Patagonian coast, which is mainly attributable to the change of soil; the northern part being a very poor clay, whilst here a schistose sub-soil is covered by a mixture of alluvium, deposited by mountain streams; and decomposed vegetable matter, which, from the thickness of the forests, is in great quantity.

Two specimens of beech (*Fagus betuloides* and *antarctica*), the former an evergreen, – and the winter's bark (*Wintera aromatica*), are the only trees of large size that we found here; but the underwood is very thick, and composed of a great variety of plants, of which *Arbutus rigida*, two or three species of *Berberis*, and a wild currant (*Ribes antarctica*, Bankes and Solander[13] MSS.), at this time in flower, and forming long clustering bunches of young fruit, were the most remarkable. The berberis produces a berry of acidulous taste, that promised to be useful to us. A species of wild celery, also, which grows abundantly near the sea-shore, was valuable as an antiscorbutic. The trees in the immediate vicinity of the shore are small, but the beach was strewed with trunks of large trees, which seemed to have been drifted there by gales and high tides. A river falls into the bay, by a very narrow channel, near its south end; but it is small, and so blocked up by trees as not to be navigable even for the smallest boat: indeed, it is merely a mountain torrent, varying in size according to the state of the weather. /

Tracks of foxes were numerous about the beach, and the footsteps of a large quadruped, probably a puma, were observed. Some teal and wild ducks were shot; and several geese were seen, but, being very wary, they escaped.

Upon Point St. Mary we noticed, for the first time, three or four huts or wigwams made by the Fuegian Indians, which had been deserted. They were not old, and merely required a slight covering of branches or skins to make them habitable. These wigwams are thus constructed: long slender branches, pointed at the end, are stuck into the ground in a circular or oval figure; their extremities are bent over, so as to form a rounded roof, and secured with ligatures of rush; leaving two apertures, one towards the sea, and the other towards the woods. The fire is made in the middle, and half fills the hut with smoke. There were no Indians in the bay when we arrived, but, on the following evening, Lieutenant Sholl, in walking towards the south end of the bay, suddenly found himself close to a party which had just arrived in two canoes from the southward. Approaching them, he found there were nine individuals – three men, and the remainder women and children. One of the women was very old, and so infirm as to require to be lifted out of the canoe and carried to the fire.[14] They seemed to have no weapons of any consequence; but, from our subsequent knowledge of their habits, and disposition, the probability is they had spears, bows, and arrows concealed close at hand. The only implement found amongst them was a sort of hatchet or knife, made of a crooked piece of wood, with part of an iron hoop tied to the end. The men were very slightly clothed, having only the back protected by a seal's skin; but the females wore large guanaco mantles, like those of the Patagonian Indians, whom our pilot told us they occasionally met for the purpose of barter. Some of the party were devouring seal's flesh, and drinking the oil extracted from its blubber, which they carried in bladders. The meat they were eating was probably part of a sea lion (*Phoca jubata*); for Mr. Sholl found amongst them a portion of the neck of one of those animals, which is / remarkable for the long hair, 'like a lion's mane,' growing upon it. They appeared to be a most miserable, squalid race, very inferior, in every respect, to the Patagonians. They did not evince the least uneasiness at Mr. Sholl's presence, or at our ships being close to them; neither did they interfere with him, but remained squatting round their fire while he staid near. This seeming indifference, and total want of curiosity, gave us no favourable opinion of their character as intellectual beings; indeed, they appeared to be very little removed from brutes; but our subsequent knowledge of them has convinced us that they are not usually deficient in intellect. This party was perhaps stupified by the unusual size of our ships, for the vessels which frequent this Strait are seldom one hundred tons in burthen.[15]

We proceeded next morning at an early hour. The Indians were already paddling across the bay in a northerly direction. Upon coming abreast of them, a thick smoke was perceived to rise suddenly from their canoes; they had probably fed the fire, which they always carry in the middle of their canoe, with green boughs and leaves, for the purpose of attracting our attention, and inviting us to communicate with them.

It was remarked that the country begins to be covered with trees at Cape Negro; but they are stunted, compared with those at Freshwater Bay. Near this place, also, the country assumes a more verdant aspect, becoming also higher, and more varied in appearance. In the neighbourhood of Rocky Point some conspicuous portions of land were noticed, which, from the regularity of their shape, and the quantity as well as size of the trees growing at the edges, bore the appearance of having been once cleared ground; and our pilot Robinson (possessing a most inventive imagination) informed us that they were fields, formerly cleared and cultivated by the Spaniards, and that ruins of buildings had been lately discovered near them. For some time his story obtained credit, but it proved to be altogether void of foundation. These apparently cleared tracts were afterwards found to be occasioned by unusual poverty of soil, and by being overrun with thick / spongy moss, the vivid green colour of which produces, from a distance, an appearance of most luxuriant pasture land. Sir John Narborough noticed, and thus describes them: 'The wood shows in many places as if there were plantations: for there were several clear places in the woods, and grass growing like fenced fields in England, the woods being so even by the sides of it.'**

The wind, after leaving Freshwater Bay, increased, with strong squalls from the S.W., at times blowing so hard as to lay the ship almost on her broadside. It was, however, so much in our favour, that we reached the entrance of Port Famine early, and after some little detention from baffling winds, which always render the approach to that bay somewhat difficult, the ships anchored in the harbour. /

* Narborough, p. 67.

CHAPTER III.

Prepare the Beagle, and our decked boat (the Hope) for surveying the Strait – Beagle sails westward, and the Hope towards the south-east – Sarmiento's voyage – and description of the colony formed by him at Port Famine – Steamer-duck – Large trees – Parroquets – Mount Tarn – Barometrical observations – Geological character – Report of the Hope's cruize.

In almost every account published of the Strait of Magalhaens, so much notice has been taken of Port Famine, that I had long considered it a suitable place for our purposes; and upon examination I found it offered so many advantages, that I did not hesitate to make it our head-quarters. As soon, therefore, as the ship was moored, tents were pitched, our decked boat was hoisted out and hauled on shore, to be coppered and equipped for the survey; – and Captain Stokes received orders to prepare the Beagle for examining the western part of the Strait; previous to which she required to be partially refitted, and supplied with fuel and water.

For several days after our arrival, we had much rain and strong south-westerly wind, with thick clouds, which concealed the high land to the southward; allowing us only now and then a partial glimpse. One evening (11th) the air was unusually clear, and many of the mountains in that direction were distinctly defined. We had assembled to take leave of our friends in the Beagle, and were watching the gradual appearance of snow-capped mountains which had previously been concealed, when, bursting upon our view, as if by magic, a lofty mountain appeared towering among them; whose snowy mantle, strongly contrasted with the dark and threatening aspect of the sky, much enhanced the grandeur of the scene.

This mountain was the 'Snowy Volcano' (*Volcan Nevado*) of Sarmiento,[1] with whose striking appearance that celebrated navigator seems to have been particularly impressed, so minute / / and excellent is his description. It is also mentioned in the account of Cordova's voyage.* The peculiar shape of its summit as seen from the north would suggest the probability of its being a volcano, but we never observed any indication of its activity. Its volcanic form is perhaps accidental, for, seen from the westward, its summit no longer resembles a cra-

* Ultimo Viage, p. 120.

THE HOPE, IN THE STRAIT OF MAGALHAENS.

THE ADELAIDE, IN HUMMING BIRD COVE.

P.P. King. S. Bull.

DISTANT VIEW OF MT SARMIENTO.

ter. From the geological character of the surrounding rocks its formation would seem to be of slate. It is in a range of mountains rising generally two or three thousand feet above the sea; but at the N.E. end of the range are some, at least four thousand feet high. The height of the 'Snowy Volcano,' or as we have called it, Mount Sarmiento,* was found, by trigonometrical measurement, to be six thousand eight hundred feet† above the level / of the sea. It is the highest land that I have seen in Tierra del Fuego; and to us, indeed, it was an object of considerable interest, because its appearance and disappearance were seldom failing weather guides. In our Meteorological Diary, a column was ruled for the insertion of its appearances.‡

This clear state of the atmosphere was followed by a heavy fall of rain, with northerly and easterly winds, which did not, however, last long.

In the vicinity of our tents erected on the low land, on the S.W. side of the bay, were several ponds of water, perfectly fit for immediate use; but, perhaps, too much impregnated with vegetable matter to keep good for any length of time. Captain Stokes, therefore, filled his tanks from the river; but as that water did not keep well, it was probably taken into the boat too near the sea. This, however, was unavoidable, except by risking the boats among a great number of sunken trees in the bed of the river.

* From an attentive perusal of the voyage of Magalhaens, I have lately been led to think that this is the mountain which Magalhaens called Roldan's Bell. Sarmiento has, however, assigned that name to a mountain at the back of his Bay of Campana, which will be noticed in its proper place. The name of Mount Sarmiento was too long, and too well established with us, or I should have restored the name bestowed upon it by Magalhaens. Herrera, in his Description de las Indias Occidentales, cap, xxiii, notices the 'Campana de Roldan' as a great mountain in the midst of the entrance of a channel; they gave it this name (Campana de Roldan) because one of Magalhaens's companions, named Roldan, an artillery officer, went to examine it. 'Y la campana de Roldan una Peña grande en medio al principio de un canal: dieron le este nombre porque la fué a reconocer uno de los compañeros de Magallanes Ilamado Roldan que era artillero.'[2]

† By angular measurement, with a theodolite, from the tent, the base being by diff, of lat. 297,863 feet, and allowing 1/12 of the intercepted arc for terrestrial refraction 6,864 feet.
By angular measure with sextant (index error, dip, and 1/12 of the intercepted arc being allowed) the base being 290,074 feet 7,237
By angular measurement, with a theodolite, from Warp Bay, by Lieuts. Skyring and Graves 6,800
Mean 6,967 feet,
but as the last observation, from the angle of elevation being greater, was more likely to be correct, 6,800 feet is considered to be its elevation.

‡ At a subsequent visit, embracing a period of 190 days, it was only seen on twenty-five, and during seven days only was it constantly visible. On the remaining eighteen, portions only were seen, and those but for a few hours at a time.

The Beagle sailed on the 15th, to survey the western entrance of the Strait, with orders to return to Port Famine by the end of March.

Our decked boat, the Hope, being ready, the command of her was given to Mr. Wickham, who was in every way qualified for the trust. We were, however, much mortified by finding that she leaked so considerably as to oblige us to unload, and again haul her on shore. When ready for sea, she sailed under the direction of my assistant-surveyor, Mr. Graves, to examine the St. Sebastian channel and the deep opening to the S.E. of Cape Valentyn. Her crew consisted of seven men, besides Mr. Wickham, and Mr. Rowlett, the purser.

Having despatched the Beagle and the Hope, I was at leisure to carry on the survey of the coast in the neighbourhood of Port Famine, and to make a plan of the port itself. The / Transit, and Altitude circle, were set up; but from the very unfavourable state of the weather, and the interference of other occupations, I was only enabled to procure a series of zenith distances of the sun, and stars, for the latitude.[3]

Port Famine, a name well known to all who have interested themselves about the Strait of Magalhaens, was selected by Sarmiento as the most convenient place for the site of an establishment formed, at his suggestion, by Philip II, King of Spain.

The voyage of Sir Francis Drake through the Strait into the Pacific, and his successes against the Spanish colonies and trade on the western side of the continent of America, induced the Viceroy of Lima to send an Expedition[4] to pursue the 'Corsair,' with orders to fight and take him, dead or alive.* This Expedition, commanded by Pedro Sarmiento de Gamboa, who had already been engaged twice with Drake, consisted of two ships, containing in all two hundred armed men, sailors and soldiers; a force which was considered sufficient to ensure the capture.†

The Strait of Magalhaens being the most likely place to meet with Drake, Sarmiento was ordered to proceed through it, and take the opportunity of exploring its coasts.

All this he performed in a manner highly creditable, as well for the excellent description handed down in his unpretending journal, as for the enterprising zeal, and steady perseverance, shown among difficulties of no trifling nature. To his accounts of various places there will be frequent occasion to refer. Our object, at present, is to give a short account of the Colony.

Sarmiento sailed from Peru (1583), and entered the Strait from the Pacific. After experiencing many serious difficulties, and escaping imminent dangers, in the western part of the Strait, where the climate is so rigorous and the country so desolate, it was not surprising that he should become enraptured with the

* Sarmiento's Voyage, p. 25.

† Id. I. c.

verdant, and picturesque appearance of the shores to the eastward of Cape Forward, and with the open country in / the neighbourhood, and to the northward of Cape Virgins.* After much opposition from the Duke of Alva†[5] and other powerful people, he succeeded in convincing the King of the expediency of fortifying the shores of the First Narrow, and forming several establishments within the Strait, to prevent the passage of strange ships, to the prejudice of the King's colonies in Chile and Peru; for at that time the passage round Cape Horn was not known. Accordingly, an Expedition was prepared, consisting of twenty-three vessels, under the joint command of Diego Florez de Valdez and Sarmiento; the former being appointed Captain-general of the fleet, and of the coast of Brazil; and the latter, Captain-general of the Strait of Magalhaens, and Governor of all the Establishments that should be formed within it.

Of the twenty-three ships which sailed from Spain, five only reached the entrance of the Strait; and these, after experiencing many difficulties from bad weather and foul winds, returned to Rio de Janeiro to refit, where Sarmiento met four vessels which had been sent from Spain to his succour. His colleague and General in chief, Florez, who had deserted the Expedition, did all in his power to impede Sarmiento, to the latest moment of his stay at the Brazils. At last, however, five ships, commanded by Ribera, and manned by five hundred and thirty men,‡ sailed; and, without encountering further loss or detention, arrived off the Strait in December (1584), and soon after reached an anchorage, between the First and Second Narrows.

Ribera would go no further; but landed about three hundred men, under Sarmiento. A city was marked out, and named Jesus,§§ in a valley well provided with water. The / ships were blown away to sea, leaving the colonists very destitute; fortunately, however, they were enabled to return, but were four times, afterwards, obliged to put to sea, from stress of weather. On the last return, one of the ships, La Trinidad, was run on shore. The ardour of Ribera being damped by repeated misfortunes, he returned to Spain, without the knowledge or consent of Sarmiento, leaving, for the use of the colony, only one ship, the Maria.

While unloading the Trinidad, the Spaniards were attacked by Indians, whom they dispersed.

* See Burney, ii. p. 45, for a fuller account; also id. 71.

† Who made a remark on the occasion, which became proverbial, 'that if a ship carried out only anchors and cables, sufficient for her security against the storms in that part of the world, she would go well laden.' Burney Coll. vol. ii. 45.

‡ Burney, ii. 51.

§ The situation of 'Jesus' must have been about half-way between the First and Second Narrow, near the point named in the chart N.S. de Valle, where some peaked elevations, dividing vallies near the coast line, are conspicuous. The Beagle anchored there, and found plenty of fresh water.

Sarmiento, after making the necessary arrangements at Jesus, set out by land with one hundred men, to go to Point St. Anna,* the ship Maria being ordered to follow. On the journey, the sufferings of the party were very great, as well from the fatiguing nature of the march, as from their being harassed by the natives, with whom they had an engagement, in which one was killed, and ten men were wounded. A mutiny among his people then broke out, which was quelled by assistance from the ship. At last they reached their destination, and founded, with the usual solemnities, the city of King Philip (or San Felipe).

At the latter end of March, while preparing habitations, the winter set in so suddenly, that for fifteen days it did not cease to snow. Sarmiento, then, after quelling a mutiny which had broken out afresh among the soldiers, embarked with thirty men to visit the first encampment at Jesus, and to superintend the erection of forts in the Narrow; but upon reaching the anchorage, a gale of wind forced him to sea, and, lasting twenty days, obliged him (with his people blinded and frost-bitten) to bear up for Rio de Janeiro.

Here his ship was stranded; upon which he chartered a vessel to convey flour to the Strait, and went himself to Pernambuco, to procure large boats for carrying supplies to his / colony, and assisting in the recovery of his stranded ship; she had, however, drifted off, and sunk near Bahia; and all his boats were destroyed. Still Sarmiento persevered in his zealous efforts to succour his friends in the Strait; and succeeded in procuring a vessel of fifty or sixty tons, which, loaded with arms and whatever he considered useful, sailed, and reached Rio de Janeiro a month after the departure of the first vessel (January 1585). He followed, but in the latitude of 39° met with a furious gale, which drove him back to Rio de Janeiro, where the vessel that had preceded him had returned in distress.

Disappointed in his attempts to carry succour to the colony, he determined to go to Spain; but on his voyage thither, to complete the catalogue of his misfortunes, his ship was captured by three English vessels, and taken to England, after which the ill-fated colony in the Strait was neglected, if not entirely forgotten.

Two months after Sarmiento's departure from the Strait of Magalhaens, in the month of August, the middle of the winter of that region, the party belonging to the first establishment at Jesus set off by land, and joined that at San Felipe, with the unwelcome tidings of their deserted state. But as the provisions at San Felipe were insufficient to support all the people, Andres de Viedma, who, after Sarmiento's departure, had assumed the command, detached two hundred soldiers, under the command of Juan Iniguez, back to Jesus, for the purpose of communicating with any ship that might make her appearance, and awaiting the expected return of Sarmiento; but the winter and following summer passed by without any relief.

* Close to Port Famine.

In this unhappy state, the colonists were obliged to think only of providing for their safety, and built two boats; in which fifty people embarked, besides Viedma, Suarez, a Franciscan friar named Antonio, and five Spanish women. They had not proceeded farther than Point Santa Brigida,* / when one of the boats struck upon a reef, and was lost, but the people were saved. The loss of this boat caused them to give up every hope of saving themselves in that way; and Viedma, with Suarez, the friar, and twenty soldiers, returned in the remaining boat to San Felipe, leaving the rest of the party, consisting of thirty men and five women, to support themselves through the approaching winter as they could. After that season had passed, Viedma sent to collect the wanderers; but fifteen men, and three women only, could be found; the rest having died of hunger and disease. The survivors then determined upon going to the first establishment at Jesus; on their way to which they passed by the skeletons of the two hundred who had been first detached. Travelling onwards, they observed three ships entering the strait, which anchored at a distance to the southward.

During the night, Viedma and his companions kept up large fires, supposing that the ships belonged to their own nation. Next morning a boat was despatched from them; and three of Viedma's party obtained permission to go and reconnoitre her. Having approached near enough, a signal was made; upon which, the people in the boat pulled towards the beach, and said they were from England, bound to Peru, and that if the Spaniards wanted a passage, they had better embark. After some hesitation, arising from the fear of trusting themselves in the power of heretics, they consented; and one was permitted to get in, but the other two were left on the beach. In the boat was the enterprizing Cavendish† himself, who, on hearing the particulars of their story, sent the other two soldiers to Viedma, offering to take him and the residue of his people on board. Cavendish returned to his ship; but, without further delay, sailed on to the Isla dos Patos (Santa Magdalena Island), where he leisurely salted down six casks of penguins; and then proceeded to San Felipe, for wood and water; he remained there four days (during which time he destroyed the houses of the Spaniards, and embarked six guns); and thence continued his voyage. The person saved / by Cavendish, whose name was Tome Hernandez, afterwards escaped from him at Quintero, near Valparaiso; and, proceeding to Peru, gave an account of the fate of this cruelly neglected colony.[6]

* From Sarmiento's description of the coast, Point Santa Brigida is the outward point of Nassau Island, (a) See Sarmiento's Voyage, p. 220.

(a) By Nassau Island is meant the land forming the south shore of the Second Narrow. – R.F.

† Formerly spelled 'Candish.'

This was the first, and perhaps will be the last, attempt made to occupy a country, offering no encouragement for a human being; a region, where the soil is swampy, cold, and unfit for cultivation, and whose climate is thoroughly cheerless.

The name, San Felipe, ceased with the colony; for Cavendish called it Port Famine, in allusion to the fate of the colonists, all of whom, except the man he took away, and one saved two years afterwards (in 1589), by Andrew Mericke,*[7] perished from hunger and its attendant diseases; and by this appellation the bay has since been universally known. To commemorate the ill-fated town, a very thickly-wooded mountain at the bottom of the bay, which forms a conspicuous and picturesque object, has been named by us Mount San Felipe.

At this port, Sarmiento, on his first voyage through the Strait, communicated with a large party of Indians, in consequence of which he called it Bahia de la Gente; and the river, which now bears the name of Sedger, he named San Juan. Of this river Sarmiento took formal possession, as well as of the whole Strait, for the 'Mui Poderoso y mui Católico Señor Phelipe Segundo,' &c. &c. It was also here that, in consequence of the miraculous preservation of his vessel on many / occasions, he attempted to change the name of the strait to Estrecho de la Madre de Dios; but it had been too long called Magalhaens, for even the influence of Sarmiento, backed by the power of Philip, to persuade the world to countenance so great an injustice.

> 'Magallanes, Señor, fué el primer hombre
> Que abriendo este camino le dió nombre.'
> Ercilla Araucana,[8] Cant. I. oct. 8.

During an excursion with Mr. Tarn to Eagle Bay,† beyond Cape San Isidro, we found many wigwams. They were then novelties to us, and we were ignorant of their being such certain indications of very sheltered places, as subsequent experience has shown them to be. We often used them, after they had been well cleaned out: a boat's sail, thrown over the hemispherical roof, was a sufficient protection from rain; – and from wind they are always well defended by their situation. Here we saw, for the first time, that most remarkable bird the Steamer-

* 'Near to Port Famine they took on board a Spaniard, who was the only one then remaining alive of the garrison left in the Strait by Sarmiento. The account given by this man, as reported by Magoths, is, that he had lived in those parts six years, and was one of the four hundred men sent thither by the King of Spain in the year 1582, to fortify and inhabit there, to hinder the passage of all strangers that way into the South Sea. But that town (San Felipe) and the other Spanish colony being destroyed by famine, he said he had lived in a house, by himself, a long time, and relieved himself with his caliver *(b)* until our coming thither.' Burney, ii. p. 96. This man died on the voyage to Europe Id. p. 97.

(b) A kind of gun – R.F.

† So named by Bougainville.

duck. Before steam-boats were in general use, this bird was denominated, from its swiftness in skimming over the surface of the water, the 'race-horse,' a name which occurs frequently in Cook's, Byron's, and other voyages. It is a gigantic duck, the largest I have met with. It has the located hind-toe, legs placed far backwards, and other characteristics of the oceanic ducks.* The principal peculiarity of this bird is, the shortness and remarkably small size of the wings, which, not having sufficient power to raise the body, serve only to propel it along, rather than through the water, and are used like the paddles of a steam-vessel. Aided by these and its strong, broad-webbed feet, it moves with astonishing velocity. / It would not be an exaggeration to state its speed at from twelve to fifteen miles an hour. The peculiar form of the wing, and the short rigid feathers which cover it, together with the power this bird possesses of remaining a considerable length of time under water, constitute it a striking link between the genera *Anas* and *Aptenodytes*. It has been noticed by many former navigators. The largest we found measured forty inches, from the extremity of the bill, to that of the tail, and weighed thirteen pounds; but Captain Cook mentions, in his second voyage, that the weight of one was twenty-nine pounds.† It is very difficult to kill them, on account of their wariness and thick coat of feathers, which is impenetrable by any thing smaller than swan shot. The flavour of their flesh is so strong and fishy, that at first we killed them solely for specimens. Five or six months, however, on salt provisions, taught many to think such food palatable, and the seamen never lost an opportunity of eating them. I have preferred these ducks to salt-beef, but more as a preventive against scurvy, than from liking their taste.

I am averse to altering names, particularly in natural history, without very good reason, but in this case I do think the name of 'steamer' much more appropriate, and descriptive of the swift paddling motion of these birds, than that of 'racehorse.' I believe, too, the name of 'steamer' is now generally given to it by those who have visited these regions.

Many shells‡ were taken from the bottom by means of a fizgig[10] which Mr. Tarn found in one of the wigwams: it was a / rough pole, eight or ten feet long,

* It belongs to the group which M. Temminek has lately named *Hylobates*, without attending to the name long since conferred upon it by Dr. Fleming. I designated it *Oidemia Patachonica*, from its large dimensions, in my communication upon the Ornithology of the Straits. Zoological Journal, vol. iv. p. 100. On my return to England, I found that M. de Freycinet had figured this bird, in the account of his last voyage in l'Uranie, where it is described by Messrs. Quoy and Gaimard under the name of *Micropterus brachypterus*.[9]

† Cook's Second Voyage, 4 to. p. 570.

‡ On the shores of Eagle Bay we procured a large collection of shells, among which were *Margarita violacea* (Nob. in Zool. Journ. v.346, No. 53), a beautiful Modiola (*M. trapesina*, Lam.[k]), a new *Pecten* (P. vitreus Nob. in Zool. Jour. v.337, No. 17), and a delicate transparent-shelled *Patella*, answering the description of *P. cymbularia*. These four species were found attached to floating leaves of the kelp (*Fucus giganteus*), and afford food

split crosswise at one end, and opened so as to form four prongs, kept apart by two small pieces of wood. Although rudely made, it was excellently adapted for a shell-gatherer, and is used by the Indians for collecting sea-eggs, which are found in the Strait of very large size, and are doubtless, to them, a great delicacy.

During our excursion we ascertained the best place to ascend the snowy mountain, since named 'Tarn;' and the surgeon, whose name it bears, set off with a party of officers to make the attempt, in which he succeeded, and obtained such an extensive view as induced me to decide upon ascending it, a few days afterwards, to procure bearings from the summit, and for the purpose of measuring its height with a barometer.

In the meantime I visited the Sedger river (Sarmiento's 'Rio de San Juan de Posesion'), and found some difficulty in entering it, because of several banks which are dry at low water. Between them, however, the stream keeps a small channel open, by which we effected our purpose. Every gale of wind causes the banks to shift, and between the times of our first, and last, visit to Port Famine, the river's mouth underwent many changes. The bed of the river is so full of fallen trees, that we could not go, with the boat, more than three miles and a half above the entrance; there it was about fifteen yards wide, bounded on each side by thickly wooded banks, of moderate height. The trees on these banks are large, chiefly the two species of Beech before-mentioned, and Winter's-bark; there are besides many shrubs, and an impenetrable underwood of Arbutus, Berberis, and currant bushes. The largest Beech tree that we saw could not have been more than thirty or forty inches in diameter, which was insignificant compared with those noticed by Commodore Byron. In describing his excursion up this river, he mentions 'trees that would supply the British navy with the best masts in the world.'* 'Some of them are of a great height, and more than eight feet in diameter, which is proportionably more than eight yards in circumference.'† / The Commodore may have been pleased by the appearance of these trees, but must have fancied their quality and dimensions such as he describes. The largest are generally rotten at the heart, and all are more or less defective. Their wood is heavy, and far too brittle for masts: we could not use it even for boat-hook staves. It makes, however, tolerable plank for boat-building, and, when seasoned, might be used in ships. For common purposes, such as houses, or fences, it is very serviceable.

We wandered about to examine the country; but, excepting the track of some quadruped, whose foot was small and cloven, rather like a pig's, we saw nothing new. The traces of foxes were numerous everywhere. We found no fish of any description in the river. Geese and wild ducks were numerous, whose young

to the steamer-duck. We also collected good specimens of *Murex Magellanicus*, Lam.[k], of *Fissurella picta*, Lam.[k], and a great number of the common patella of the Strait, which forms a considerable article of food for the Natives.

* Byron's Voyage round the World, 4to. p. 38.

† I. c.

were at this time scarcely fledged, and an easy prey. We also observed here, for the first time, the parroquet, which Bougainville described to be common in the Strait. He carried specimens home with him; but some naturalists of those days decided that there must have been a mistake, because, as they averred, parroquets did not exist in so high a latitude. Bougainville, however, made no mistake, for the species* is very abundant in the neighbourhood of Port Famine, and has been seen by us in all parts of the Strait. It feeds principally upon the seeds of the Winter's-bark. The existence of this bird in Tierra del Fuego is also mentioned by Cook and Narborough.†

All accounts of Port Famine informed us of its abounding in fish, but as yet we had taken none excepting with hook and line, although the seine had been frequently shot. At last, however, in the first week of February, we had a successful haul of mullet and smelts, many of the former weighing eight pounds, and the latter measuring fifteen inches in length. After this we were often very fortunate, and on one occasion caught, at one haul of the seine, sixteen hundred-weight of smelts, some weighing two pounds, and measuring twenty inches in length. A few days previously we had a draught of mullet, which served the crews of both Adventure and Beagle for three days. Geese, wild ducks and teal, snipe, and now and then woodcocks, were to be found by taking a short walk; there were, however, no quadrupeds fit for food which we could take. Foxes and wild cats were occasionally seen, and a foot-mark of some large animal of the feline race, probably a puma, was once observed upon the beach. We found many traces of horses, which showed that the Patagonian Indians sometimes come thus far south. Had we been so fortunate as to meet them here, we might have procured, perhaps, a regular supply of guanaco meat.

One the 9th of February, as the weather seemed favourable for ascending Mount Tarn,‡ Lieutenant Cooke, the Surgeon, and Anderson, the botanical collector, set off in advance to select a convenient place for passing the night, carrying with them a tent and provisions. I followed later in the day, and, while the boat's crew were arranging their loads, made some observations with a barometer on the beach.

* *Psittacus smaragdinus*, Gmel. I have no doubt that the bird we saw is the same as Bougainville procured, and from which a description has been given in the Ency. Méth., art. Ornith. 139; although a material error is made, for they are not *splendidè viridis*, nor is the *uropygium* red; in other points, however, the description is correct. See Buffon's Hist. Nat. des Oiseaux, vi.262. Pl. enl. n. 85, Perruche des Terres Magellaniques.

† Bougainville says, 'we have likewise perceived some perrokeets: the latter are not afraid of the cold.' To which the English translator, T.R. Forster,[11] who is incredulous of the correctness of Bougainville's assertion, appends the following note: 'Perruches, probably sea-parrots, or auks.' Buffon also doubted the fact, and the author of Histoire Naturelle, art. Oiseaux, tom. ii. p. 322, suggests the possibility of a specimen having been obtained in some other part of the world, and put, by mistake, amongst those collected in the Strait.[12] /

‡ So named because Mr. Tarn, the surgeon of the Adventure, was the first person who reached its summit.

Our way led through thick underwood, and then, with a gradual ascent, among fallen trees, covered with so thick a coating of moss, that at every step we sunk up to the knees / before firm footing could be found. It was very laborious work, and the ground being saturated, and each tree dripping with moisture, we were soon wet through. We proceeded along the same sort of road up a steep ascent; some one of the party constantly falling into deep holes covered by moss, or stumbling over fallen trunks of trees. As I carried a barometer, I was obliged to proceed with caution, and succeeded in emerging from this jungle without accident. After about three quarters of an hour spent in this way, we reached an open space, where we rested, and I set up the barometer. Here we found a cypress of very stunted growth.

Our road hence was rather more varied: always steep, but sometimes free from impediment. Here and there we observed the boggy soil was faced with a small plant (*Chamitissp.*) of a harsh character, growing so thick and close as to form large tufts, over which we walked as on hard ground. We struggled through several thickets of stunted beech-trees, with a thick jungle of Berberis underneath, whose strong and sharp thorns penetrated our clothes at every step; and began to find the fatigue very oppressive: some of my boat's crew suffered much, being unused to such exercise. At last we approached the place where Mr. Cooke and his party had established themselves, and upon hailing, were invigorated by a cheer in reply. We reached the bivouac in a very way-worn condition, and found, to our great comfort, the tent pitched, and a good fire burning.*

The ground was so exceedingly wet, that although we slept upon branches, forming a layer at least a foot thick, we found ourselves, in the night, lying as if in a morass, and suffering from cold, even with a large fire blazing at our feet. At daylight next morning, just as we were starting, a boat was seen sailing round Cape San Isidro, which, by the aid of a telescope, I made out to be the Hope.

We resumed the ascent, and passed over, rather than through, thickets of the crumply-leaved beech, which, from / their exposure to the prevailing winds, rose no higher than twelve or fourteen inches from the ground, with widely-spreading branches, so closely interwoven, as to form a platform that bore our weight in walking. We next traversed an extent of table-land,† much intersected by ponds of water. Mr. Tarn shot two plovers of a new species (*Charadrius rubecola,* Zool. Jour. vol. iv. p. 96), and a snipe. We then ascended three or four hundred feet, and crossed a deep ravine. The bottom of the ravine was clay-slate in a decomposing state, but the surface of the ground was strewed with pebbles of granite. Another plain, with many ponds, succeeded; the intervening spaces being cov-

* The height of this place, as shown by the barometer, on the ascent, was 941 feet, and, on the descent, 973 feet.

† On this table-land the barometer stood at 27,767. Temperature of the air 46°, 5, and of the mercury 47°,5, which gave the elevation 1,327 feet.

ered with tufts of chamitis, and studded here and there with small clusters of dwarf beech; but the ground was so hard, and firm, that we proceeded rapidly, without fatigue, until we attained the height of 1,800 feet, when the ascent became very steep. Near the summit lay a large mass of snow, rapidly melting away. We reached the highest pinnacle of the mount at seven o'clock (having left our resting place at four), and immediately set up the instruments. I was obliged to avail myself of Mr. Tarn's assistance to hold the barometer, whilst two of my boat's crew held the legs of the theodolite-stand, for the wind was blowing very strongly, and the edge of a precipice was close to us, perpendicular for many hundred feet, and thence downwards so steep, that any body going over would fall at least a thousand feet. The theodolite-stand was unavoidably placed within a very few inches of the edge, and I took a round of angles, suffering, however, intense pain from the piercing coldness of the wind, which, heated as we were by the ascent, was much felt, though the temperature was not lower than 39°. I was lightly clothed, and should have fared badly, had not one of the party lent me his Flushing jacket, while he descended under the lee of the mountain-top to make a fire. The barometer stood at 26, 618, the temperature of the air being 40°, and of / the mercury 43°.* Unfortunately the day was very cloudy, and many squalls of sleet and rain, which obscured the hills, passed whilst I was taking bearings. To the N.E., towards the supposed Sebastian Channel, the horizon was too hazy to allow much view. A deep inlet was seen in that direction; but whether the land closed round, or whether a channel was at the bottom, we could not distinguish. A considerable body of water was observed to the southward of Cape St. Valentyn, behind Lomas Bay, but its extent was screened from our view by the intervention of the Lomas hills. It appeared to be a channel, the opposite or eastern side of it being formed by the high ranges previously seen from Point St. Mary. Cordova's Ports San Antonio and Valdez were distinctly made out; but, to the southward, every thing was enveloped in mist.

By angular measurement from Observation Cove, Port Famine, with theodolite, allowing 1/12 of the intercepted arc for terrestrial refraction, the height is 2,850 feet.

Another observation, with the sextant, made it 2,855 feet. The mean 2,852 I consider more correct, from the difficulty of obtaining a correct reading of the barometer on the summit.

* The result of the barometeric observation for the height of Mount Tarn is as follows:

Height by one barometer	ascent	2,567.7	mean	2,596.5 feet.
	descent	2,625.4		
Do. two do	ascent	2,619.3		
	descent	2,596.7		2,608.0
				2,602.2

The bearings and observations, which occupied me nearly two hours, being completed, we all adjourned to a sheltered cleft in the rock close to our station, where we soon recovered the use of our fingers.*

Having accomplished our object, we began the descent. In a comparatively mild and agreeable spot, I again set up the theodolite and barometer, while some of the party employed themselves in fruitless attempts to kindle a fire. The height, by the barometer, proved to be 1,845 feet above the sea; and the bearings from this station were much better than those I had taken from the exposed summit.

We reached our tent at noon, having been absent seven hours. At three we reached the beach, where the barometer stood at 29,312 (air 61°3,† and mercury 62,°5).

Excepting near the sea, where clay-slate (very similar to that of Point St. Anna, but with an opposite dip) showed itself, the side of the hill is clothed with trees and underwood, and no rock is visible until one arrives at the ravine. Around the summit of Mount Tarn the ground is bare, but so covered with small decomposed fragments, that the solid rock only appears occasionally: it is very hard, and breaks with a conchoidal fracture: some of the specimens which we detached bore indistinct impressions of organic remains. We also found, projecting from the rock in which they were embedded, nodules, or small rounded masses of stone, in an advanced state of decomposition, mouldering away in laminar forms somewhat resembling the inner leaves of a cabbage. Several were brought away carefully, but before we arrived on board they had crumbled to pieces: the nucleus was quite hard, but was surrounded by concentric laminæ, more brittle the nearer they approached to the outer surface. It seemed as if the face of the summit / above-mentioned was covered with the decomposing fragments of these nodules.

* By Daniell's hygrometer, used in this sheltered spot, I found the temperature of the air to be 48°; dew point 41°: but upon exposing the instrument to the wind, the air was 39 1/2°, and the dew point 36°: the difference in the former being 7°; and the latter 3 1/2°; from which the following results are obtained:

	air.	dew pt.	diff.	exp.	dryness.	weight of a cubic. foot of air.
In the ravine	48	41	7	292	776	3.323
Exposed to wind	39 ½	36	3 ½	248	898	2.871
Difference	8 ½	5	3 ½	44	122	0.452

The above being the difference in the short space of three feet apart; the instrument, in the first case, being just under the lee of the rocky summit of the mountain, and in the last, above it, exposed to the wind. /

† The air was so dry this afternoon that I failed to procure a deposit of dew upon Daniell's hygrometer, although the internal temperature was lowered from 61° to 37°. One of Jones's portable hygrometers[14] was also tried, and the temperature was lowered to 31° 1/2 without a deposit; so that, the difference being more than thirty degrees, the expansive force of the air must have been less than 212, the dryness, on the thermometric scale, less than 367, and the weight of vapour, in a cubic foot of air, less than 2,355 grains.

The highest parts of the Mount form a ridge extending S. E. and N.W., being a succession of strata of slaty rock, dipping to the eastward, at an angle of 15° or 20° from the horizon. The strata are very narrow, and separated from each other by a vein of quartz, much of which is in a crystallized state. We reached the ship about seven o'clock, and found that the Hope and her party had done well. Her cruize proved interesting, with regard to the geography of the Strait, and a summary of it is subjoined.

Mr. Graves's orders were to survey the Sebastian Channel; but in the event of his seeing any thing more interesting to the S.E., he was allowed to defer that service to another opportunity. The Hope crossed the Strait, and anchored in a small bay, formed between the two projecting points of Cape Valentyn, where some few defects in the vessel were remedied, and a good round of angles obtained from the summit of the Cape, whence there was a fine view. The country was low, undulating, and destitute of trees. From a station about two miles overland, to the eastward, a large body of water was observed to the southward, forming a channel, or deep sound, and it was determined to follow up its examination, rather than risk the crew in the deep bay that was supposed to communicate with the San Sebastian Channel, on board a vessel whose capabilities were unknown. Several fire-places and remains of wigwams were seen; the latter were, however, very different, both in shape and material, from those at Port Famine, for the country being destitute of trees, they were built of driftwood, piled up in a conical form.

Passing round Cape Valentyn, the Hope hauled to the southward, keeping the land on board. At night she anchored in Philip Gidley Cove, at the bottom of Willes Bay, where she was weather-bound until the 29th of January. The shores of Willes Bay are thickly clothed with wood, growing to the water's edge, except at the S.W. side. The great abundance of muscles and limpets attracts the Indians, whose wigwams / were found standing, and from the green appearance of the branches with which they were formed, seemed to have been lately erected. After leaving Willes Bay, the Hope visited Fox Bay, and Sir Edward Owen's Sound, which, it was thought, would lead into Lomas Bay, opposite to Port Famine; but, after running ten miles up, they got into shoal water, and as there was no current, or stream of tide, they landed, and found that a mile and a half farther on, the sound was terminated by low land. Another day, while proceeding along the south side of Brenton Sound, the smoke of Indians' fires was noticed near the beach. As this was the first time the Natives of this part had been seen, the course was shaped towards them, until the Hope anchored. Three Indians then approached, holding up the skins of some animal, and inviting them to land. The small boat was hoisted out, and Messrs. Wickham and Rowlett, with Robinson the pilot, went on shore. The Fuegians presented a fox skin to each of the party, who in return gave them some trifles. After a short interview the boat left them, and no further communication was held that night. The following morning a canoe came off to the vessel, containing three young men, two women, and three children, the youngest not more than four months old. They were no sooner alongside than the men went on board, and commenced an active traffic with all the valuables they possessed; and for a few buttons, a glass bottle, or an empty preserved-meat canister, many of their goods were bartered. They had several fox-skins with them, but no other kind of peltry, except their clothing, obtained from the seal or guanaco: and though many of them wore a penguin skin suspended from their girdle, some were without even that covering. This canoe was followed by another, containing an old man, sixty or seventy years of age, with a grey beard; an elderly woman, and two children. Before they came alongside they put their dogs on shore.

Although the visit from these Indians did not last very long, they had time enough to pilfer. One of the young men, who was seen going into a canoe, excited, by his manner, a suspicion / of his having stolen something, and a tin pot was found concealed under his mantle. As there was every probability of their soon separating, and Mr. Graves feared that punishment would cause a rupture, he only turned him out of the vessel: the rest soon followed him, and landed. Having made a fire, the men squatted round it; while the women were despatched to collect shell-fish.

As soon as the Natives had finished their meal, they embarked, and proceeded eastward. Next day they again visited the Hope, but in consequence, perhaps, of the occurrence the day before, did not venture alongside, until invited by the words, 'ho-say, ho-say,' which mean, 'come, come.' In a few minutes confidence was restored, and they began to barter. The trade was opened by one of the women making a peace-offering of a shell necklace, in return for which, red caps and medals[15] were given to each of the women and children. The Hope went thence to Soapsuds Cove, where the crew washed their clothes, and replaced a broken spar.

In a S.E. direction from this cove there appeared to be a considerable channel leading to the S.E., and to the southward was a deep sound, towards which they were proceeding the next morning; but having advanced about two miles, the land of Cape Expectation trended suddenly round to the eastward, and a long narrow channel presented itself, which seemed likely to communicate with the Strait, to the southward of Port San Antonio. They proceeded through this channel, which takes a very straight course, and gradually narrows from Port Waterfall, where it is two miles and a half wide, to Passage Cove, where it is scarcely three quarters of a mile; and there they anchored.

Between Port Waterfall and Passage Cove, a party of Natives was seen; but, being probably the same who were met at Indian Cove, no attention was paid to their hallooings and fires of invitation.* The Hope came into the Strait, eastward of an opening then called Magdalen Sound; her passage / must therefore have been through Sarmiento's 'San Gabriel' Channel.

At night, when between Cape Forward and Port San Antonio, a heavy squall from S.W. carried the little vessel rapidly towards Cape San Isidro, and, at daylight the next morning, she was in the position observed by us, while ascending Mount Tarn. /

* Fires made to attract attention, and invite strangers to land.

CHAPTER IV.

Deer seen – Hope sails again – Eagle Bay – Gabriel Channel – 'Williwaws' – Port Waterfall – Natives – Admiralty Sound – Gabriel Channel – Magdalen Channel – Hope returns to Port Famine – San Antonio – Lomas Bay – Loss of boat – Master and two seamen drowned.

FROM Mr. Graves's report of the appearance of the channel to the S.E. of Dawson Island, I decided to proceed there as soon as the Hope was ready, for she required some alteration, and repairs.

A deer having been seen on Point St. Anna, Mr. Tarn landed, very early in the morning, eager for the prize, but could only get an ineffectual shot. At another time a few deer were seen by our party, near the river; but instead of returning with the information, they fired their guns, loaded with small shot only, which served but to scare them away. As the animal was new to us, and we had evidence of its being equally new to Science, I was anxious to procure a specimen, but never afterwards had an opportunity. Here Sarmiento saw the only deer which he mentions in his journal.

The morning of the 16th seeming more favourable, I set out in the Hope. The heights were covered with snow which had fallen the preceding night, the thermometer had been at freezing point, and much ice had formed; but the appearance of the weather deceived us: we had scarcely left the ship, when it began to rain, and by the time we reached Cape San Isidro the wind had freshened to a gale, which obliged me to anchor in Eagle Bay.

Having landed, a tent was pitched, and a blazing fire made to dry our clothes. In the evening the gale blew with great / violence from S.W., and the Hope, at her anchor, sheered about by the squalls, was occasionally laid over so as to dip her gunwale under water.

The following day (17th), although the rain had ceased, the wind was still strong. Towards evening it fell, and early on the 18th we left Eagle Bay with a fresh breeze from E.N.E., and passed close to Port San Antonio; but were then delayed by calms and squalls. At noon a westerly wind sprung up, and we proceeded down the Gabriel Channel, with the wind aft, and the tide in our favour. Port Waterfall sheltered us for the night.

The apparently artificial formation of this channel is very striking. It seems to have been formerly a valley between two ridges of the range, in the direction of the strata (of which there are frequent instances, such as the valley in the Lomas Range, opposite Cape San Isidro, the valley of Valdez Bay, and one immediately to the north of the channel itself, besides many others), and that at some remote period the sea had forced its way through, effecting a communication between the Strait and the waters behind Dawson Island: as if one of those great 'northern waves,' of which we once heard so much,[1] had rolled down the wide reach of the Strait (the parallelism of whose shores is also remarkable) from the north-west, towards Cape Froward; and finding itself opposed by the Lomas Range, had forced a passage through the valley until stopped by the mountains at Fitton Bay. Having imagined such a wave in motion, the reader may fancy it uniting with another northern roller from Cape San Valentyn, attacking the hills and carrying all before it, until Mount Hope, at the bottom of Admiralty Sound, arrested its course. I have already noticed the remarkably straight direction in which this curious channel trends. At both extremities the width may be from two to three miles; but the shores gradually approach each other midway, and the coast on each side rises abruptly to the height of fifteen hundred feet. The south shore, sheltered from the prevailing and strongest winds, is thickly covered with trees and luxuriant underwood, which, being chiefly evergreen, improve / the scenery greatly, particularly in the winter season: the north shore is also well wooded for about two-thirds up; but the summit is barren and the outline very much serrated, as is usual in slate formations.

On the north shore we noticed some extraordinary effects of the whirlwinds which so frequently occur in Tierra del Fuego. The crews of sealing vessels call them 'williwaws,' or 'hurricane-squalls,' and they are most violent. The south-west gales, which blow upon the coast with extreme fury, are pent up and impeded in passing over the high lands; when, increasing in power, they rush violently over the edges of precipices, expand, as it were, and descending perpendicularly, destroy every thing moveable. The surface of the water, when struck by these gusts, is so agitated, as to be covered with foam, which is taken up by them, and flies before their fury until dispersed in vapour. Ships at anchor under high land are sometimes suddenly thrown over on their beam-ends, and the next moment recover their equilibrium, as if nothing had occurred. Again a squall strikes them, perhaps on the other side, and over they heel before its rage: the cable becomes strained, and checks the ship with a jerk, that causes her to start a-head through the water, until again stopped by the cable, or driven astern by another gust of wind.

At all these anchorages, under high land, there are some parts more exposed than others; and by watching for those places which are least troubled by these squalls, a more secure, or rather a more quiet, spot may be selected. I do not consider ships so anchored to be in danger if their ground tackle be good; but every thing that offers a stiff resistance must suffer from the fury of these blasts. In many parts of this country trees are torn up by the roots, or rent asunder by the wind; and in the Gabriel Channel the 'williwaws' bursting over the mountainous ridge, which forms the south side of the channel, descend, and striking against the base of the opposite shore, rush up the steep, and carry all before them. I know of nothing to which I can better compare the bared track left by one of these squalls than to a bad broad road. After / having made such an opening, the wind frequently sweeping through prevents the growth of vegetation. Confused masses of up-rooted trees lie at the lower ends of these bared tracks, and show plainly what power has been exerted.

The southern shore of the channel is formed by the base of that range of hills, which extends, from the eastern side of the Magdalen Channel, towards the E.S.E. It is the highest part of Tierra del Fuego, and on it are several remarkable mountains, besides Sarmiento, towering over all.

Close to the east end of the Gabriel Channel is Mount Buckland, a tall obelisk-like hill, terminating in a sharp needlepoint, and lifting its head above a chaotic mass of 'reliquiæ diluvianæ,'[2] covered with perpetual snow, by the melting of which an enormous glacier on the leeward, or north-eastern side, has been gradually formed. This icy domain is twelve or fourteen miles long, and extends from near the end of the channel to Port Waterfall, feeding, in the intermediate space, many magnificent cascades, which, for number and height, are not perhaps to be exceeded in an equal space of any part of the world. Within an extent of nine or ten miles, there are upwards of a hundred and fifty waterfalls, dashing into the channel from a height of fifteen hundred, or two thousand feet. The course of many is concealed, at first, by intervening trees, and, when half-way down the descent, they burst upon the view, leaping, as it were, out of the wood. Some unite as they fall, and together are precipitated into the sea, in a cloud of foam; so varied, indeed, are the forms of these cascades, and so great their contrast with the dark foliage of the trees, which thickly cover the sides of the mountain, that it is impossible adequately to describe the scene. I have met with nothing exceeding the picturesque grandeur of this part of the Strait.

There are several coves on the south shore, but opposite to them there is no shelter until you reach a deep bay in which are several islets; and where, I think, there is a communication with Brenton Sound, but we did not enter it.

Port Waterfall may easily be known by a large flat-topped bare rock, lying across the summit of the eastern head, and / by a magnificent cascade formed by the union of two torrents.

All the plants of the Strait grow here: a sweet-scented Callixene (*C. marginata*, Lam[k.]) filled the air with its odour; and a beautiful flower we had not previously seen, was found by Mr. Graves: it was pendulous, tubular, about two inches long (Class. Hexand. Monóg. Cal. 2. Pet. 3. pointed), and of a rich carnation colour.

The trees are small and stunted; they are of the usual species, Beech and Winter's-bark. Here we first noticed a large fern,** having a stem two or three feet long, and five or six inches in diameter, very similar to the Zamia of New Holland. We saw very few birds, and no quadrupeds. Among the former was a king-fisher, which at the time was new to us; but it is distributed over a large tract of South America, and I have since seen a specimen said to have been shot at Rio de Janeiro.

Fitton Harbour is a deep inlet, surrounded on all sides by precipitous land, rising to the height of three, or four thousand feet, and terminated by peaks, of most fantastic shape, covered with ice and snow.

Between Fitton Harbour and Cape Rowlett are high mountains, two of which, more conspicuous than the rest, we called 'Mount Sherrard,' and 'Curious Peak.'

Card Point proved to be clay-slate, and I think the projection of Cape Rowlett, and the mountains, are also of this rock.

While crossing over towards Cape Rowlett, (the south head of a deep sound, trending to the S.E., which it was my intention to examine), we were met by three canoes, containing, together, about twenty-four people, and ten or twelve dogs. Mr. Wickham recognised them to be the same party who had visited the Hope on her last cruize; the thief, however, was not amongst them, fearing probably he might be known.

These natives conducted themselves very quietly, and, except one of the women, who wished to keep a tin-pot in which some / / water had been given her, made no attempt to pilfer. One of the party, who seemed more than half an idiot, spit in my face; but as it was not apparently done angrily, and he was reproved by his companions, his uncourteous conduct was forgiven.

If possessed of any furs, they had left them, perhaps concealed, near their wigwams: only a few arrows, a necklace of shells, and a fillet for the head, made of ostrich feathers, were obtained by barter. Their canoes were paddled by the women, occasionally helped by the men. One or two of the former were young, and well-featured, but the rest were hideous; and all were filthy and most disagreeable, from the quantity of sealoil and blubber, with which they had covered their bodies. After we had obtained, by barter, all the articles they had to dispose of, I presented them with red caps and medals, of which they were very proud: the latter they requested might have a hole drilled through them, that they might be suspended by a string round their necks. Their astonishment was much excited, and they were pleased by hearing a watch tick; but I believe I had

* This fern we found at the island of Juan Fernandez also.

PORT FAMINE.
(Looking Northwards)

PORT FAMINE.
(Looking Southwards)

P.P. King. CURIOUS PEAK, ADMIRALTY SOUND. S. Bull.

very nearly, though unintentionally, given great offence, by cutting off a lock of hair, from the head of one of the men. Assuming a grave look, he very carefully wrapped the hair up, and handed it to a woman in the canoe, who, as carefully, stowed it away in a basket, in which she kept her beads and paint: the man then turned round, requesting me, very seriously, to put away the scissors, and my compliance restored him to good-humour.

The features of these people bore a great resemblance to those of the Patagonian Indians, but in person they were considerably shorter and smaller. The elderly people of both sexes had hideous figures; the children, however, and young men, were well-formed; particularly one of the boys, whom they called 'Yaāl-lă-bă,' which, I believe, meant a youth, or a young warrior. The word 'Shērrōo' was used to denote a canoe, or vessel.

They were ill-clothed, with mantles made of guanaco, or otter skins, but not so neatly as those of the Patagonians. / Their bodies were smeared over with a mixture of earth, charcoal, or red-ccre, and seal-oil; which, combined with the filth of their persons, produced a most offensive smell. Some were partially painted with a white argillaceous earth; others were blackened with charcoal; one of the men was daubed all over with a white pigment. Their hair was bound by a fillet of plaited twine, made perhaps with strips of bark, and a few of them had it turned up; but to none did it appear to be an object of attention, except one of the young women, who repeatedly combed and arranged hers with the well-toothed jaw of a porpoise.

During a remarkably calm night, we were frequently startled by the loud blowing of whales, between us and the shore. We had noticed several of those monsters on the previous day, but had never heard them blow in so still a place.

At dawn, a light air carried us towards some broken land to the S.E. of Cape Rowlett, between the eastern trend of which, and the projecting point of an island, we found a secure and land-locked harbour, with two entrances, one to the north and the other to the south of High Islet. The south side of the port, which I called Port Cooke,* is a narrow strip of land, forming the head of a deep inlet or sound, called† Brook Harbour. It seemed to extend to the base of the high mountainous range, and to be separated only by a narrow isthmus from Fitton Harbour.

We had scarcely been at anchor half an hour when the same party of Fuegians was seen arriving. The men hastened to us in their canoes, as soon as the women had landed, to cover or thatch the wigwams, which they found standing, and to light fires.

We afterwards went ashore, and, sitting down near them, commenced a brisk trade for arrows, skins, necklaces, and other commodities. The furs which cov-

* After the lieutenant of the Adventure.

† At Mr. Tarn's request.

ered their backs they parted with, for a few beads, and went quite naked the whole evening.

Among them was a young man, who appeared to be treated / with some deference by the others; he was one of the best-looking of the party; and there was a good-natured smile on his countenance during our communication, while the rest frequently manifested displeasure, even about trifles. He was, at least, the master of one of the two families; his wigwam contained his wife, and two children, his, or his wife's father, and mother, as well as the idiot, and his wife, who, from her appearance, must have been a Patagonian, or else a woman of unusual size among these people. The old woman was very inquisitive, and the man, in a long speech, described to her all the wonders I had shown him, applying to me, from time to time, to point out to her the articles he was trying to describe.

Their dexterity with the sling is extraordinary; and, I should think, when used as a weapon of offence, it must be very formidable. Upon asking the same man to show us its use, he picked up a pebble, about the size of a pigeon's egg, and placed it in the sling; then intimating that he was going to strike a canoe, he turned his back to the mark, and threw the stone in an opposite direction, against the trunk of a tree, whence it rebounded over his head, and fell close to the canoe.

I have seen them strike a cap, placed upon the stump of a tree, fifty or sixty yards off, with a stone from a sling. In using the bow and arrow, also, with which they kill birds, they are very dexterous. The spear is principally for striking porpoises and seals, but is also used in war; and from the nature of the barb, must be an efficient weapon. For close quarters, they use clubs, stones held in the hand, and short wooden daggers, pointed with very sharp-edged quartz, pitch-stone, or flint.

The next morning, seeing us underweigh, they came alongside and tried to induce us to anchor again. The young man, of whom I have spoken, was very importunate, and at last offered us his wife, as a bribe, who used all her fancied allurements to second his proposal.

So highly did they esteem beads and buttons, that a few of each would have purchased the canoe, the wife, and children, / their dogs, and all the furniture. Seeing us proceed to the southward, with the apparent intention of sailing down the inlet, they motioned to us to go to the north, repeatedly calling out 'Sheroo, sherroo,' and pointing to the northward; which we thought intimated that there was no passage in the direction we were taking.

At noon, I landed to observe the latitude, and take bearings down the Sound to the S.E., at the bottom of which was a hill, standing by itself, as it were, in mid-channel. The view certainly excited hopes of its being a channel; and as we had begun to calculate upon reaching Nassau Bay in a few days, we named this hill, Mount Hope.

The point on which we landed was at the foot of a high snow-capped hill, called by us Mount. Seymour; whence, had not the Indians been near, I should have taken bearings.

We sailed south-eastward, close to the south shore, until the evening; when from the summit of some hills, about three hundred feet above the sea, we had a view down the Sound, which almost convinced us it would prove to be a channel. The rock at this place differed from any we had seen in the Strait. The mountains are high, and evidently of clay-slate; but the point, near which we anchored, is a mass of hard, and very quartzose sand-stone, much resembling the old red sandstone formation of Europe, and precisely like the rock of Goulburn Island, on the north coast of New Holland.*

The following morning (23d), we proceeded towards Mount Hope, while running down to which some squalls passed over, clouding the south shore, and as we passed Parry Harbour it bore so much the appearance of a channel, that we stood into it; but the clouds clearing away soon exposed the bottom to our view, where there seemed to be two arms or inlets. In the south-eastern arm, the shores were covered with thick ice (like the bottom of Ainsworth Harbour, to the west of Parry Harbour, where an immense glacier slopes down to the water's edge). The south-west arm appeared to be well sheltered, and if it affords a moderate depth of water, would be an excellent harbour. /

After satisfying ourselves that there was no channel here, we bore up on our original course; but, before long, found ourselves within two miles of the bottom of the Sound; which is shallow, and appears to receive two rivers. The great quantity of ice water, which mingles here with the sea, changed its colour to so pale a blue, that we thought ourselves in fresh water.

Mount Hope proved to be an isolated mass of hills, lying like the rest N.W. and S.E., having low land to the southward, over which nothing was visible except one hill, thirty or forty miles distant, covered with snow, to which the rays of the sun gave the appearance of a sheet of gold. Finding ourselves embayed, we hastened out of the scrape, and, after beating for some hours, anchored in Parry Harbour.

Our entrance into a little cove in Parry Harbour disturbed a quantity of ducks, steamers, shags, and geese. Their numbers showed that Indians had not lately visited it.

Next day we reached Ainsworth Harbour, which is of the same character as Parry Harbour, and affords perfect security for small vessels: by dint of sweeping, we reached a secure anchorage in a cove at the south-east corner.

The bottom of the port is formed, as I before said, by an immense glacier, from which, during the night, large masses broke off and fell into the sea with

* King's Australia,' vol. i. p. 70; also vol. ii. pp. 573, 582, and 613.

a loud crash,* thus explaining the nocturnal noises we had often heard at Port Famine, and which at the time were thought to arise from the eruption of volcanoes. Such were also, probably, the sounds heard by the Spanish officers during their exploration of the Straits, whilst in the port of Santa Monica, where they had taken refuge from a violent gale of wind.†

The harbour was full of fragments of ice, the succeeding morning, drifting into the Sound, where the sea-water, being at a higher temperature than the air, rapidly melted them.

Since our departure from Port Waterfall, the weather had been mild, clear, and settled; but as it wanted only three days of the change of the moon, at which period, as well as at the full, it always blew a gale, I wished to reach a place of security in the Gabriel Channel or Magdalen Sound.

Near the islands of Ainsworth Harbour, three canoes passed us, steering across the Sound, each with a seal-skin fixed up in the bow for a sail; and we recognised in them the party left at Port Cooke, among whom was the Indian who had been detected in stealing a tin pot. They did not come along-side; but as we went by, pointed to the north, apparently urging us to go in that direction.

We had noticed several wigwams at Parry and Ainsworth Harbours, which shows that they are much frequented by Indians, perhaps on their way to the open low country east of Mount Hope, where numerous herds of guanacoes may be found.

Porpoises and seal were not scarce in this inlet, and in the entrance there were many whales. The presence of seal and whales made me think it probable there was a channel; but I believe every person with me was satisfied of its being a sound, terminating under Mount Hope. Since my later experience of the deceptive character of some passages in Tierra del Fuego (the Barbara Channel, for example), I have felt less certain that there may not be a communication with the low land, behind Mount Hope, round its northern base. The improbability was, however, so great, – from the bottom of the sound / being shoal, – from the very slight tide-stream, – and from the information of the Natives; who evidently

* At high tide the sea-water undermines, by thawing, large masses of ice, which, when the tide falls, want support, and, consequently, break off, bringing after them huge fragments of the glacier, and falling into the still basin with a noise like thunder.

† 'En los dias 24, y 25, oimos un ruido sordo, y de corta duracion, que, por el pronto, nos parecio trueno; pero habiendo reflexîonado, nos inclinamos a creer que fue efecto de alguna explosion subterranea, formado en el seno de alguna de las montañas inmediatas, en que parece haber algunos minerales, y aun volcanes, que estan del todo 6 casi apagados, moviéndonos a hacer este juicio, el haberse encontrado, en la cima de una de ellas, porcion de materia compuesta de tierra y metal, que en su peso, color, y demas caracteres, tenia impreso el sello del fuego activo en que habia tomado aquel estado, pues era una perfecta imagèn de las escorias del hierro que se ven en nuestras ferrerías. – *Apendice al Viage de Cordova al Magallanes,* p.65.[3]/

intended to tell us we could not get out to sea, – that we did not consider it worth while to make another examination.

I have before observed that the strata of the slate rocks, in the Strait, dip to the S.E.; and I found that they dip similarly all the way to the bottom of this inlet, which I named Admiralty Sound.

The north side, like that of the Gabriel Channel, is steep, without indentations, excepting where there is a break in the hills; but on the south shore there are many coves, and bights, the cause of which is shown in the accompanying imaginary section of the Gabriel Channel. The same cause operates on the outline of the north shore of the reach of Cape Froward, westward as far as Cape Holland, where the rock assumes a still more primitive form. Its general character, however, is micaceous slate, with broad veins of quartz; the latter being particularly conspicuous at Port Gallant.

The following slight sketch, intended to represent an imaginary section of such an opening as the Gabriel Channel, may also serve to give a general idea of many Fuegian anchorages; – of deep water passages existing between the almost innumerable islands of Tierra del Fuego; – and of the effects of those sudden, and violent gusts of wind, – so frequent and dangerous, – commonly called hurricane-squalls,* or williwaws.

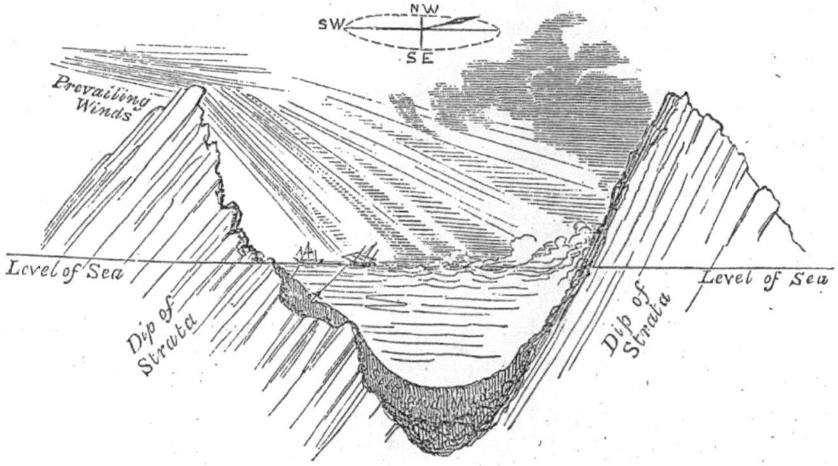

* No canvas could withstand some of these squalls, which carry spray, leaves, and dirt before them, in a dense cloud, reaching from the water to the height of a ship's lower yards, or even lower mast-heads. Happily their duration is so short, that the cable of a vessel, at anchor, is scarcely strained to the utmost, before the furious blast is over. Persons who have been some time in Tierra del Fuego, but fortunate enough not to have experienced the extreme violence of such squalls, may incline to think their force exaggerated in this description: but it ought to be considered, that their utmost fury is only felt during unusually heavy gales, and in particular situations; so that a ship might pass through the Strait of Magalhaens many times, without encountering one such blast as has occasionally been witnessed there. – R.F. /

The rock, of course, decomposes equally on both sides; but on that exposed to the south wind, it breaks off in flakes parallel to the direction of the strata, and therefore does not make the course of the beach more irregular; while on the other side it moulders away transversely to the direction of the dip, leaving holes, in which water lodges, and hastens decomposition by entering deeply into the interstices. Water, air, and frost decompose the rock, and form a soil, which, if not too much exposed to the wind, is soon occupied by vegetation.

The rugged faces of the cliffs, on the southern shore, caused by the rock decomposing across the grain, collect sand and mud; and hence it happens that anchorages are frequently found on one side, whilst, on the other, the anchor will not hold, from the steepness of the ground; there being nothing upon the smooth declivity to retain mud and sand before it gets to the bottom; which, in most cases known to me, lies far beyond the reach of the anchor.

After a tedious and difficult passage through the Gabriel Channel, we anchored in a snug harbour within the entrance of Magdalen Channel, on the west side, under a peaked hill called by Sarmiento 'El Vernal,' – in our plan, the 'Sugarloaf.' The entrance is about a quarter of a mile wide; but after a few hundred yards the harbour opens, extending in for nearly a mile. It is of easy depth; seven fathoms in the entrance, and four, five, and six fathoms within; so that it is / very convenient for a small vessel: to us, indeed, it was a most welcome discovery. The land rises, around this cove, to the height of two or three thousand feet. It is covered with Beech, and Winter's-bark, and near the water is adorned with large groves of Fuchsia, Berberis, and the common shrubs of Port Famine,[4] growing so thickly as to form an almost impenetrable jungle; but, notwithstanding the picturesque character of its scenery, the towering height of the hills, which exclude the sun's rays for the whole day, during the greater portion of the year, renders it a gloomy and melancholy spot.*

We found a family of Fuegians in the inner harbour. Three canoes were hauled up on the beach, but their owners were not at first visible. At last, after our repeatedly calling out 'Ho-say, ho-say,'[6] they appeared, and, rather reluctantly, invited us, by signs, to land. There seemed to be fourteen or fifteen people, and seven or eight dogs. Mr. Wickham and Mr. Tarn went on shore to these natives, who exhibited some timidity, until a hideous old woman began to chatter, and soon made them understand that the young men (Lā-ā-pas) were absent on a hunting excursion, but were every moment expected to return. There were only three men with the women and children. To inspire them with confidence in our good intentions, Mr. Wickham gave each man a red cap, and some other trifles. One of them complained of being sick, but I rather imagine his illness was feigned, and the others did not at all seem to like our visit. By degrees their fears subsided, and, restraint being laid aside, an active trade began; in which several

* 'sub rupe cavatâ Arboribus clausam circum atque horrentibus umbris.'[5]

otter skins, shell-necklaces, spears, and other trifles, were obtained from them in exchange for beads, buttons, medals, &c. The otters are caught by the help of dogs, on which account, principally, the latter are so valuable.

These people were slightly clothed with skins of the seal and otter, but some had pieces of guanaco mantles over their shoulders, whence we supposed that they were either of the same tribe, or at peace, with the Indians of Admiralty Sound:[7] / unless, indeed, they trade with the Patagonia Indians; but such is the poverty of the Fuegians, they can scarcely possess any thing of value sufficient to exchange with the good of their northern neighbours, unless it be iron pyrites, which I think is not found in the open country inhabited by the Patagonian Indians, and, from the facility with which it yields sparks of fire, must be an object of importance.

We were not a little amused by the surprise which these natives showed at the things in our possession, and by the effect produced in their countenances when they saw any thing extraordinary: the expression was not that of joy or surprise, but a sort of vacant, stupified, stare at each other. They must have been very suspicious of our intentions, or very much excited by what they had seen during the day, as throughout the night an incessant chattering of voices was heard on shore, interrupted only by the barking of their dogs.

Looking down the Magdalen Inlet, we saw two openings, which, while the hills were enveloped in mist, had the appearance of being channels. We proceeded for some distance into the more westerly of the two, but found that it was merely a sound, terminated by high land. The boat was then steered under a steep mass of black mountainous land,* the summit of which is divided into three peaks, which Sarmiento called 'El Pan de Azucar de los Boquerones' (the Sugar-loaf of the Openings). We ran southward, fifteen miles down this sound, and reached the Labyrinth Islands; but finding there no suitable anchorage, resumed our course towards the bottom of what we thought another sound, terminated by mountains. At noon, the furthest point, on the west shore, which we called Cape Turn, was within three miles of us, and we should soon have discovered the continuation of the channel (as it has since been proved); but a breeze set in from the S.W., and in a short time it blew so strong as to oblige us to turn back. 'Williwaws' and baffling eddy winds kept us seven hours under Mount Boqueron. These squalls were at first alarming, but by taking in all sail, before they passed, we sustained no injury. At sunset / we were abreast of Hope Harbour, in which we purposed taking shelter from the gale. Our late neighbours, the Indians, had lighted a fire at the entrance to invite our return; but wind and tide were against us, and as we knew of no port to leeward, our only resource was to run out of the sound. Furious squalls carried us into the true, or

* Mount Boqueron.

steady, wind, which we found very strong; and as Port San Antonio was on the lee-bow, we had to carry such a press of sail, that our excellent boat had nearly half the lee side of her deck under water. By daylight we got into smooth water, and, with less wind and better weather, steered for Port Famine. The smoother water enabled us to light a fire and cook a meal, not an unimportant affair, as we had eaten nothing since six o'clock on the preceding morning.

In our absence Mr. Graves had surveyed Lomas Bay, and, after his return, Mr. Ainsworth had crossed the Strait with the gig and cutter to survey Port San Antonio. They were victualled for five days; the gig was manned by my own boat's crew, and the cutter by volunteers: but although they had not come back, we felt no anxiety about their safety, being assured that Mr. Ainsworth would not run the risk of crossing the Strait during bad weather. The tempestuous state of the two following days, however, made us uneasy, and on the third morning, when the wind moderated much, we looked out anxiously for their arrival. In the evening the cutter returned; but, alas! with the melancholy information of the loss of Mr. Ainsworth, and two seamen, drowned by the upsetting of the gig. One of the latter was my excellent coxswain, John Corkhill. The remainder of the gig's crew were only rescued from drowning by the strenuous exertions of those in the cutter.

Mr. Ainsworth, anxious to return to the ship, thought too little of the difficulty and danger of crossing the Strait during unsettled weather. He set out from Port San Antonio under sail, and, while sheltered by the land, did very well; but as soon as they got into the offing, both wind and sea increased so much that the gig was in great danger, although under only a small close-reefed sail. /

The people in the cutter were anxiously watching her labouring movements, when she disappeared! They hastened to the spot – saved three men; but the other two had gone down. Poor Ainsworth was still clinging to the gig's gunwale when his shipmates eagerly approached; but letting go his hold from extreme exhaustion, and being heavily clothed, he sunk from their sight to rise no more.

He had been cheering the drowning crew, and trying to save his companions, till the moment his grasp relaxed. Just before Ainsworth himself let go, Mr. Hodgskin lost his hold, exclaiming, 'Ainsworth, save me!' when, exhausted as he was, with one hand he rescued his friend, and, directly afterwards, his strength failing, sunk.

This addition of three people to the already loaded cutter, made her cargo more than was safe, therefore Mr. Williams, who commanded her, very prudently bore up for the first convenient landing-place, and happily succeeded in reaching the only part of the beach, between Lomas Bay and Cape Valentyn, where a boat could land.

The following morning, the weather being more favourable, they crossed under sail to Freshwater Bay, and thence pulled to Port Famine.

This melancholy disaster was much felt by every one. Ainsworth was a deserving officer, and highly esteemed. Corkhill was captain of the forecastle, and had served in the Polar voyages under Sir Edward Parry.[8] On the Sunday following, the colours were hoisted half-mast high, and the funeral service was read after morning prayers: for although to recover the bodies was impossible, their watery grave was before our eyes; and the performance of this last sad duty was a melancholy satisfaction.

> 'Ours are the tears, tho' few, sincerely shed,
> When ocean shrouds and sepulchers our dead.'[9]

A tablet was subsequently erected, on Point St. Anna, to record this fatal accident. /

CHAPTER V.

Lieutenant Sholl arrives – Beagle returns – Loss of the Saxe Cobourg sealer – Captain Stokes goes to Fury Harbour to save her crew – Beagle's proceedings – Bougainville's memorial – Cordova's memorial – Beagle's danger – Difficulties – Captain Stokes's boat-cruize – Passages – Natives – Dangerous service – Western entrance of the Strait of Magalhaens – Hope's cruize – Prepare to return to Monte Video.

THE Beagle's time of absence had expired on the 1st of April, and our anxiety, more excited by our recent loss, was becoming painful. I detained the Hope from going upon a service for which she was prepared, in case she might be required to search for our consort: but on the 6th a strange whale-boat was descried pulling towards us from the southward, in which we soon distinguished Lieut. Sholl. His appearance, under such circumstances, of course raised fears for the Beagle's safety; but, on approaching, his gratifying shout, 'all's well!' at once removed anxiety.

Mr. Sholl informed me, that the Beagle had picked up a boat, belonging to the schooner Prince of Saxe Cobourg, wrecked in Fury Harbour, at the south entrance of the Barbara Channel; and that she had put into Port Gallant, whence Captain Stokes had gone with the boats to assist the Sealers, leaving Lieut. Skyring on board.

The safety of the Beagle being established, I despatched Mr. Graves, in the Hope, to examine some openings between the Magdalen Channel and the Dos Hermanos of Bougainville.

Several days earlier than I expected, the Beagle made her appearance, and Captain Stokes soon gave me the agreeable intelligence of having succeeded in saving the Prince of Saxe Cobourg's crew. Favoured by the weather, though delayed by his guide having forgotten the way, Captain Stokes reached Fury Harbour in two days, and embarked the master and / crew of the wrecked vessel, with all their personal property, and the greater part of the seal-skins which they had cured. He reached Port Gallant again on the fourth day; sailed immediately in the Beagle, and two days afterwards anchored in Port Famine.

The Prince of Saxe Cobourg, belonging to Mr. Weddel (whose voyage towards the South Pole is so well known), and commanded by Mr. Matthew Brisbane, who accompanied Weddel on that occasion, sailed from England in the summer of 1826, on a sealing voyage.[1] At South Shetland she encountered a continuance of bad weather, was beset by a large body of ice for several days, and received so much damage as to oblige her to run for the Fuegian coast, and anchor in Fury Harbour, at the entrance of the Barbara Channel. There (December 16th, 1826) she was driven on shore by the furious strength of the williwaws, and wrecked. The crew were, however, enabled to save most of-the provisions and stores, as well as their three boats. Having made tents, and established themselves on shore, they remained in anxious expectation of the arrival of some vessel which might relieve them; day after day however passed, without succour.

Two boats were despatched to look for any sealing vessel that might be in the vicinity, but after fifteen day's absence they returned unsuccessful. In this interval one of the crew, who had long been sickly, died; and another, in carelessly discharging a musket, exploded twenty pounds of gunpowder, by which he was very much burned. Three of the people being mutinous, were punished by being sent, each to a different island, with only a week's provisions.

Soon afterwards another boat was sent away, which reached Hope Harbour, but found no vessel there. Seven of the people then obtained permission from the master (who kept up a very proper state of discipline), to take the largest whale-boat, and go towards the River Negro. Previous to their departure they drew up articles of agreement for their general conduct, a breach of which was to be punished by the offender being left upon the coast, wherever they might happen to / be. The boat eventually arrived safely at the place of her destination, and the crew entered as volunteers on board of the Buenos Ayrean squadron, at that time engaged in the war with Brazil.

Again a boat was despatched, directed to go westward through the Strait in search of vessels. She had only reached as far as Playa Parda, when the Beagle fell in with her (March 3d, 1827). While passing through the small channels, before entering the Strait, she met several canoes, with Indians, who endeavoured to stop her, and shot arrows at the crew; but, happily, without doing any mischief.

After the last boat's departure, Mr. Brisbane began to build a small vessel, and, while so employed, was visited by a party of natives, who conducted themselves very peaceably, and went away. Their visit, however, gave the shipwrecked people, now much reduced in number,* reason to apprehend the return of a larger body, who might try to possess themselves of the property which was lying about on the shore; they therefore buried a great deal, and took means to preserve the rest by making preparations to repel attack. When Captain Stokes appeared with his two boats, the Sealers flew to their arms, calling out 'the Indians, the Indians!' but in a very few minutes excess of joy succeeded to their sudden alarm.

Captain Stokes found the vessel lying on the rocks, bilged, and an utter wreck. The master and crew were extremely anxious to get away, he therefore

* Including the master, there were on board, when cast away, twenty-two persons.

embarked them, with as much of the property as could be carried, and succeeded (after another night in the boats, and a long pull of eighty miles,) in conveying them safely to the Beagle.

The following is an abstract of Captain Stokes's journal of his cruise to the western entrance of the Strait.

The Beagle sailed from Port Famine on the 15th of January, to explore the Strait westward of Cape Froward, and to fix particularly the positions of Cape Pillar, the rock called Westminster / Hall, and the Islands of Direction, at the western entrance of the Strait.

For the first night Captain Stokes anchored in San Nicolas Bay, and in the evening examined a harbour* behind Nassau Island, which Bougainville, in the year 1765, visited for the purpose of procuring wood for the French settlement at the Falkland Islands.

On the second night, after a day nearly calm, the Beagle was anchored in a cove to the eastward of Cape Froward, and the next day (17th) passed round the Cape, carrying a heavy press of sail against a dead foul wind. Captain Stokes's account of this day's beat to wind ward will give the reader an idea of the sort of navigation.

'Our little bay had screened us so completely from the wind, that though, when (at five A.M.) we weighed, the breeze was so light as scarcely to enable us, with all sail set, to clear its entrance; no sooner were we outside, than we were obliged to treble reef the topsails. We continued to beat to windward under a heavy press of sail; our object being to double Cape Froward, and secure, if possible, an anchorage ere night-fall under Cape Holland, six leagues further to the westward. At first we made 'boards' right across the Straits to within a third of a mile of each shore, gaining, however, but little. We then tried whether, by confining our tacks to either coast, we could discover a tide by which we might profit; and for that purpose I began with the north shore, for though we were there more exposed to violent squalls which came down the valleys, I thought it advisable to avoid the indraught of various channels intersecting the Fuegian coast; but having made several boards without any perceptible advantage, we tried the south shore, with such success that I was induced to keep on that side during the remainder of the day.

'And here let me remark, that in consequence of the westerly winds which blow through the western parts of the Straits of Magalhaens, with almost the constancy (as regards / direction, not force) of a trade-wind; a current setting to the eastward, commonly at the rate of a knot and three quarters an hour, will be found in mid-channel. The tides exert scarcely any influence, except near either

* Bougainville Harbour, better known to Sealers by the name of 'Jack's Harbour.'

shore; and sometimes appear to set, up one side of the Straits, and down the other: the weather tide is generally shown by a rippling. (*c*)

'Heavy squalls off Cape Froward repeatedly obliged us to clew all up. By day their approach is announced, in time for the necessary precautions, by their curling up and covering with foam the surface of the water, and driving the spray in clouds before them.

'At last we doubled Cape Froward. This Cape (called by the Spaniards El Morro de Santa Agueda), the southern-most point of all America, is a bold promontory, composed of dark coloured slaty rock; its outer face is nearly perpendicular, and whether coming from the eastward or westward, it 'makes' as a high round-topped bluff hill ('Morro').

'Bougainville observes, that 'Cape Froward has always been much dreaded by navigators.'* To double it, and gain an anchorage under Cape Holland, certainly cost the Beagle as tough a sixteen hours' beat as I have ever witnessed: we made thirty-one tacks, which, with the squalls, kept us constantly on the alert, and scarcely allowed the crew to have the ropes out of their hands throughout the day. But what there is to inspire a navigator with 'dread' I cannot tell, for the coast on both sides is perfectly clear, and a vessel may work from shore to shore.'

From Cape Holland, the Beagle proceeded to Port Gallant, and during her stay there, Mr. Bowen ascended the Mountain de la Cruz. Upon the summit he found some remains of a glass bottle, and a roll of papers, which proved to be the memorials stated to have been left by Don Antonio de Cordova, / and a copy of a document that had previously been deposited there by M. de Bougainville. With these papers was found a Spanish two-rial piece of Carlos III., which had been bent to admit of its being put into the bottle. It was with considerable difficulty that any of the writing could be decyphered, for the papers, having been doubled up, were torn, and the words defaced at the foldings, and edges.

Bougainville's memorial was in Latin. Cordova's, besides a document in Latin, was accompanied by an account of his voyage, written in four languages, Spanish, French, Italian, and English. The legible part of the former† was as follows:–

(*c*) While the 'current' runs eastward for many days in mid-channel, or along one shore, it often happens that the 'stream of tide' either sets in a contrary direction, along each side of the Strait, or that it follows only the shore opposite to that washed by the 'current.' – R.F.

* 'Voyage autour du Monde.' 1767.[2]

† *part of the former:* the Bougainville text, dated 1766, was indeed fragmentary. It referred to a hazardous journey, uncertain harbours and closed with what seemed to be a list of officers and scientific staff, possibly those who made the ascent and left the memorial: the astronomer Pierre-Antoine Veron; the physician and naturalist Philibert Commerçon; the hydrographer and engineer Charles Routier de Romainville; the surgeon Francois Vives; and Landais, Lavery –de-Roi and Fontaine, who were officers

Viatori Benevolo salus
_______ que a periculose admodum naviga _____
_______ Brasilie Bonarve et insularum __________
____
__________ incertis freti Magellanici portubus __________
_____________________ historia astronomia
____ Boug __________
____ Boug _____ Duelos et de la Giranda 2 navium _____
______________ Primaris
____ Comerson _____ Doct med naturalista Regio __________
accu _____ m. Veron astronomo de Romainville hidrographio
____________ a rege Christianissimo demandans
____________ Landais Lavan Fontaine navium
Loco tenentibus-et Vexillariis __________
________itineris locus DD Dervi Lemoyne
___________ Riouffe voluntariis. __________
_______________ vives __________ scriba
Anno MDCCLXVI.

The Latin inscription of Cordova was as follows:–

Benevolo Navigatori
Salutem
Anno Domini MDCCLXXXVIII Vir celeberrimus
DD Antonius de Cordova Laso de Vega navibus duabus *(quarum*
nomina SS Casilda et Eulalia *erant ad scrutamen* Magellanici
freti subsequendum *unâque* littorum, portuum aliorumque notabilium
___________________ iter iterum fecit.
____ e Gadibus classis tertio nonas Octobris habenas *immittit*
quarto idus *ejusdem* Nova __________ vidit /
A Boreali ad Austra __________ *miserium* postridie Kalendæ
Novembris emigravit.
Decimo quarto Kalendas Januarii Patagonicis recognitis
litoribus ad ostium appulit freti.
Tandem ingentibus periculis et horroribus tam in mari quam
in freto magnanime et constanter superatis et omnibus
portubus atque navium *fundamentis* utriusque litoris
correctissime cognitis ad hune portum Divini Jose vel
Galante septimo idu Januarii pervenit ubi ad
perpetuam rei memoriam in monte sanctissimæ crucis hoc
monumentum reliquit.
Tertio et excelso Carolo regnante potente
Regali jussu facta fuere suo.
Colocatum fuit nono Kalendæ Februarii Anno MDCCLXXXIX.[3]

---

of the *Etoile*, one of the two ships on the voyage. Bougainville was detained for weeks in January 1768 in this bay, but does not mention the memorial.

together with a list of the officers of both vessels, and enclosing a memorial of Cordova's former voyage in the Santa Maria de la Cabeza. The originals are placed in the British Museum; but before we finally left the Strait, copies were made on vellum, and deposited on the same spot.

The Beagle left Port Gallant* with a fair wind, which carried her to Swallow Harbour.

The next stopping place was Marian's Cove, a very snug anchorage on the north shore, a few miles beyond Playa Parda. Proceeding thence to the westward, with the wind 'in their teeth,' and such bad weather, that they could only see the land of either coast at intervals, and failing in an attempt to find anchorage under Cape Upright, the Beagle was kept under weigh during a squally dark night.

In that very place, Commodore Byron, with the Dolphin and Tamar, passed the anxious night, which he thus describes:–

'Our situation was now very alarming; the storm increased every minute, the weather was extremely thick, and the rain seemed to threaten another deluge; we had a long dark night before us, we were in a narrow channel, and surrounded on / every side by rocks and breakers.'† The Beagle was under similar circumstances, but the land being known to be high and bold, her danger was not considered so imminent.

Eastward of Cape Upright the water was smooth; but between it and Cape Providence a heavy breaking sea was caused by the deep swell of the Pacific. Captain Stokes found an anchorage the next night in a bay under Cape Tamar; and the following evening very nearly reached another under Cape Phillip; but the darkness of a rainy night, and strong squalls, prevented their attempting to anchor in an unknown place, and the only resource was to bear up for shelter under Cape Tamar, where the previous night had been passed. Even this was a dangerous attempt; they could hardly discern any part of the high land, and when before the wind could not avoid the ship's going much too fast. While running about eight knots, a violent shock – a lift forward – heel over – and downward plunge – electrified every one; but before they could look round, she was scudding along, as before, having fairly leaped over the rock.

It was afterwards found that a great part of the gripe and false keel[4] were knocked away. Captain Stokes's account of this day's beat will give an idea of the difficulties which the Beagle's crew encountered, in working out of the Strait.

January 31st. 'The hands were turned up at daylight 'up anchor;' but the heavy squalls that came off the high land of the harbour, rendered it too hazardous to weigh, until a temporary lull enabled us to make sail, and re-commence

* One of the feathered tribe, which a naturalist would not expect to find here, a 'humming bird,' was shot near the beach by a young midshipman. – Stokes MS.

† Hawkesworth's Coll. of Voyages, vol. i. p. 76.

beating to the westward against a dead foul wind, much rain, hard squalls, and a turbulent cross sea.

'The squalls became more frequent and more violent after noon; but they gave, in daylight, sufficient warning, being preceded by dark clouds gradually expanding upwards, until their upper line attained the altitude of about fifty degrees: then came heavy rain, and perhaps hail; immediately after followed the squall in all its fury, and generally lasted fifteen or twenty minutes. /

'In working to windward we frequently extended our 'boards' to the south shore (not without risk considering the state of the weather), with the hope of making out Tuesday Bay, or some anchorage thereabout; but the coast was covered with so thick a mist, that not a single point, mentioned by preceding navigators, could be recognised.

'About seven in the evening we were assailed by a squall, which burst upon the ship with fury far surpassing all that preceded it; had not sail been shortened in time, not a stick would have been left standing, or she must have capsized. As it was, the squall hove her so much over on her broadside, that the boat which was hanging at the starboard quarter was washed away. I then stood over to the north shore, to look for anchorage under the lee of a cape, about three leagues to the north-west of Cape Tamar. On closing it, the weather became so thick that at times we could scarcely see two ships' lengths a-head.

'These circumstances were not in favour of exploring unknown bays, and to think of passing such a night as was in prospect, under sail in the Straits, would have been a desperate risk; I was obliged therefore to yield the hard-gained advantage of this day's beat, and run for the anchorage whence we had started in the morning.

'It was nearly dark ere we reached it; and in entering, desirous to keep well up to windward, in order to gain the best anchorage, I went too close to the outer islet, and the ship struck violently on a rocky ledge. However, she did not hang a moment, and was soon anchored in safety.'

Finding so much danger and difficulty, in proceeding with the ship, without first knowing where to run for anchorages, Captain Stokes left her in Tamar Bay, under the charge of Lieutenant Skyring; and, accompanied by Mr. Flinn, set out in the cutter, with a week's provisions, to examine the south coast.

In a very arduous and dangerous cruize he discovered several well-sheltered anchorages, but experienced a 'constant heavy gale from W.N.W., with thick weather and incessant drenching rain.' /

Captain Stokes says, 'Our discomfort in an open boat was very great, since we were all constantly wet to the skin. In trying to double the various headlands, we were repeatedly obliged (after hours of ineffectual struggle against sea and wind) to desist from useless labour, and take refuge in the nearest cove which lay to leeward.'

From the Harbour of Mercy, Captain Stokes attempted to cross the Strait, on his return to the Beagle; but the sea ran too high, and obliged him to defer his daring purpose until the weather was more favourable.

During his absence, Lieutenant Skyring surveyed Tamar Bay and its vicinity.

Again the Beagle weighed, and tried hard to make some progress to the westward, but was obliged a third time to return to Tamar Bay. After another delay she just reached Sholl Bay, under Cape Phillip, and remained there one day, to make a plan of the anchorage, and take observations to fix its position.

The Beagle reached the Harbour of Mercy (Separation Harbour of Wallis and Carteret),* after a thirty days' passage from Port Famine, on the 15th, having visited several anchorages on the south shore in her way. But tedious and harassing as her progress had been, the accounts of Byron, Wallis, Carteret, and Bougainville show that they found more difficulty, and took more time, in their passages from Port Famine to the western entrance of the Strait. Byron, in 1764, was forty-two days; Wallis, in 1766, eighty-two; Carteret, in the same year, eighty-four; and Bougainville, in 1768, forty days, in going that short distance.

Five days were passed at this place, during which they communicated with a few natives, of whom Captain Stokes remarks; 'As might be expected from the unkindly climate in which they dwell, the personal appearance of these Indians does not / exhibit, either in male or female, any indications of activity or strength. Their average height is five feet five inches; their habit of body is spare; the limbs are badly turned, and deficient in muscle; the hair of their head is black, straight, and coarse; their beards, whiskers, and eyebrows, naturally exceedingly scanty, are carefully plucked out; their forehead is low; the nose rather prominent, with dilated nostrils; their eyes are dark, and of a moderate size; the mouth is large, and the under-lip thick; their teeth are small and regular, but of bad colour. They are of a dirty copper colour; their countenance is dull, and devoid of expression. For protection against the rigours of these inclement regions, their clothing is miserably suited; being only the skin of a seal, or sea-otter, thrown over the shoulders, with the hairy side outward.

'The two upper corners of this skin are tied together across the breast with a strip of sinew or skin, and a similar thong secures it round the waist; the skirts are brought forward so as to be a partial covering. Their comb is a portion of the jaw of a porpoise, and they anoint their hair with seal or whale blubber; for removing the beard and eyebrows they employ a very primitive kind of tweezers, namely, two muscle shells. They daub their bodies with a red earth, like the ruddle used in England for marking sheep. The women, and children, wear necklaces, formed of small shells, neatly attached by a plaiting of the fine fibres of seal's intestines.

* It was here that Commodore Wallis and Captain Carteret separated, the Dolphin going round the world; the Swallow returning to England. Sarmiento's name of Puerto de la Misericordia, or 'Harbour of Mercy,' being of prior date, ought doubtless to be retained.

'The tracts they inhabit are altogether destitute of four-footed animals; they have not domesticated the geese or ducks which abound here; of tillage they are utterly ignorant; and the only vegetable productions they eat are a few wild berries and a kind of sea-weed. Their principal food consists of muscles, limpets, and sea-eggs, and, as often as possible, seal, sea-otter, porpoise, and whale: we often found in their deserted dwellings bones of these animals, which had undergone the action of fire.

'Former voyagers have noticed the avidity with which they swallowed the most offensive offal, such as decaying seal-skins, / rancid seal, and whale blubber, &c. When on board my ship, they ate or drank greedily whatever was offered to them, salt-beef, salt-pork, preserved meat, pudding, pea-soup, tea, coffee, wine, or brandy – nothing came amiss. One little instance, however, happened, which showed what they preferred. As they were going ashore, a lump of the tallow used for arming the lead[5] was given to them, and received with particular delight. It was scrupulously divided, and placed in the little baskets which they form of rushes, to be reserved for eating last, as the richest treat.

'To their dwellings have been given, in various books of voyages, the names of huts, wigwams, &c.; but, with reference to their structure, I think old Sir John Narborough's term for them will convey the best idea to an English reader; he calls them 'arbours.'[6] They are formed of about a couple of dozen branches, pointed at the larger ends, and stuck into the ground round a circular or elliptical space, about ten feet by six; the upper ends are brought together, and secured by tyers of grass, over which is thrown a thatching of grass and seal-skins, a hole being left at the side as a door, and another at the top as a vent for the smoke. A fire is kept burning within, over which the natives are constantly cowering; hence, when seen abroad, instead of appearing to be hardy savages, inured to wet and cold, you see wretched creatures shivering at every breeze. I never met people so sensible of cold as these Fuegian Indians.

'The nature of their domestic ties we had no opportunity of discovering; their manner towards their children is affectionate and caressing. I often witnessed the tenderness with which they tried to quiet the alarms our presence at first occasioned, and the pleasure which they showed when we bestowed upon the little ones any trifling trinkets. It appeared that they allow their children to possess property, and consult their little whims and wishes, with respect to its disposal; for lying in a boat, alongside one of the canoes, bargaining for various articles, spears, arrows, baskets, &c., I took a fancy to a dog lying near one of the women, and offered a price for it; one of my seamen, supposing the bargain concluded, laid / hands on the dog, at which the woman set up a dismal yell; so bidding him desist, I increased my offers. She declined to part with it, but would give two others. At last, my offers became so considerable, that she called a little boy out of the thick jungle (into which he had fled at our approach), who was the owner of the dog. The goods were shown to him, and all his party urged him to sell it, but the little urchin would not consent. He offered to let me have his necklace, and what he received in exchange was put away in his own little basket.

'These people never evinced any thankfulness for our presents. Whatever was offered they 'clutched at,' doubtful of getting it, although held out to them; and when in their own hand, it was instantly stowed away, as if they feared it would be recalled.

'I sometimes tried to discover whether they preferred any particular colour, and for that purpose held out three strings of beads, black, white, and red; they clutched at all three, in their usual manner, without showing any preference.

'Their pronunciation is exceedingly harsh and guttural; not more than two words, whose signification was at all ascertained, could be made out, 'sherroo,' a ship, boat, or canoe, and 'peteet,' a child. They have a wonderful aptitude for imitating the sounds of strange languages: let a sentence, of even a dozen words, be distinctly pronounced, and they will repeat it with the utmost precision.[7]

'Their only articles of traffic, besides such implements and weapons as they use, are seal and otter skins; and I should say that the quantity of peltry to be procured from them would be insignificant towards completing the cargo of a sealing vessel.'

During the next few days the Beagle was employed in the most exposed, the least known, and the most dangerous part of the Strait. Fortunately, she was favoured by weather, and effected her purpose without injury or loss; but I never reflect upon this piece of service without an inward tribute of admiration to the daring, skill, and seamanship of Captain Stokes, Lieutenant Skyring, and Mr. Flinn. /

In his journal Captain Stokes says:

'Incessant rain and thick clouds prevented my completing, until this day (19th), the observations necessary for making an island, just outside the Harbour of Mercy, the southern end of my base, for the trigonometrical connection of the coasts and islands near the western entrance of this weather-beaten Strait.

'On the 20th, I weighed and beat to windward, intending to search for anchorage on the north shore, where I might land and fix the northern end of our base line. In the evening we anchored in an archipelago of islands, the real danger of whose vicinity was much increased to the eye by rocks, scattered in every direction, and high breakers, occasioned doubtless by reefs under water. We observed that most of the larger islands have small banks of sand at their eastern sides, on which anchorage may be found; but for ordinary purposes of navigation, this cluster of islands* need only be pointed out to be avoided. The number and contiguity of the rocks, below as well as above water, render it a most hazardous place for any square rigged vessel: nothing but the particular duty on which I was ordered would have induced me to venture among them.

Fore-and-aft vessels might work with far less risk; and as the rocks are frequented by vast numbers of fur seal, a season or two might be profitably passed here by a sealing vessel so rigged.

This morning (21st) I landed on one of the larger islands, with Lieutenant Skyring, and having ascended an eminence (Observation Mount) with the necessary instruments, fixed its position, and made it the northern end of our base.

It was a beautiful, and clear day; the Isles of Direction (or Evangelists), as well as every point of importance on the adjacent coast, were seen distinctly during several hours.

My next object was to fix the position of Cape Victory, and ascertain whether anchorage could be found in its neighbourhood. Accordingly, we weighed early next morning (22d,) and after extricating ourselves from this labyrinth (not without much difficulty and danger), we beat to the westward. Violent squalls, a heavy sea, and thick weather, which came on about / noon, obliged me to choose the least evil, and run for the Harbour of Mercy.

On the 23d, we went out again, and beat towards the Isles of Direction, off which we passed a night under sail.

The morning of the 24th was very fine, and the wind moderate. Leaving the Beagle to sound about the Isles of Direction, I set out in my boat, with two days' provisions, towards Cape Victory. As we rowed along these rocky shores, threading the mazes of the labyrinth of islets which fringe them, we saw vast numbers of black whales, and the rocks were quite covered with fur seal and brant geese.

After pulling, in earnest, for six hours, we landed upon Cape Victory, the north-western limit of the Strait of Magalhaens, and there, with a sextant, artificial horizon, and chronometer, ascertained the position of this remarkable promontory. From an eminence, eight hundred feet above the sea, we had a commanding view of the adjacent coasts, as well as of the vast Pacific, which enabled us to rectify former material errors. Late in the evening we were fortunate enough to get safely on board again, which, considering the usual weather here and the heavy sea, was unexpected success. This night was passed under sail in the Pacific, and next morning we commenced our return to Port Famine.

When within four or five miles of Cape Pillar, and to the westward of it, a current was found to set southward, at about two knots an hour. As we neared the Cape the wind fell, and the Beagle was set rapidly towards those dangerous rocks, called the Apostles. Fortunately, a commanding breeze sprung up, and we extricated ourselves from the difficulty. While passing Cape Pillar, I landed in a cove near it, and determined its position. By sunset we had arrived near the Harbour of Mercy; and being becalmed, towed the ship in, with her boats, until an anchor was dropped at the proper place.

On the 26th, we went to Tuesday Bay, and on the 27th crossed the Strait, and anchored under Cape Parker. I have rarely witnessed such a high, cross, and irregular sea as we this day passed through, near the strange mass of rock, called by / Narborough, 'Westminster Hall.' The coast about our unsafe anchorage was as barren and dismal-looking as any part of this country, which, as the old navigator above-mentioned said, is 'so desolate land to behold.'[8]

Next day (March 1st) we ran down to Cape Upright, and there remained until the 3d, collecting the required data for our survey.

While standing towards the bay called Playa Parda (on the 3d), a boat under sail was seen making towards us from the southern coast. I fired several guns, to show our position, before we became shut in by the land, and soon after anchoring a whale-boat came alongside, with the second mate and five men belonging to the sealing-vessel Prince of Saxe Cobourg.

'Anxious not to lose a moment in hastening to the relief of our shipwrecked countrymen, I ran down next day to Port Gallant, and thence proceeded with two ten-oared boats (on the 5th) through the Barbara Channel, and the following evening reached Fury Harbour.'

Having already given a short account of the Saxe Cobourg's loss, and the rescue of her crew by Captain Stokes, I will not repeat the story by extracting more from his journal.

Mr. Graves returned from his cruize in the Hope on the 17th, after suffering much from stormy weather and incessant rain; but having made a survey of the openings in the land to the west of Magdalen Channel as far as the Sugar Loaf Point, at the west head of Lyell Sound, which he found to be deep inlets, affording no anchorages of value to navigation.

The time having arrived for our return to Monte Video, preparations were made for sailing, and in the mean time I went to the northward, in the Hope, to survey the coast between Port Famine and Elizabeth Island, including Shoal Haven.

At the bottom of Shoal Haven we were stopped by the water shoaling to five feet, so that we were obliged to haul out till we could anchor in more than two fathoms. During the night the wind shifted to N.E., and blew right in, obliging us to weigh, and work under the S.W. end of Elizabeth Island into a bay close to that shore. From the summit of the S.W. / point I afterwards took angles, among which the most important gave Mount Sarmiento bearing S. 1 1/2° W. (true). Its distance must have been (by recent observations) ninety-four miles.

Elizabeth Island is a long, low strip of land, lying parallel to the shores of the Strait, which here take a N.N.E. direction. Compared with the land to the southward it is very low, no part being more than two or three hundred feet high. It is composed of narrow ranges of hills, extending in ridges in the direction of its length, over which are strewed boulders of the various rocks, which have been noticed before as forming the shingle beaches of Point St. Mary and Point St. Anna; two kinds of rock, greenstone and hornblende, being the most common. The vallies which divide the hilly ridges were well clothed with grass, and in many places were seen hollows, that had contained fresh water, but now were entirely dried up. These spots were marked by a white crust, apparently caused by the saline quality of the soil.

Geese and wild ducks, and the red-bill (*Hæmatopus*), seem to be the only inhabitants of this island. The Indians sometimes visit it, for at the S.W. end we found remains of wigwams and shell-fish. Perhaps it is a place whence they communicate with the Patagonian natives, or they, may in the season frequent it for eggs.

We anchored in Laredo Bay, and visited a lake about a mile from the beach, distinguished on the chart by the name of Duck Lagoon: it is very extensive, and covered with large flights of gulls, ducks, and widgeons. We shot one widgeon, which was a most beautiful bird, and of a species we had not before seen.*

Here the country begins to be clothed with the deciduous leaved Beech tree (*Fagus Antarctica*), which is stunted in growth, but very convenient for fuel. Though the hardiest tree of this region, it is never found of large size, the larger trees being the evergreen Beech (*Fagus betuloides*). We also met with several small plants common to Cape Gregory. One / may consider Cape Negro to be the boundary of two countries, as entirely different from each other in geological structure and appearance, as they are in climate, to which last difference may be attributed the dissimilarity of their botanical productions.

Hence we returned to Port Famine. In our absence, a boat from the Beagle had crossed the Strait to Lomas Bay, where a party of natives had kindled fires of invitation.

The weather, since the sun crossed the equator, had been unusually fine; and, with the exception of one day's heavy rain, the sky was so clear (the wind being moderate from the N.E.) that all the heights were exposed to our view, and amongst them Mount Sarmiento stood pre-eminent.

Our preparations for sailing being nearly completed, the Hope was unrigged and hoisted in, and our temporary settlement on shore abandoned. It consisted of a marquee and a large bell tent.[9] In the former was Mr. Harrison (mate), who had charge of the party, and of the meteorological instruments: the bell tent held the crew. Near them were the observatory, a sawpit, and a cooking place, where a cheerful fire was always blazing. The carpenter's shop, cooper's bench, and armourer's forge had each its place, as well as a rope-walk, close to which our rigging was refitted, and the sails were repaired. After working-hours the shore party roamed about the woods with guns, or at low water picked up shell fish,†[10] by which they usually procured a fresh meal twice, but always once, a week.

* 'Anas Rafflesii,' Zool. Journ., vol. iv., and Tab. Supp., xxix.

† Of these a species of mactra (*M. edulis Nob*). was most abundant.

Meanwhile the ship was kept carefully clean and in order. The officers not immediately employed in active duty made excursions with their guns; and although the immediate vicinity of our tents was pretty well thinned of game, yet a walk of a few miles was always rewarded by ample sport. When opportunities offered, some of the men were permitted to amuse themselves on shore with their guns, for which many had provided themselves with powder and shot. Every Sunday, after divine service, which was performed as regularly as possible under our circumstances, such of the ship's company as desired permission / to land obtained it. On one occasion, however, we had nearly suffered for this indulgence, which was conducive to the men's health, and seldom abused: for one of them having made a fire at a little distance from the tents, the flames spread, and the exertions of all hands, for three hours, only just prevented it from communicating to the tents. On another occasion, two men set out on a shooting excursion, intending to cross the river Sedger, against doing which there had been no particular orders, as such a proceeding was scarcely contemplated. Having reached the bank near its mouth, and searched for a fordable place unsuccessfully, they launched a log of wood, and sitting astride, without providing themselves with a pole or paddle, pushed off from the shore, supposing it would go across; but, on reaching the middle of the stream, it was soon carried, by the current, out of the river, into the bay. One man, Gilly, seeing that the log was still floating away with the ebb tide, plunged in, and just reached the shore south of the river, in a very exhausted state; the other, Rix, unable to swim, kept his place, and was carried out to sea on a voyage that might have been fatal, had he not been seen from the ship, and saved by a boat.

Before leaving Port Famine we hauled one of our boats ashore, and left her (as we thought) securely hidden among the trees.

Being now ready to sail, and only waiting for wind, the officers of both ships, twenty-seven in number, dined together on shore. /

CHAPTER VI.

Trees – Leave Port Famine – Patagonians – Gregory Bay – Bysante – Maria – Falkner's account of the Natives – Indians seen on the borders of the Otway Water, in 1829 – Maria visits the Adventure – Religious Ceremony – Patagonian Encampment – Tomb of a Child – Women's employment – Children – Gratitude of a Native – Size of Patagonians – Former accounts of their gigantic height – Character – Articles for barter – Fuegians living with Patagonians – Ships sail – Arrive at Monte Video and Rio de Janeiro.

WHILE detained by northerly winds, the carpenter and a party of people were employed in the woods selecting and cutting down trees to be ready for our next visit. After felling thirteen trees, from twenty-four to thirty-six inches in diameter, eight were found to be rotten at the heart; but by afterwards taking the precaution of boring the trees with an augur, while standing, much trouble was saved, and fifteen sound sticks of considerable diameter were cut down. We found one tree, an evergreen beech, too large for any of our saws: it measured twenty-one feet in girth at the base, and from the height of six feet to twenty it was seventeen feet in circumference; above this height, three large arms (each from thirty to forty inches in diameter), branched off from the trunk. It is, perhaps, the very tree described by Byron in his account of this place.[1] We only once saw it equalled in size, and that was by a prostrate trunk, very much decayed.

In this interval of fine weather and northerly wind, we had the thermometer as high as 58°, and the barometer ranging between 29.80 and 30.00; but for two days before the wind shifted, the alteration was predicated by a gradual descent of the mercurial column, and a considerable increase of cold. On the 7th May, as there was some appearance of a change, we got under weigh; but were hardly outside the port, when a northerly wind again set in, and prevented our going farther than Fresh-water / Bay, where we passed the night. At last, on the 8th, accompanied by the Beagle, we proceeded on our course with a strong south-westerly breeze, which carried us quickly up to Cape Negro, when it blew so hard that I anchored off Laredo Bay. At this anchorage we certainly felt the air much colder and sharper than at Port Famine, arising from our being in a more exposed situation, and from the approach of winter, as well as from the severe south-west gale which was blowing.

After the gale had abated, we proceeded with fair weather and a light breeze to the Second Narrow, when the wind fell; but the tide being in our favour, we passed rapidly through. On a hill near us we observed three or four Patagonian Indians standing together, and their horses feeding close to them. A fire was soon kindled, to attract our notice, to which signal we replied by showing our colours; and had we not already communicated with these people, we should certainly have thought them giants, for they 'loomed very large' as they stood on the summit of the hill. This optical deception must doubtless have been caused by mirage: the haze has always been observed to be very great during fine weather and a hot day, arising from rapid evaporation of the moisture so abundantly deposited, on the surface of the ground, in all parts of the Strait.

As soon as the Patagonians found they were noticed, they mounted and rode along the shore abreast of us, being joined by other parties, until the whole number could not have been less than forty. Several foals and dogs were with them. Having anchored in Gregory Bay, where I intended remaining for two days to communicate with them, I sent up a rocket, burnt a blue-light, and despatched Lieutenant Cooke on shore to ask for a large supply of guanaco meat, for which we would pay in knives and beads. The boat returned on board immediately, bringing off four natives, three men and 'Maria.'[2] This rather remarkable woman must have been, judging by her appearance, about forty years old: she is said to have been born at Assuncion, in Paraguay, but I think the place of her birth was nearer Buenos Ayres. She spoke broken, but intelligible, / Spanish, and stated herself to be sister of Bysante, the cacique of a tribe near the Santa Cruz River, who is an important personage, on account of his size (which Maria described to be immense), and his riches. In speaking of him, she said he was *very* rich; he had many mantles, and also many hides ('*muy* rico, tiene muchas mantas y tambien muchos cueros').[3] One of Maria's companions, a brother of Bysante, was the tallest and largest man of this tribe; and though he only measured six feet in height, his body was large enough for a much taller man. He was in great affliction: his daughter had died only two days before our arrival; but, notwithstanding his sad story, which soon found him friends, it was not long before he became quite intoxicated, and began to sing and roar on the subject of his misfortunes, with a sound more like the bellowing of a bull than the voice of a human being. Upon applying to Maria, who was not quite so tipsy as her brother, to prevent him from making such hideous noises, she laughed and said, 'Oh, never mind, he's drunk; poor fellow, his daughter is dead' (Es boracho, povrecito, murio su hija); and then, assuming a serious tone, she looked towards the sky, and muttered in her own language a sort of prayer or invocation to their chief demon, or ruling spirit, whom Pigafetta, the companion and historian of Magalhaens, called *Setebos*, which Admiral Burney supposes to have been the original of one of Shakspear's names in the 'Tempest' –

> ' – his art is of such power
> He would controul my dam's god Setebos.'*[4]

Maria's dress was similar to that of other females of the tribe; but she wore ear-rings, made of medals stamped with a figure of the Virgin Mary, which, with the brass-pin that secured her mantle across her breast, were given to her by one Lewis, who had passed by in an American sealing-vessel,[5] and who, we understood from her, had made them 'Christians.' The Jesuit Falkner, who lived among them for many years, has written a long and, apparently, a very authentic account / of the inhabitants of the countries south of the River Plata, and he describes those who inhabit the borders of the Strait and sea-coast to be, 'Yacana-cunnees, which signifies foot- people, for they have no horses in their country; to the north they border on the Sehuau-cunnees, to the west on the Keyyus, or Key-yuhues, from whom they are divided by a ridge of mountains; to the east they are bounded by the ocean; and to the south by the islands of Tierra del Fuego, or the South Sea. These Indians live near the sea on both sides of the Strait, and often make war with one another. They make use of light floats, like those of Chilóe, in order to pass the Straits, and are sometimes attacked by the Huilliches and other Tehuelhets, who carry them away for slaves, as they have nothing to lose but their liberty and their lives. They subsist chiefly on fish, which they catch either by diving, or striking them with their darts. They are very nimble afoot, and catch guanacoes and ostriches with their bowls. Their stature is much the same as that of the other Tehuelhets, rarely exceeding seven feet, and oftentimes not six feet. They are an innocent, harmless people.'†

To the north of this race, Falkner describes 'the Sehuaucunnees, the most southern Indians who travel on horseback; Sehuau signifies in the Tehuel dialect a species of black rabbit, about the size of a field rat; and as their country abounds in these animals, their name may be derived from thence: cunnee signifying 'people.'[6]

With the exception of their mode of killing the guanaco by bowls, or balls, the description of the Key-yus would apply better to the Fuegian Indians; and if so, they have been driven across the Strait, and confined to the Fuegian shores by the Sehuau-cunnees, who must be no other than Maria's tribe. The Key-yus, who are described to inhabit the northern shore of the Strait, between Peckett's Harbour and Madre de Dios, are probably the tribe found about the south-western islands, and now called Alikhoolip; whilst the eastern Fuegians,

* Burney,i. 35 and 37.

† Falkner's Patagonia, pp. 110, 111.

or Yacana-cunnees, who have also been turned off the continent / by their powerful neighbours, are now called Tekeenikas. Our knowledge of the names of these two tribes, Alikhoolip and Tekeenika, results from Captain Fitz-Roy's subsequent examination of the outer coast of Tierra del Fuego in the Beagle (1830). A Cacique, belonging to the nation of the Key-yus, told Falkner that he had been in a house made of wood, that travelled on the water. A party of the Indians, in four canoes, were met on the borders of the Otway Water by Captain Fitz-Roy in 1829, whose arms, implements, and every thing they had, were precisely like the Fuegian Indians, excepting that they had a quiver made from the skin of a deer, and were in form a superior race, being both stronger and stouter.

For want of better information upon the subject, we must be content to separate the natives into Patagonians and Fuegians. The sealing vessels' crews distinguish them as Horse Indians, and Canoe Indians.[7]

These people have had considerable communication with the sealers who frequent this neighbourhood, bartering their guanaco skins and meat, their mantles, and furs, for beads, knives, brass ornaments, and other articles; but they are equally anxious to get sugar, flour, and, more than all, 'aqua ardiente,' or spirits.[8] Upon the arrival of a boat from any vessel, Maria, with as many as she can persuade the boat's crew to take, goes on board, and, if permitted, passes the night. As soon as our boat landed, Maria and her friends took their seats as if it had been sent purposely for them. Not expecting such a visit, I had given no order to the contrary, and the novelty of such companions overcame the scruples of the officer, who was sent on shore to communicate with them. Their noisy behaviour becoming disagreeable, they were soon conducted from below to the deck, where they passed the night. Maria slept with her head on the windlass; and was so intoxicated, that the noise and concussion produced by veering eighty fathoms of cable round it did not awake her. The following morning, whilst I was at breakfast, she very unceremoniously introduced herself, with one of her companions, and / seating herself at table, asked for tea and bread, and made a hearty meal. I took the precaution of having all the knives, and articles that I thought likely to be stolen, removed from the table; but neither then, nor at any time, did I detect Maria in trying to steal, although her companions never lost an opportunity of pilfering.

After breakfast the Indians were landed, and as many of the officers as could be spared went on shore, and passed the whole day with the tribe, during which a very active trade was carried on. There were about one hundred and twenty Indians collected together, with horses and dogs. It is probable that, with the exception of five or six individuals left to take care of the encampment, and such as were absent on hunting excursions, the whole of the tribe was mustered on the beach, each family in a separate knot, with all their riches displayed to the best advantage for sale.

I accompanied Maria to the shore. On landing, she conducted me to the place where her family were seated round their property. They consisted of Manuel, her husband, and three children, the eldest being known by the appellation of Capitan Chico, or 'little chief.' A skin being spread out for me to sit on, the family and the greater part of the tribe collected around. Maria then presented me with several mantles and skins, for which I gave in return a sword, remnants of red baize, knives, scissors, looking-glasses, and beads: of the latter I afterwards distributed bunches to all the children, a present which caused evident satisfaction to the mothers, many of whom also obtained a share. The receivers were selected by Maria, who directed me to the youngest children first, then to the elder ones, and lastly to the girls and women. It was curious and amusing, to witness the order with which this scene was conducted, and the remarkable patience of the children, who, with the greatest anxiety to possess their trinkets, neither opened their lips, nor held out a hand, until she pointed to them in succession.

Having told Maria that I had more things to dispose of for guanaco meat she dismissed the tribe from around me, and, / saying she was going for meat (carne), mounted her horse, and rode off at a brisk pace. Upon her departure a most active trade commenced: at first, a mantle was purchased for a string of beads; but as the demand increased, so the Indians increased their price, till it rose to a knife, then to tobacco, then to a sword, at last nothing would satisfy them but 'aqua ardiente,' for which they asked repeatedly, saying 'bueno es boracho – bueno es – bueno es boracho;'* – but I would not permit spirits to be brought on shore.

At Maria's return with a very small quantity of guanaco meat, her husband told her that I had been very inquisitive about a red baize bundle, which he told me contained 'Cristo,' upon which she said to me 'Quiere mirar mi Cristo' (do you wish to see my Christ), and then, upon my nodding assent, called around her a number of the tribe, who immediately obeyed her summons. Many of the women, however, remained to take care of their valuables. A ceremony then took place. Maria, who, by the lead she took in the proceedings, appeared to be high priestess† as well as cacique of the tribe, began by pulverising some whitish earth in the hollow of her hand, and then taking a mouthful of water, spit from time to time upon it, until she had formed a sort of pigment, which she distributed to the rest, reserving only sufficient to mark her face, eyelids, arms, and hair with

* It is good to be drunk, it is pleasant to be drunk.

† Two Portuguese seamen, however, who had resided some months with them, having been left behind by a sealing vessel, and taken off by us at a subsequent period of the voyage at their own request, informed us that Maria is not the leader of religious ceremonies. Each family possesses its own household god, a small wooden image, about three inches in length, the rough imitation of a man's head and shoulders, which they consider as the representative of a superior being, attributing to it all the good or evil that happens to them.

the figure of the cross. The manner in which this was done was peculiar. After rubbing the paint in her left hand smooth with the palm of the right, she scored marks across the paint, and again others at right angles, leaving the impression of as many crosses, which she / stamped upon different parts of her body, rubbing the paint, and marking the crosses afresh, after every stamp was made.

The men, after having marked themselves in a similar manner (to do which some stripped to the waist and covered all their body with impressions), proceeded to do the same to the boys, who were not permitted to perform this part of the ceremony themselves. Manuel, Maria's husband, who seemed to be her chief assistant on the occasion, then took from the folds of the sacred wrapper an awl, and with it pierced either the arms or ears of all the party; each of whom presented in turn, pinched up between the finger and thumb, that portion of flesh which was to be perforated. The object evidently was to lose blood, and those from whom the blood flowed freely showed marks of satisfaction, while some whose wounds bled but little underwent the operation a second time.

When Manuel had finished, he gave the awl to Maria, who pierced his arm, and then, with great solemnity and care, muttering and talking to herself in Spanish (not two words of which could I catch, although I knelt down close to her and listened with the greatest attention), she removed two or three wrappers, and exposed to our view a small figure, carved in wood, representing a dead person, stretched out. After exposing the image, to which all paid the greatest attention, and contemplating it for some moments in silence, Maria began to descant upon the virtues of her Christ, telling us it had a good heart ('buen corazon'), and that it was very fond of tobacco. 'Mucho quiere mi Cristo tabaco, da me mas,' (my Christ loves tobacco very much, give me some). Such an appeal, on such an occasion, I could not refuse; and after agreeing with her in praise of the figure, I said I would send on board for some. Having gained her point, she began to talk to herself for some minutes, during which she looked up, after repeating the words 'muy bueno es mi Cristo, muy bueno corazon tiene,' and slowly and solemnly packed up the figure, depositing it in the place whence it had been taken. This ceremony ended, the traffic, which had been suspended, recommenced with redoubled activity. /

According to my promise, I sent on board for some tobacco, and my servant brought a larger quantity than I thought necessary for the occasion, which he injudiciously exposed to view. Maria, having seen the treasure, made up her mind to have the whole, and upon my selecting three or four pounds of it, and presenting them to her, looked very much disappointed, and grumbled forth her discontent: I taxed her with greediness, and spoke rather sharply, which had a good effect, for she went away and returned with a guanaco mantle, which she presented to me.

During this days' barter we procured guanaco meat, sufficient for two days' supply of all hands, for a few pounds of tobacco. It had been killed in the morning, and was brought on horseback cut up into large pieces, for each of which we had to bargain. Directly an animal is killed, it is skinned and cut up, or torn as under, for the convenience of carrying. The operation is done in haste, and therefore the meat looks bad; but it is well tasted, excellent food, and although never fat, yields abundance of gravy, which compensates for its leanness. It improves very much by keeping, and proved to be valuable and wholesome meat.

Captain Stokes, and several of the officers; upon our, first reaching the beach, had obtained horses, and rode to their 'toldos,' or principal encampment. On their return, I learned that, at a short distance from the dwellings, they had seen the tomb of the child who had lately died. As soon, therefore, as Maria returned, I procured a horse from her, and, accompanied by her husband and brother, the father of the deceased, and herself, visited these toldos, situated in a valley extending north and south between two ridges of hills, through which ran a stream, falling into the Strait within the Second Narrow, about a mile to the westward of Cape Gregory.

We found eight or ten huts arranged in a row; the sides and backs were covered with skins, but the fronts, which faced the east, were open; even these, however, were very much screened from wind by the ridge of hills, eastward of the plain. Near them the ground was rather bare, but a little / farther back there was a luxuriant growth of grass, affording rich and plentiful pasture for the horses, among which we observed several mares in foal, and colts feeding and frisking by the side of their dams: the scene was lively and pleasing, and, for the moment, reminded me of distant climes, and days gone by.

The 'toldos' are all alike. In form they are rectangular, about ten or twelve feet long, ten deep, seven feet high in front, and six feet in the rear. The frame of the building is formed by poles stuck in the ground, having forked tops to hold cross pieces, on which are laid poles for rafters, to support the covering, which is made of skins of animals sewn together so as to be almost impervious to rain or wind. The posts and rafters, which are not easily procured, are carried from place to place in all their travelling excursions. Having reached their bivouac, and marked out a place with due regard to shelter from the wind, they dig holes with an iron bar or piece of pointed hard wood, to receive the posts; and all the frame and cover being ready, it takes but a short time to erect a dwelling. Their goods and furniture are placed on horseback under the charge of the females, who are mounted aloft upon them. The men carry nothing but the lasso and bolas, to be ready for the capture of animals, or for defence.

Maria's toldo was nearly in the middle, and next to it was her brother's. All the huts seemed well stored with skins and provisions, the former being rolled up and placed at the back, and the latter suspended from the supporters of the roof; the greater part was in that state well known in South America by the name of charque (jerked beef); but this was principally horse-flesh, which these people esteem superior to other food. The fresh meat was almost all guanaco. The only vessels they use for carrying water are bladders, and sufficiently disagreeable substitutes for drinking utensils they make: the Fuegian basket, although sometimes dirty, is less offensive.

About two hundred yards from the village the tomb was erected, to which, while Maria was arranging her skins and / mantles for sale, the father of the deceased conducted me and a few other officers.

It was a conical pile of dried twigs and branches of bushes, about ten feet high and twenty-five in circumference at the base, the whole bound round with thongs of hide, and the top covered with a piece of red cloth, ornamented with brass studs, and surmounted by two poles, bearing red flags and a string of bells, which, moved by the wind, kept up a continual tinkling.

A ditch, about two feet wide and one foot deep, was dug round the tomb, except at the entrance, which had been filled up with bushes. In front of this entrance stood the stuffed skins of two horses, recently killed, each placed upon four poles for legs. The horses heads were ornamented with brass studs, similar to those on the top of the tomb; and on the outer margin of the ditch were six poles, each carrying two flags, one over the other.

The father, who wept much when he visited the tomb, with the party of officers who first went with him, although now evidently distressed, entered into, what we supposed to be, a long account of the illness of his child, and explained to us that her death was caused by a bad cough. No watch was kept over the tomb; but it was in sight of, and not very far from their toldos, so that the approach of any one could immediately be known. They evidently placed extreme confidence in us, and therefore it would have been as unjust as impolitic to attempt an examination of its contents, or to ascertain what had been done with the body.

The Patagonian women are treated far more kindly by their husbands than the Fuegian; who are little better than slaves, subject to be beaten, and obliged to perform all the laborious offices of the family. The Patagonian females sit at home, grinding paint, drying and stretching skins, making and painting mantles. In travelling, however, they have the baggage and provisions in their charge, and, of course, their children. These women probably have employments of a more laborious nature than what we saw; but they cannot be compared with / / those of the Fuegians, who, excepting in the fight and chace, do every thing. They paddle the canoes, dive for shells and sea-eggs, build their wigwams, and keep up the fire; and if they neglect any of these duties, or incur the displeasure of their husbands in any way, they are struck or kicked most severely. Byron, in his narrative of the loss of the Wager, describes the brutal conduct of one of these Indians, who actually killed his child for a most trifling offence.[9] The Patagonians are devotedly attached to their offspring. In infancy they are carried behind the sad-

P. P. King. T. Landseer.

PATAGONIAN 'TOLDO' AND TOMB.

dle of the mother, within a sort of cradle, in which they are securely fixed. The cradle is made of wickerwork, about four feet long and one foot wide, roofed over with twigs like the frame of a tilted waggon. The child is swaddled up in skins, with the fur inwards or outwards according to the weather. At night, or when it rains, the cradle is covered with a skin that effectually keeps out the cold or rain. Seeing one of these cradles near a woman, I began to make a sketch of it, upon which the mother called the father, who watched me most attentively, and held the cradle in the position which I considered most advantageous for my sketch. The completion of the drawing gave them both great pleasure, and during the afternoon the father reminded me repeatedly of having painted his child ('pintado su hijo.')

One circumstance deserves to be noticed, as a proof of their good feeling towards us. It will be recollected that three Indians, of the party with whom we first communicated, accompanied us as far as Cape Negro, where they landed. Upon our arrival on this occasion, I was met, on landing, by one of them, who asked for my son,[10] to whom they had taken a great fancy; upon my saying he was on board, the native presented me with a bunch of nine ostrich feathers, and then gave a similar present to every one in the boat. He still carried a large quantity under his arm, tied up in bunches, containing nine feathers in each; and soon afterwards, when a boat from the Beagle landed with Captain Stokes and others, he went to meet them; but finding strangers, he withdrew without making them any present. /

In the evening my son landed, when the same Indian came down to meet him, appeared delighted to see him, and presented him with a bunch of feathers, of the same size as those which he had distributed in the morning. At this, our second visit, there were about fifty Patagonian men assembled, not one of whom looked more than fifty-five years of age. They were generally between five feet ten and six feet in height: one man only exceeded six feet – whose dimensions, measured by Captain Stokes, were as follows:-

	ft.	in.
Height	6	1 ¾
Round the chest	4	1 ⅛
Do. loins	3	4 ¾

I had before remarked the disproportionate largeness of head, and length of body of these people, as compared with the diminutive size of their extremities; and, on this visit, my opinion was further confirmed, for such appeared to be the general character of the whole tribe; and to this, perhaps, may be attributed the mistakes of some former navigators.[11] Magalhaens, or rather Pigafetta, was the first who described the inhabitants of the southern extremity of America as giants. He met some at Port San Julian, of whom one is described to be 'so tall, that our heads scarcely came up to his waist, and his voice was like that of a bull.'

Herrera,* however, gives a less extravagant account of them: he says, 'the least of the men was larger and taller than the stoutest man of Castile;' and Maxim. Transylvanus says they were 'in height ten palms or spans; or seven feet six inches.'[12]

In Loyasa's voyage (1526), Herrera mentions an interview with the natives, who came in two canoes, 'the sides of which were formed of the ribs of whales.' The people in them were of large size 'some called them giants; but there is so little conformity between the accounts given concerning them, that I shall be silent on the subject.'†[13]

As Loyasa's voyage was undertaken immediately after the return of Magalhaens' expedition, it is probable that, from the / impressions received from Pigafetta's narrative, many thought the Indians whom they met must be giants, whilst others, not finding them so large as they expected, spoke more cautiously on the subject; but the people seen by them must have been Fuegians, and not those whom we now recognise by the name of Patagonians.

Sir Francis Drake's fleet put into Port San Julian, where they found natives 'of large stature;' and the author of the 'World Encompassed,'[14] in which the above voyage is detailed, speaking of their size and height, supposes the name given them to have been *Pentagones,* to denote a stature of 'five cubits, viz. seven feet and a half,' and remarks that it described, the full height, if not somewhat more, of the tallest of them.‡ They spoke of the Indians whom they met within the Strait as small in stature.§

The next navigator who passed through the Strait was Sarmiento; whose narrative says little in proof of the very superior size of the Patagonians. He merely calls them 'Gente Grande,'¶ and 'los Gigantes;' but this might have originated from the account of Magalhaens' voyage. He particularises but one Indian, whom they made prisoner, and only says 'his limbs are of large size:' ('Es crecido de miembros.') This man was a native of the land near Cape Monmouth, and, therefore, a Fuegian. Sarmiento was afterwards in the neighbourhood of Gregory Bay, and had an encounter with the Indians, in which he and others were wounded; but he does not speak of them as being unusually tall.[15]

After the establishment, called 'Jesus,' was formed by Sarmiento, in the very spot where 'giants' had been seen, no people of large stature are mentioned, in the account of the colony; but Tomé Hernandez, when examined before the Vice-Roy of Peru, stated, 'that the Indians of the plains, who are giants, communicate with the natives of Tierra del Fuego, who are like them.**

* Burney, i.p.33.

† Ibid. p. 135.

‡ Burney, i. 318.

§ Ibid, i. 324.

¶ Sarmiento, p. 244.

** Sarmiento's Appendix, xxix.

Anthony Knyvet's account* of Cavendish's second voyage / (which is contained in Purchas), is not considered credible.[16] He describes the Patagonians to be fifteen or sixteen spans in height; and that of these cannibals, there came to them at one time above a thousand! The Indians at Port Famine, in the same narrative, are mentioned as a kind of strange cannibals, short of body, not above five or six spans high, very strong, and thick made.†

The natives, who were so inhumanly murdered by Oliver Van Noort,[17] on the Island of Santa Marta (near Elizabeth Island), were described to be nearly of the same stature as the common people in Holland, and were remarked to be broad and high-chested. Some captives were taken on board, and one, a boy, informed the crew that there was a tribe living farther in-land, named 'Tiremenen,' and their territory 'Coin;' that they were 'great people, like giants, being from ten to twelve feet high, and that they came to make war against the other tribes,‡ whom they reproached for being eaters of ostriches!'§‡

Spilbergen (1615)[18] says he 'saw a man of extraordinary stature, who kept on the higher grounds to observe the ships; and on an island, near the entrance of the Strait, were found the dead bodies of two natives, wrapped in the skins of penguins, and very lightly covered with earth; one of them, was of the common human stature, the other, the journal says, was two feet and a half longer.¶§ The gigantic appearance of the man on the hills may perhaps be explained by the optical deception we ourselves experienced.

Le Maire and Schouten,[19] whose accounts of the graves of the Patagonians agree precisely with what we noticed at Sea Bear Bay, of the body being laid on the ground covered with / a heap of stones, describe the skeletons as measuring ten or eleven feet in length, 'the skulls of which we could put on our heads in the manner of helmets!'

The Nodales did not see any people on the northern side of the Strait; those with whom they communicated were natives of Tierra del Fuego, of whose form no particular notice is taken.

Sir John Narborough saw Indians at Port San Julian, and describes them as 'people of a middling stature: well-shaped. * * * Mr. Wood[20] was taller than any

* Purchas, iv. ch. 6 and 7.

† Burney, ii. p. 106.

‡ The tribes described by this boy are the
1. Kemenites, inhabiting a place called Karay.
2. Kennekas Karamay.
3. Karaike Morine.
4. Enoo, the tribe to which the Indians, whom they murdered, belonged.

§ Burney, ii. 215.

¶ Ibid. ii. 334.

of them.' He also had an interview with nineteen natives upon Elizabeth Island, but they were Fuegians.

In the year 1741, Patagonian Indians were seen by Bulkley and his companions. They were mounted on horses, or mules, which is the first notice we have of their possessing those animals.

Duclos de Guyot, in the year 1766, had an interview with seven Patagonian Indians, who were mounted on horses equipped with saddles, bridles, and stirrups. The shortest of the men measured five feet eleven inches and a quarter English. The others were considerably taller. Their chief or leader they called 'Capitan.'

Bougainville, in 1767, landed amongst the Patagonians. Of their size he remarks: 'They have a fine shape; among those whom we saw, not one was below five feet ten inches and a quarter (English), nor above six feet two inches and a half in height. Their gigantic appearance arises from their prodigiously broad shoulders, the size of their heads, and the thickness of all their limbs. They are robust and well fed: their nerves are braced and their muscles strong, and sufficiently hard, &c.' This is an excellent account; but how different is that of Commodore Byron, who says, 'One of them, who afterwards appeared to be chief, came towards me; he was of gigantic stature, and seemed to realise the tales of monsters in a human shape: he had the skin of some wild beast thrown over his shoulders, as a Scotch Highlander wears his plaid, and was painted so as to make the most hideous appearance I / ever beheld: round one eye was a large circle of white, a circle of black surrounded the other, and the rest of his body was streaked with paint of different colours. I did not measure him; but if I may judge of his height by the proportion of his stature to my own, it could not be less than seven feet. When this frightful colossus came up, we muttered somewhat to each other as a salutation, &c.'** After this he mentions a woman 'of most enormous size;' and again, when Mr. Cumming, the lieutenant, joined him, the commodore says, 'Before the song was finished, Mr. Cumming came up with the tobacco, and I could not but smile at the astonishment which I saw expressed in his countenance upon perceiving himself, though six feet two inches high, become at once a pigmy among giants, for these people may, indeed, more properly be called giants than tall men: of the few among us who are full six feet high, scarcely any are broad and muscular, in proportion to their stature, but look rather like men of the common bulk grown up accidentally to an unusual height; and a man who should measure only six feet two inches, and equally exceed a stout well-set man of the common stature in breadth and muscle, would strike us rather as being of a gigantic race, than as an individual accidentally anomalous;

* Hawksworth's Coll. i. 28.

our sensations, therefore, upon seeing five hundred people, the shortest of whom were at least four inches taller, and bulky in proportion, may be easily imagined.'*

This account was published only seven years after the voyage, and the exaggeration, if any, might have been exposed by numbers. There can be no doubt, that among five hundred persons several were of a large size; but that all were four inches taller than six feet must have been a mistake. The commodore says, that he 'caused them all to be seated,' and in that position, from the length of their bodies, they would certainly appear to be of very large stature.†

Shortly afterwards, Wallis, in the neighbourhood of Cape Virgins, communicated with the same people, and as the story of the Patagonian giants had been spread abroad, and was very much discredited, he carried two measuring rods with him; and says, in his narrative, 'We went round and measured those that appeared to be the tallest. One was six feet seven inches high, several more were six feet five, and six feet six inches; but the stature of the greatest part of them was from five feet ten to six feet.'

In the voyage of the Santa Maria de la Cabeza,‡ 1786, it is related that the height of one or two Patagonians, with whom the officers had an interview, was six feet eleven inches and a half (of Burgos),[22] which is equal to six feet four inches and a half (English). This man wore a sword, on which was engraved 'Por el Rey Carlos III.,' and spoke a few words in Spanish, proofs of his having had communication with some of the Spanish settlements. It does not, however, appear from the account that there were many others, if any, of that height.

Of all the above accounts, I think those by Bougainville and Wallis the most accurate. It is true, that of the number we saw, none measured more than six feet two inches; but it is possible that the preceding generation may have been a larger race of people, for none that we saw could have been alive at the time of Wallis's or Byron's voyage. The oldest certainly were the tallest; but, without discrediting the accounts of Byron, or any other of the modern voyagers, I think it probable that, by a different mode of life, or a mixture by marriage with the southern or Fuegian tribes, which we know has taken place, they have degenerated into a smaller race, and have lost all right to the title of giants; yet their bulky, / muscular forms, and length of body, in some measure bear out the

* Ibid.

† See a letter from Mr. Charles Clarke, an officer on board the Dolphin, to Mr. Maly, M.D., secretary of the Royal Society, dated Nov. 3, 1766, read before the Royal Society on 12th April 1767, and published in the fifty-seventh volume of the Phil. Trans., part i. p. 75, in which an exaggerated account is given of this meeting. The men are described to be eight feet high, and the women seven and a half to eight feet. 'They are prodigious stout, and as well and proportionably made as ever I saw people in my life.'[21] This communication was probably intended to corroborate the commodore's account. /

‡ Ultimo Viage, p. 21.

above accounts; for had the present generation proportionate limbs, they might, without any exaggeration, justify the account of Commodore Byron. The Jesuit Missionary Falkner,* who, from an intercourse of forty years with the Indians of South America, must be considered as one of the best authorities, says, speaking of a Patagonian named Cangapol, 'This chief, who was called by the Spaniards the Cacique Bravo,† was tall and well-proportioned; he must have been seven feet and some inches in height, because on tiptoe I could not reach the top of his head: I was very well acquainted with him, and went some journeys in his company: I do not recollect ever to have seen an Indian that was above an inch or two taller than Cangapol. His brother Sausimian was but about six feet high. The Patagonians or Puelches are a large-bodied people; but I never heard of that gigantic race which others have mentioned, though I have seen persons of all the different tribes of the Southern Indians.'

This is an account in 1746, only twenty years before that of Bougainville. Taking all the evidence together, it may be considered, that the medium height of the males of these southern tribes is about five feet eleven inches. The women are not so tall, but are in proportion broader and stouter: they are generally plain-featured. The head is long, broad and flat, and the forehead low, with the hair growing within an inch of the eyebrows, which are bare. The eyes are often placed obliquely, and have but little expression, the nose is generally rather flat, and turned up; but we noticed several with that feature / straight, and sometimes aquiline: the mouth is wide, with prominent lips, and the chin is rather large; the jaws are broad, and give the face a square appearance; the neck is short and thick; the shoulders are broad; the chest is broad, and very full; but the arm, particularly the fore-arm, is small, as are also the foot and leg; the body long, large and fat, but not corpulent. Such was the appearance of those who came under my observation.

As to their character, the Patagonians are friendly, without that disposition to quarrel, after the novelty of first acquaintance has worn off, which is so common among savages in general. This probably arises from interested motives, certainly not from fear, unless it be the fear of being avoided instead of visited by the ships which pass by, and from which they procure many useful articles, and many temporary gratifications.

* Falkner, according to Dean Funes, was originally engaged in the slave trade at Buenos Ayres; but afterwards became a Jesuit, and studied in the college at Cordova, where, to an eminent knowledge of medicine, he added that of theology. He is the author of a description of Patagonia, published in London after the expulsion of the Jesuits. – (*Ensayo de la Historia Civil del Paraguay, Buenos Ayres, y Tucuman, por el Doctor Don Gregorio Funes, iii. p. 23, note. Published at Buenos Ayres. 8vo.* 1817.)

† See Dean Funes's account of Buenos Ayres, and of the Indian tribes, vol. ii. 394.

A. Earle. T. A. Prior.

MONTE VIDEO, MOLE.

Swords, long knives, tobacco, Paraguay tea, bits, saddles, guns, lead for balls, red cloth, beads (particularly of a sky-blue colour), flour, sugar, and spirits, are much desired in exchange for their peltry and guanaco meat; but they have no idea beyond that of satisfying the wants of the moment.

After a few pounds of tobacco had been distributed amongst them, although they are very fond of smoking, it became quite a drug, and it was necessary to produce something new to excite their attention. From Maria's influence, and the reference so constantly made to her, it would seem that she was considered as cacique of the tribe; but her apparent superiority may arise from her connexion with Bysante, of whom they all spoke as 'El Cacique Grande,' or from the attention paid to her by ships with whom they communicate.

The people of this tribe seemed to live together harmoniously; no bickerings or jealous feelings were observed, and certainly none were expressed by any one of our bulky friends on witnessing another receiving a valuable present, or a good exchange for his property.

At sunset our people were ordered to embark, upon which the price of Patagonian goods immediately fell, at least, a thousand per cent., though many held back in expectation of / the next day. Maria put into the boat, after my refusal to let her go on board to pass the night, two bags, and asked me to send her flour and sugar. She was most importunate for aqua ardiente, which, however, I refused. Her constant cry was 'It is very good to be drunk; I like drinking very much; rum is very good. – Give me some?' ('Muy bueno es boracho, mucho mi gusta, mucho mi gusta de beber, muy bueno es aqua ardiente. – Da me no mas?')

Among them was a Fuegian Indian; but it did not appear clearly whether he was living with them permanently, or only on a visit. Some of us thought we understood the account of one of the Patagonians, who seemed to be the most interested about him, to be, that a master of a sealer had left him amongst them. We knew him instantly by his squalid and comparatively diminutive appearance, and were confirmed in our ideas by his recognition of the words 'Hosay' and Sherroo.' The Patagonian name for a ship is 'Carro grande,' and for a boat 'Carro chico,' a mixture of their own and the Spanish language. All that I could understand of his history was, that he was Cacique of some Indian tribes at a distance: he was evidently a great favourite, and although Maria spoke generally with much contempt of the Fuegian Indians, she had patronised this stranger, for he lived in her toldo, and shared all the presents that were made to her.

The following morning it rained hard, and blew so fresh a gale, from the westward, that it would have been dangerous to send a boat on shore: and I was obliged to weigh without landing the things which I had promised. After we were under weigh, the weather cleared partially, when we observed Maria on the beach, mounted on her white horse, with others watching our departure, and when it was evident that we were really gone, she rode slowly back to her toldo, no doubt considerably vexed. I was very sorry to treat them in this way, for their conduct towards us had been open and friendly. All I could hope to do, to make amends, was to give something of value at my return.

We steered across the Bay of St. Philip, accompanied by / / the Beagle,* left the Strait of Magalhaens with a fair wind, and, after a favourable passage, reached Monte Video on the 24th April 1827.

From Monte Video we went to Rio de Janeiro, to procure stores, and prepare for another voyage to the Strait. On our arrival I received the Commander-in-chief's leave to apply to the Lord High Admiral for permission to employ a tender, to facilitate the surveys of the sounds and deep channels, in the neighbourhood of the Strait, and the inner sounds on the west coast; for which, neither the Adventure, nor the Beagle, were adapted; and I thought it best to delay our departure until an answer to my application was received. /

* We left Gregory Bay in the morning, and passed Cape Virgins in the evening of the same day.

CHAPTER VII.

Leave Rio de Janeiro – Santos – St. Catherine's – Monte Video – Purchase the Adelaide schooner, for a Tender to the Adventure – Leave Monte Video – Beagle goes to Port Desire – Shoals off Cape Blanco – Bellaco Rock – Cape Virgins – Possession Bay – First Narrow – Race – Gregory Bay – View – Tomb – Traffic with Natives – Cordial meeting – Maria goes on board – Natives intoxicated – Laredo Bay – Port Famine.

WE were ready to resume our voyage early in September (1827); but not having received any communication by the packet, from the Admiralty, relative to the purchase of a tender, I determined to await the arrival to the next, early in October. I was again disappointed, and very reluctantly left Rio de Janeiro, on the 16th, for Monte Video; but that I might still benefit by the orders which were sure to be in the following packet, I determined upon calling at Santos, and St. Catherine's, for chronometrical observations; leaving the Beagle to wait for letters conveying the decision of his Royal Highness the Lord High Admiral.

We reached Santos on the 18th, and staid there until the 28th. In this interval I paid a short visit to St. Paul's, for the purpose of making barometrical observations.* At St. Catherine's / / we remained eight days, and during the interval necessary for ascertaining the rates of the chronometers, I obtained magnetic observations.

* On our passage from Santos to St. Catherine's, in latitude 28° south, we caught a 'dolphin' (*Coryphena*), the maw of which I found filled with shells, of *Argonauta tuberculosa*, and all containing the '*Octopus Ocythöe*' that has been always found as its inhabitant. Most of the specimens were crushed by the narrow passage into the stomach, but the smaller ones were quite perfect, and had been so recently swallowed that I was enabled to preserve several of various sizes containing the animal. To some of them was attached a nidus of eggs, which was deposited between the animal and the spire. The shells varied in size from two-thirds of an inch to two and a half inches in length; each contained an octopus, the bulk and shape of which was so completely adapted to that of the shell, that it seemed as if the shell increased with the animal's growth. When so many learned naturalists have differed so materially as to the character of the inhabitants of the argonauta,[1] it would be presumption in me to express even an opinion; I therefore merely mention the fact, and state that in no one specimen did there appear to be any connexion between the animal and the shell.

A. Earle. T. Heir.

MOLE PALACE AND CATHEDRAL, RIO DE JANEIRO.

After a tedious voyage of nineteen days from St. Catherine's, I arrived at Monte Video, and there received intelligence that the long-wished permission from the Lord High Admiral, to procure a tender, had been obtained. I accordingly purchased a schooner, which I named the Adelaide, and appointed Lieutenant Graves to the command. Five months' additional provisions for both vessels were purchased, and put into her; and on the 2d December, after running up the river to complete our water, we sailed out by the southern entrance, passing to the westward of the Archimedes' Shoal, and proceeded without farther detention to the southward.

On the 1st of January (in latitude 43° 17' and long. 61° 9'), I was informed that we were close to a rock. Upon going on deck, I saw the object; but in a very short time I perceived it was a dead whale, upon whose half-putrid body large flocks of birds were feeding. Many on board were, however, sceptical, until, on passing to leeward, the strong odour testified the fact. Its appearance certainly was very like the summit of a dark brown rock, covered with weeds and barnacles, and the myriads of birds which surrounded it added to the deception. It could, however, be distinguished by its buoyancy; for the water did not break over it, as of course it would have done had it been a fixed body. Such is probably the origin of half the 'vigias' that are found on the charts.[2] Whales, when struck by the fishers, frequently escape and perish; the carcass then floats on the surface of the sea, until decomposed or eaten by birds and fishes. A small vessel striking against such a mass, would probably be severely injured; and at night, the / body, from its buoyancy and the sea not breaking against it, would not be readily seen.

On the 4th, being about one hundred miles to the N.E. of Cape Blanco, I communicated with Captain Stokes, and gave him directions to proceed to Port Desire for chronometrical observations, and then follow me immediately to Cape Fairweather or Cape Virgins. We had light winds during the night, so that the Beagle made very little progress. In the afternoon, Cape Blanco, a long level-topped ridge, came in sight, of which good views are given in Lord Anson's voyage. We steered towards the land, and at six o'clock, were in eighteen fathoms, the rocky hill at the extremity of the Cape bearing S. 10° E. thirteen miles; at seven o'clock, the same hill was six miles and a half off, bearing S. 3° E., when we observed a line of rippling water, extending from east to as far as we could see on the south horizon. The depth was seventeen fathoms, but as we proceeded it gradually decreased to twelve and ten, and soon afterwards to seven fathoms, when the Beagle was observed to be firing guns; but whether they were intended to warn us of danger, or as signals of her own distress, we could not determine, and I hauled to the wind to cross where the ripple appeared least violent. In passing through it we had not less than seven fathoms, and then it deepened to twelve and fifteen fathoms. We had now leisure to attend to the Beagle, and soon saw that her signals were only to warn us, for she had resumed her course under a press of sail.

After steering four miles to the S.E., we again found ourselves in the midst of ripplings, in which the water shoaled to six fathoms. It being then dark, and not knowing how to proceed, we shortened sail and brought to the wind, in order that if the ship struck it might be with less force; but happily we passed on without any further decrease of soundings. In going through the ripple, the Adelaide, though deeply laden, behaved well.

Commodore Byron passed over these shoals, which he describes as lying at a greater distance from the shore: it was to avoid them that we passed so near the land. /

During the following evening there was a very heavy dew, the never-failing prognostic of a northerly wind; the horizon, also, was very hazy, and the water perfectly smooth. We were not more than ten miles off shore, yet the land was completely distorted in appearance by mirage.

Next morning we were very close to the position assigned to the Bellaco, or St. Estevan's Shoal, the existence of which has been very much doubted. It was discovered by the Nodales, and in the diary of their voyage is thus described: 'At five o'clock, or later in the evening, we discovered a rock a-wash ('una baxa que lababa la mar en ella') about five leagues from the shore, more or less. It is a very deceitful rock ('Es muy bellaco baxo'), because it is under water, over which, in fine weather and smooth water, the sea breaks. We sounded near it, and found twenty-six fathoms stony bottom. Its latitude is 48 1/2°., according to our noon observation, and the course and distance we have since run.'**

The late Don Felipe Bauza, one of the companions of Malespina,[3] informed me, that on the voyage of the Descubierta and Atrevida, their boats were sent to look for it, but were unsuccessful.

At noon we were in lat. 48° 40' S., long. 66° 6', depth forty-two fathoms, but without any signs of the Bellaco. Sailing on, the coast was seen in the neighbourhood of Beachy Head (so named from its resemblance to the well-known promontory). Afterwards, Cape Fairweather came in sight, and on the 10th Cape Virgins, which we passed in the evening, and, half an hour afterwards, rounding Dungeness, we again entered the Strait of Magalhaens; and anchored near the northern shore.

In Possession Bay we were detained several days, although repeated attempts to pass the First Narrow were anxiously made.

One evening, clouds gathered, and the weather assumed such a threatening appearance, that I expected to be obliged to run to sea; but to our surprise, when the cloudy mass seemed on / the point of bursting over us with a deluge of rain, it

* Nodales, p. 48.

suddenly vanished, and was succeeded by a beautifully clear and fine night. This favourable appearance gave us hopes of being able to make good our entrance on the following day; but a fresh gale set in, and kept us at our anchorage.

Early on the 14th we made another fruitless attempt to pass the First Narrow. As the Adelaide sailed under our stern, Lieutenant Graves informed me that he had lost an anchor, and had only one left, to which he had bent his chain-cable; and that she had shipped so much water in attempting to beat through, that he was on the point of asking permission to bear up when we ourselves gave up the attempt. It blew too hard to give any assistance to the Adelaide, but next morning, when the weather was more moderate, I seized an opportunity of sending our two kedge anchors; and in the afternoon we supplied her with some water and other necessaries, so that she was comparatively well off, and my anxiety on her account much relieved.

Fires on the Fuegian side had been kept up since our arrival, but we could not distinguish any inhabitants; on the Patagonian shores we saw a great number of guanacoes feeding quietly, a proof of there being no Indians near them.

On the 16th, the weather appearing favourable, our anchor was weighed, and, with the Adelaide, we soon entered the sluice of the Narrow, proceeding rapidly, though the wind blew hard against us. The tide carried us to an anchorage, about four miles beyond the western entrance, and it was slack water when the anchor was dropped; but, no sooner had the stream turned, than we found ourselves in the midst of a 'race,' and during the whole tide, the water broke furiously over the ship. At slack water we got underweigh, but the Adelaide not being able (from the strength of the tide), to purchase her anchor, was obliged to slip the cable: it was fortunate that we had supplied her with our kedges, or she would then have been without an anchor. The night was tempestuous, and although we reached a much quieter birth, the Adelaide drifted considerably; had she remained at the morning's anchorage, / in order to save her anchor and cable, we should probably never have seen her again.

The succeeding morning, after a hard beat to windward, both vessels anchored in Gregory Bay. No Indians were in the neighbourhood, or we should have seen their fires. In the afternoon the wind moderated, and as there was every appearance of fine weather, I remained to survey the coast.

On the summit of the land, about half a mile northward of the extremity of the Cape, while Lieutenant Graves and I were taking bearings, and making observations, two guanacoes came up and stood neighing at us; the observation, however, was of consequence, and as they were not disturbed, they remained watching us for some minutes before they took alarm and fled.

Lieutenant Wickham and Mr. Tarn made an excursion to the summit of the Table Land, previously described as extending from the low land behind the Second Narrow to the N.E., in the direction of Mount Aymond, and were amply repaid for a fatiguing walk, with the thermometer at 81°, by a magnificent view: Cape Possession to the eastward, and to the south the mountains near Mount Tarn, eighty miles distant, were plainly distinguished. The view to the westward, stretching over a large extent of grassy plains, was bounded by lofty ranges of snow-capped mountains; but to the north it was intercepted by another summit of the mountain upon which they stood. The country they passed over was covered with short grass, through which a mass of granite occasionally protruded. Neither trees nor shrubs were observed, excepting a few herbaceous plants, and the berberis; a goose, some ducks, snipe, and plovers were shot; and guanacoes were seen at a distance, but no ostriches, nor did they meet any Indians. Large fires were, however, kindled on both shores of the Strait, in answer to the fire which they made for cooking. In consequence of those on the Patagonian coast appearing so close to us, we expected a visit from the natives before night, but none made their appearance.

Next morning, Mr. Graves accompanied me in a boat to a / station three miles within the Second Narrow on the north side, and in our way we found the geological structure of the cliffs to be of a decomposed clay-slate, arranged in strata, much distorted by the violent action of the water, and dispersed in vertical and inclined directions in very thin laminæ.

These cliffs are about one hundred feet high, the soil a sandy alluvium, of a sterile character, scantily covered with a wiry, stunted grass, and here and there a berberis bush, loaded with ripe fruit, which, from the poverty of the soil, was tasteless and dry; the ground was also, in many parts, over-run to a considerable extent with an insipid cranberry, scarcely worth the trouble of gathering.

We struck across the country, with the view of examining the place where the Indians were residing at our last visit, and the tomb which had then been erected. Grass had grown up, and effaced the traces of feet; but the tomb had suffered no farther alteration than the weather might have effected. We found that the place had been recently visited by the natives, for within a few yards of the entrance were strewed the ashes of a large fire, containing vestiges of the former decorations of the tomb, and the end of one of the flag-staffs, with the unburnt corner of one of the banners. Amongst the ashes, also, we found calcined bones; but whether they were human or not, we could not ascertain.

The discovery of the bones impressed us with the idea that the body had been burnt, and determined me to examine the tomb. The bushes that filled up the entrance appeared to be placed exactly as when we first saw them, and indeed the whole pile seemed to have remained quite undisturbed; but there was no appearance of the brass ornaments, or of the effigies of the horses.

Having effected an opening in the bushes, we found an inner covering, made of horse-skins. Having cut two holes opposite each other, for the admission of light, we saw nothing but two parallel rows of stones, three in each row, probably intended as a bier for the body or a covering for the grave; but the ground around and between them bore no appearance / of having been disturbed for burial.** As we hourly expected the Indians would arrive (the place being in the direct line of their journey to the ships), and were unwilling to let them know we had disturbed the sanctuaries of their dead, we restored the former appearance of the tomb; and it was fortunate we did so, for three women on horseback, carrying their children in cradles, with a quantity of skins, provisions, and other merchandise, evidently the harbingers of the tribe, made their appearance, and immediately began to erect their tents.

When we next went on shore we found several Indians arrived, and divided into three groups, with mantles, ostrich-feathers, skins, and joints of guanaco meat displayed for sale.

As the meat appeared fresh, it is probable that, on seeing us, the women were despatched to place the toldos, while the men set out to provide guanaco meat, for they knew our partiality for this excellent food. When we landed, an active barter began.

From the haste and avidity shown in offering their goods, and closing the bargains, it seemed as if they were anxious to monopolize our articles of barter before the rest of their party, or tribe arrived. One old man attempted to cheat; but my interdiction of all farther traffic with him brought him to a sense of his error, and I then made him a present of some tobacco and allowed him to trade, which he afterwards did, with cheerfulness and honesty.

One of the party was the Fuegian chief, whom I previously noticed, as a squalid, meagre-looking man; but he was now enlarged to Patagonian dimensions, by his improved diet and more cheerful mode of life. The appearance of bad weather obliged us to suspend the barter and get on board. After we had reached the ship, successive parties of the tribe arrived, / and formed the encampment. Among them, mounted on her white horse, was Maria, who, duly

* Falkner says, in his account of the burial ceremonies of the southern Patagonians[4] – that, after a certain interval, the bodies are taken out of the tomb, and skeletons are made of them by the women – the flesh and entrails having been burnt. It is possible that in this case the body had been so treated, and that the fire near it was for the purpose of burning the flesh, and perhaps with it all the flags and ornaments of the tomb.

escorted, paraded on the beach to challenge our recognition. In the centre of the encampment, a large flag suspended from a pole was a signal to us, and showed the position of her toldo.

The next morning being fine, we landed near the encampment, and were most cordially received. Maria was particularly attentive, and embraced me closely, while her companions chaunted in chorus a song of delight at our arrival.

When we reached her toldo, a mat was spread out for me to sit on. Maria and her family placed themselves in front of me, while the rest sat round. Almost the first question was an inquiry for my son Philip, whom they called Felipe,** and two or three skins were given to me for him. They then asked for our pilot on the former voyage, and were much disappointed to find he had left the ship. After a short conversation I returned the two bags (which I had so unwillingly carried away at our last visit), having filled them with flour and sugar, and then proceeded to deliver our presents. As each article was delivered into her hands, she repeated, in Spanish, 'I'll pay for this;' but upon a bit for her horse being presented, a general burst of admiration followed, and it was handed round the tents, whilst each individual, as it passed on, looked, I thought, anxious to be its possessor.

Maria then began to consider what adequate requital she could possibly make me. The result was, a present of two mantles, one new, of guanaco skin, and the other well worn, of zorillo skin, besides two or three skins of the puma. She then produced a piece of paper, carefully wrapped up in canvas, containing a letter, or memorandum, left by Mr. Low, master of the Uxbridge sealer,[5] addressed to any shipmaster passing through the Strait, apprising him 'of the friendly disposition of the Indians, and impressing him with the necessity of treating them well, and not deceiving them; for they had good memories, and would seriously resent it.'

The advice, no doubt, was good; but I think the fear of / forfeiting advantages and comforts to be derived from traffic would induce them to restrain their resentment.

I brought no spirits; for which, after a short time, Maria asked, complaining that she was very ill, and had sore eyes, and for some time past had nothing but water to drink, and wood to smoke. Her illness was evidently assumed, but her eyes seemed highly inflamed; and no wonder, for the upper part of her face was smeared over with an ochrous red pigment, even to the very edge of her eyelids: indeed, the whole tribe had ornamented themselves similarly, in compliment, I suppose, to our visit.

As I prepared to return on board, Maria's importunity induced me to allow her to accompany me; upon which she began to muster up all her empty bags, old mantles, and skins, and, attended by her husband, her brother-in-law, his

* He was a great favourite with them.

wife and daughter, got into the boat. While going on board, the spray washed the painted countenances of our visitors, much to their regret.

Upon reaching the ship, I ordered them to be regaled with meat and biscuit, of which they partook very sparingly, but took care to put what remained into their bags. Some spirits and water, too, which I thought would be soon dispatched, and which had been plentifully diluted to prevent their being made tipsy, they emptied into bottles to take on shore 'for the evening,' when, as Maria said, they would be 'very drunk.'

Among various things shown to amuse them was a musical snuff-box, which I had procured for the express purpose of exciting their astonishment; but I was surprised to find, that a penny-whistle produced a ten-fold greater effect upon their senses. This indifference to musical sounds I should not have suspected, because they frequently sing, though certainly in a monotonous manner.

As soon as their repast was concluded, the party, except Maria and the girls, commenced bartering their mantles and skins, and, by the time their stock was expended, they had amassed a large quantity of biscuit, and a bundle of various / trifles, some of which they had attempted to get by pilfering.

They made themselves so contented, that it was not without much difficulty we could persuade them to go on shore. Maria had made her mind up to pass the night on board, and so anxious were they all to remain, that it was only by giving Maria two bottles of spirits (which had been well diluted) that they were induced to get into the boat, and accompany me ashore. Being a lee-tide, and low water, the boat grounded at a considerable distance from the beach; seeing this, some of the Indians rode into the water, and taking us up behind them, conveyed us to the encampment, my place being behind Maria, the smell of whose zorillo-skin mantle was hardly bearable; but it was necessary to conceal our dislike of our companions as much as possible, for they are very sensitive, and easily offended.

While waiting for the tide, we witnessed a drunken scene at Maria's toldo. Fifteen persons, seated around her, shared the spirits she had obtained on board, until all were intoxicated. Some were screaming, others laughing, some stupefied, and some bellowing. The uproar drew all the other Indians round the tent, who tendered their assistance to compose their friends, and we returned to the ship. When we visited them the next day, they were quite recovered, and gave us some guanaco meat, which had been brought in that morning. On communicating my intention of proceeding on the voyage, Maria wished to know when we should finish our 'seal-killing,' and come back. I told her 'in five moons,' upon which she endeavoured to persuade me to return in four, because she would then have plenty of skins to barter.

I wrote a few lines to Captain Stokes, who, I expected, would arrive in a day or two, communicating my desire that he should follow, as soon as possible, to

Port Famine, and committed the letter to Maria's care, who promised to deliver it to him; then, taking leave of her and her companions, I embarked, and proceeded through the Second Narrow to an anchorage off Cape Negro.

Our visit to Gregory Bay, and communication with the / Indians, furnished us with many additions to our zoological collection; among them was a tiger-cat, which seemed, from the description, to be the *Felis pajaros* of the Encyclopédie Méthodique (the 'Chat de Pampa' of D'Azara). Maria gave me a very large bezoar stone, that was taken from the stomach of a guanaco. It is used medicinally by the Indians, as a remedy for bowel complaints.*

Whilst we were at the anchorage before Cape Negro, Mr. Tarn and Mr. Wickham visited the lake at the back of Laredo Bay, and saw two swans, which, from the colour of their plumage, seemed to be the black-necked swan of the River Plata and of the Falkland Islands† (Dom Pernettey, ii. p. 148). They brought on board with them a new species of duck, which is described in the proceedings of the Zoological Society as *Anas specularis* (Nob.), and a small burrowing animal, of the rat tribe, that, from the character of its teeth, is probably of a genus not hitherto noted: it approaches nearest to F. Cuvier's *Helamys*.

We next anchored in Port Famine, where the tents, &c. were replaced in their former positions, the ship was unrigged and secured for the winter, and all hands set to work, preparing the Adelaide for service. /

* The medicinal property of this intestinal concretion is well known wherever the animal is found. Maregrave, in his 'Tractatus topographicus et meteorologicus Brasiliae,' folio, p. 36, says:– 'Hæe animalia (guanacoes) generant lapides Bezoares in sinu quodam ventriculi, quimaximi æstimantur contra venena et febres malignos ad roborandum et refocillandum cor, aliosque affectus. Materia è qua generantur sunt herbæ insignis virtutis, quibus veseuntur naturæ instinctu ad sanitatem tuendum, aut morbos et venena superandum. Hi lapides inveniuntur in adultioribus hisee animalibus atque interdum tam grandes, ut unum in Italiam attulerim qui pendet uncias duas supra triginta.' – Mr. Thompson, on Intestinal Concretions. See his Syn. of Chemistry, iv. 576.[6]

† *Anser nigrocollis*. Eneye. Méthod., art. Ornithol. 108.

CHAPTER VIII.

Find that the Cutter had been burned – Anxiety for the Beagle – Uxbridge Sealer – Beagle arrives – Her cruize – Bellaco Rock – San Julian – Santa Cruz – Gallegos – Adeona – Death of Lieutenant Sholl – Adelaide Sails – Supposed Channel of San Sebastian – Useless Bay – Natives – Port San Antonio – Humming-birds – Fuegians – Beagle sailed – Sarmiento – Roldan – Pond – Whales – Structure – Scenery – Port Gallant.

PORT Famine bore evident marks of having been visited in our absence by the Indians, for a large fire, apparently recent, had over-run the grass, and burned the trees upon Point Santa Anna, particularly in that part where our boat had been so carefully concealed. Eager to know whether she had escaped the fire, I lost no time in hastening to the spot, directly after the Adventure anchored, and found, as our fears had anticipated, that she had been completely destroyed, scarcely a vestige of her wood remaining, and most of the iron-work having been carried away; for which, doubtless, the Indians had set her on fire.

The sheds for the cooper and armourer, which had been erected with some pains, were also entirely consumed, and every thing portable had been carried away. Those things which were of no use to them were either broken or burnt; but some of our station poles on Point Santa Anna were left uninjured; as well as the tablet erected to the memory of Mr. Ainsworth and the boat's crew; which was singular, because it was secured by iron hoops – of great value, in their eyes.

From the fresh traces of horses in the neighbourhood, we at first suspected the conflagration to have been caused by the Patagonians; but we soon found we owed our loss to the Fuegians, for in two new wigwams were strewed some remains of our boat.

The last winter appeared to have been milder than that preceding it, for last January, Mount Sarmiento and the hills to / the southward, over Fitton Bay, were so covered with snow, that not a particle of the rock could be seen; but this year many bare spots were visible. Every thing else, however, indicated a bad season, and the berberis bushes and arbutus shrubs had scarcely any show of fruit; which was rather a disappointment, as the berries of the former plant proved an agreeable addition to our food last year. However, there was no scarcity of birds, and with the seine we procured plenty of fish.

The Beagle's long and unexpected absence caused us much uneasiness, and some apprehension for her safety. Her visit to Port Desire ought not to have occupied more than three days, and her superior sailing should have enabled Captain Stokes to rejoin us in the entrance of the Strait. People were sent daily to look out for her, and every succeeding day increased our anxiety.

A long succession of blowing and rainy weather much impeded our progress with the Adelaide; but the Hope was hoisted out, and prepared for service.

Before daylight on the 14th I was informed that the Beagle was seen in the offing. Blue lights were burnt, and lanterns immediately shown to guide her to the anchorage; but our disappointment was great when the stranger proved to be Mr. W. Low's schooner, the Uxbridge. He had been sealing since November in the neighbourhood of Noir Island, near the outer entrance of the Barbara Channel, and was on his way to Cape Gregory to meet his elder brother, who had been collecting sea-elephant oil at South Shetland. The Uxbridge had entered the Strait from the Pacific, by the Magdalen 'Channel,' which last year we thought a Sound, and had attempted to explore in the Hope, but had been deceived by the abrupt change in the direction of the Channel at Cape Turn.

At last (on the 28th), after the Beagle's absence had been protracted to more than a month beyond the time intended, we were relieved from painful anxiety, and much rejoiced, by Mr. Tarn's telling us he had just seen her, and in two hours afterwards she arrived.

Captain Stokes, to my great surprise, told me that he had / been examining the whole coast between Port Desire and Cape Virgins, and for the last ten days had been detained in the Gallegos River by heavy gales of wind. He had sounded round, and fixed the position of the Bellaco Rock, or St. Estevan's Shoal, the existence of which had been so long doubted. He had also visited and partially surveyed, the harbours of Port San Julian and Santa Cruz, besides Coy Bay, and had made almost a complete survey of the River Gallegos, which he found to be a large and rapid river, whose entrance forms a spacious port: instead of being blocked up by a mound of shingle four or five feet above the level of the sea, and having so small a stream as to escape the notice of Mr. Weddell as he walked along the beach.*[1] Cape Fairweather is so remarkable, and so correctly placed upon the chart, that Mr. Weddell, in his search for the river, must have very much deceived himself. I should think he must have mistaken the ravine described upon my former visit, since that is the only part which answers his description: it could not be Coy Bay, because that opening, although of minor importance, has a broad boat communication with the sea.

Captain Stokes described the tide at the anchorage, within the mouth of the Gallegos, as running at the rate of five knots, and rising forty-six feet. From Mr.

* Weddell's Voyage.

Weddell's account, he was on the point of passing by without examining it; but the weather being fine, he determined to go in his boat and ascertain the truth of that description. It was soon evident that the river was large, and, returning to his ship, he lost no time in anchoring her within the entrance, where she rode out a heavy gale from S.W.

The Beagle left the Gallegos on the 23d, and reached Port Famine on the 28th, a very short passage, since she remained for a night and the greater part of a day at Gregory Bay, to communicate with the natives. When approaching the First Narrow, Captain Stokes observed a brig, apparently at anchor, under Cape Orange, and supposing her either to have found a good anchorage, or to be in distress, steered towards her. / Before he had reached within two miles of her, the Beagle touched the ground, but was extricated from the danger most fortunately, because it was nearly high water; and had she remained a-ground during the tide, the consequences might have been serious – at least, she could not have been got off without lightening her considerably. The brig proved to be the Adeona (Mr. Low's vessel), on her way to meet the Uxbridge. In attempting to enter the narrow, she grounded on the shoals, and had been left dry. The following tide again floated her, and she was on the point of getting underweigh, when the Beagle hove in sight. Captain Stokes finding that the Adeona had received no damage, proceeded to Gregory Bay.

By the Beagle's arrival we were informed of the death of Lieutenant Robert H. Sholl, after an illness of ten days. His remains were interred at Port San Julian, where a tablet was erected to his memory.

This excellent young man's death was sincerely regretted by all his friends, and by none more than by me. He was appointed to the expedition, as a midshipman, solely on account of his high character.

During our voyage from England, he made himself conspicuously useful in saving the cargo of a vessel, which was stranded in Port Praya; and on our arrival at Rio de Janeiro, the Commander-in-chief appointed him to a vacant lieutenantcy on board the Beagle: an appointment which, up to the period of his lamented death, he filled zealously and most creditably.*

On the 1st of March we were surprised by the appearance of three Europeans, walking round Point St. Anna. A boat / was sent for them, and we found they were deserters from the Uxbridge, who had come to volunteer for our ships.

* I cannot avoid noticing here the considerate conduct of the Commander-in-chief (Sir George Eyre) with respect to this appointment. By the tenor of my instructions the Adventure and Beagle were placed under the Admiral's orders; and the vacancy, had he wished to exercise his prerogative, might have been filled by one of his own followers. It was, however, given, at my request, to Mr. Sholl, as being more conversant with the duties of this peculiar service than any of the midshipmen of the flag-ship. The Admiral's conduct, on this occasion, calls for my warmest thanks.[2]

The following day the Adeona and Uxbridge arrived, on their way to Port San Antonio, to boil their oil; but I recommended Bougainville, or (as the sealers call it) Jack's Harbour, as more convenient for their purpose, and more secure from storms, as well as from troublesome visits of the natives.

Upon my offering to restore the three deserters to the Uxbridge, Mr. Low requested me to keep them, and another, also, who was anxious to join the Adventure, to which I consented, as the Adelaide wanted men.

A few days after Mr. Low's departure, he returned in a whale-boat to ask assistance in repairing the Uxbridge's rudder. By our help it was soon made serviceable, and she was enabled to prosecute her voyage, which could not otherwise have been continued.

The Adelaide being ready for sea: her first service was to be an examination of the St. Sebastian Channel, which, from its delineation on the old charts, would seem to penetrate through the large eastern island of Tierra del Fuego. In the voyage of the Nodales (in the year 1618), an opening on the eastern coast, supposed to be the mouth of a channel, communicating with the Strait of Magalhaens, was discovered. After describing the coast to the south of Cape Espiritu Santo, the journal of that voyage states: 'We found, in the channel of St. Sebastian, twenty fathoms clear ground. The north shore is a beach of white sand, five leagues in extent, stretching out from the high land that terminates at Cape Espiritu Santo, and giving the coast here the appearance of a deep bay; but, on a nearer approach, a projecting tract of low shore is observed. The south extremity of this low beach is a sandy point, round which the channel trends; the mouth is a league and a half wide. The south shore is higher than the land to the northward, and in the middle of the bay the depth is from fifteen to twenty fathoms clear ground, and a good bottom; but from mid-channel to the south shore the bottom is stony, and the water, of little depth, there being only six and seven fathoms. From / hence the channel shows itself, and continues, as far as we could see, of the same breadth. It seemed to be a large sea. The latitude was observed to be 53° 16'.'*

From the above account, and from the chart that accompanies it, in which this inlet is made to communicate with the Strait of Magalhaens by the opening round Cape Monmouth, our knowledge of the supposed St. Sebastian Channel was derived. That there is a deep bay, in the latitude of 53° 16', not only appears from the account of the Nodales, who were within the heads, although it seems they did not proceed beyond the stony ground on the south side of the entrance; but also from the accounts of vessels who have lately seen it; and of one ship-master who was deterred from entering, by the formidable notice on our charts

* Relacion del Viage, &c. que hicieron los Capitanes B. G. de Nodales y Gonzalo de Nodales, p. 59.

of its being 'only navigable for small vessels,' whence he conjectured that the tides would be very strong, and the channel occasionally narrow, as well as and shoal.

Sarmiento, Narborough, Byron, Wallis, Bougainville, and Cordova, have severally noticed an opening, which corresponds to this supposed channel, namely, that between Capes Monmouth and Valentyn; but the object of those voyagers having been to make the passage through the known Strait, to explore this opening was, in all probability, considered a waste of time; yet, that such a channel was supposed to exist, we must conclude from the conspicuous figure it makes in the charts of Tierra del Fuego.

Had there been a knowledge of its affording any communication with the sea, surely Sarmiento and Narborough, as well as the Nodales, who navigated the Strait from west to east, would have been induced to attempt to pass through; and avoid the dangers, as well as difficulties, of the channels to the northward.

Anxious to set the question at rest, I gave Captain Stokes orders to proceed to survey the western coasts, between the Strait of Magalhaens and latitude 47° south, or as much of / those dangerous and exposed shores as he could examine, with the means at his disposal, and sailed myself, in the Adelaide, to explore the supposed St. Sebastian Channel. Every discretionary power was given to Captain Stokes to act as he pleased, for the benefit of the service; but he had strict orders to return to Port Famine by the 24th of July, when I hoped to move the Adventure to some other part of the Strait, and to recommence operations with the earliest days of spring, if the winter should be unfit for our work.

Having crossed over to the southward of Point Boqueron, we proceeded, on the 13th of March, to the N.E. (in which direction the opening trended), at no great distance from the northern shore; behind which the country seemed to rise gradually to the summit of a long ridge of table-land, terminating near the First Narrow, and appearing like that in the neighbourhood of Cape Gregory. It was inhabited; for here and there we observed the smoke of fires, perhaps intended as invitations for us to land.

The south side of the opening seemed (after forming a small bay under Nose Peak) to extend in a direction parallel to the northern coast of the bay, for three or four leagues, when it dipped beneath the horizon. Neither shore had any opening or indenture in its coast line, of sufficient size to shelter even a boat; so that a vessel caught here, with a south-westerly gale, would have little chance of escape; unless a channel should exist, of which, from the stillness of the water and the total absence of tide, we had very little hope. The soundings were variable between twenty and thirty fathoms, and the bottom seemed to be of shells, probably covering a substratum of clay or sand. As we stood on, a small rocky lump came in sight, which appeared to be the termination of the northern shore, and again we flattered ourselves with the expectation of finding a passage; but in less than half an hour afterwards, the bay was distinctly seen to be closed by low land, and the rocky lump proved to be an isolated mass of rock, about two miles inland. As every person on board was then satisfied of the non-existence of any channel, we put about to return, and / by bearings of Mount Tarn, crossed by angles from Mount Graves, Nose Peak, and Point Boqueron, our position, and the extent of this bay, were determined. As it affords neither anchorage nor shelter, nor any other advantage for the navigator, we have named it Useless Bay. It was too much exposed to the prevailing winds to allow of our landing to examine the country, and its productions, or to communicate with the Indians; and as there was not much likelihood of finding anything of novel character, we lost no time in retreating from so exposed a place. Abreast of Point Boqueron the patent log gave for our run twenty-six miles, precisely the same distance which it had given in the morning; so that from five o'clock in the morning until ten, and from ten o'clock until four in the afternoon, we had not experienced the least tide, which of itself is a fact confirmatory of the non-existence of a channel.

From the fires of the natives in this part having been noticed at a distance from the beach, it would seem that they derive their subsistence from hunting rather than fishing; and as there are guanacoes on the south shore of the First Narrow, it is probable the people's habits resemble those of the Patagonians, rather than the Fuegians; but as they have no horses, the chase of so shy and swift an animal as the guanaco must be fatiguing and very precarious.*

Sarmiento is the only person on record who has communicated with the natives in the neighbourhood of Cape Monmouth. He calls them in his narrative a large race (Gente grande). There it was that he was attacked by the Indians, whom he repulsed, and one of whom he made prisoner.[3]

We remained a night in Port Famine, and again set out in the Adelaide to survey some of the western parts of the Strait. / Bad weather forced us into Port San Antonio; of which Cordova gives so favourable an account, that we were surprised to find it small and inconvenient, even for the Adelaide.

He describes the port to be a mile and a halflong, and three quarters of a mile broad: we found the length a mile and a quarter, and the mean breadth scarcely a quarter of a mile. It possesses no one advantage that is not common to almost every other harbour and cove in the Strait; and for a ship, or square-rigged vessel of any kind, it is both difficult to enter, and dangerous to leave. Besides the local disadvantages of Port San Antonio, the weather in it is seldom fair, even when the day is fine elsewhere. It lies at the base of the Lomas Range, which rises

* Falkner describes the Indians who inhabit the eastern islands of Tierra del Fuego, to be 'Yacana-cunnees,' and as he designates those who inhabit the Patagonian shore of the Strait by the same name, it might be inferred that they are of the same race; but however closely connected they may have been formerly, they certainly are not so now, for Maria (the Patagonian) spoke very contemptuously of them, and disclaimed their alliance; calling them 'zapallios,' which means slaves.

almost perpendicularly to the height of three thousand feet, fronting the great western channel of the Strait, whence it receives upon its cold surface the western winds, and is covered by the vapour, which is condensed from them, while in all other parts the sun may be shining brightly.

This port is formed by a channel, a quarter of a mile wide, separating two islands from the shore. The best anchorage is off a picturesque little bay on the south island, which is thickly wooded to the water's edge with the holly leaved berberis,* fuchsia, and veronica, growing to the height of twenty feet; overtopped and sheltered by large beech, and Winter's-bark trees, rooted under a thick mossy carpet, through which a narrow Indian path winds between arbutus and currant bushes, and round prostrate stems of dead trees, leading to the seaward side of the island. Upon the beach, just within the bushes, and sheltered by a large and wide-spreading fuchsia bush, in full flower, stood two Indian wigwams, which, apparently, had not been inhabited since the visit of poor Ainsworth. He had occupied these very wigwams for two days, having covered them over with the boat's sail; and remains of the ropeyarns that tied it down were still there: a melancholy memento.

In no part of the Strait did we find the vegetation so luxuriant as in this little cove. Some of the Winter's-bark and currant trees had shoots more than five feet long, and many of the / / / Winter's-bark trees were two feet in diameter. The veronica (I believe *V. decussata*) grows in the sheltered parts to the height of twenty feet, with a stem six inches in diameter. It was found too on the windward side of the island in abundance, and of large size, rooted in the very wash of the sea-beach, and exposed to the full force of the cold winds and hail-storms, which rush down the wide western reach of the Strait.

The fuchsia also grows to a large size; but it is a more delicate plant than the veronica, and thrives only in sheltered places. Many were observed six inches in diameter; the stems of the two last plants were used by us, during our stay, for fuel.

The day after our arrival, the gale subsided, and the weather became very fine indeed. The stillness of the air may be imagined, when the chirping of humming-birds, and buzzing of large bees, were heard at a considerable distance. A humming-bird had been seen at Port Gallant last year, and was brought to me by Captain Stokes, since which none had been noticed. Here, however, we saw, and procured several; but of only one species.†* It is the same as that found on

* Berberis ilicifolia. – Banks and Solander MSS.

† The specimen that was found at Port Gallant was sent by me to Mr. Vigors, who considering it, although well known to ornithologists, as never having yet been named, describes it in the Zoological Journal (vol. iii.p. 432, Aug. 1827), as Mellisuga Kingii. Shortly afterwards M. Lesson[4] published it in his Manuel d'Ornithologie (vol.ii. p. 80.), as *Ornismya sephaniodes*, as a discovery belonging to La Coquille's voyage, in the illustra-

P. P. King S. Bull.

FUEGIAN WIGWAMS AT HOPE HARBOUR IN THE MAGDALEN CHANNEL.

the western coast, as high as Lima; so that it has a range of 41° of latitude, the southern limit being 53 1/2°, if not farther south.

The islets, at the north part of the port, were well stocked with geese and other birds, which supplied our people with fresh meals. The steamer duck we found difficult to shoot, from its excessive wariness, and power of remaining, for a great length of time, under water.

Our fine weather lasted but a few hours, and (no unusual occurrence in these regions) was succeeded by a week's rain and wind, during which we were confined to the small space / of the Adelaide; and for some days had three anchors down, owing to the violent squalls. Farenheit's thermometer ranged between thirty-six and forty-six degrees, and we had several snow storms, but the snow did not lie on the low grounds.

On the 28th the gale began to subside, and there was a change for the better; but we were again disappointed, and not until the 31st could we effect our departure from this dreary and confined little place.

The day before we sailed, three canoes, containing in all sixteen persons, of whom six only were men, came alongside.

For about an hour they had hesitated to approach; but when once near us, very little invitation was necessary to induce them to come on board. One was clothed in a duck shirt, which was recognised by one of our people, who had joined us from the Uxbridge, as having been given to them a few weeks before, when that vessel passed through Magdalen Channel: another wore a red flannel shirt, and in the canoe we observed an European boarding-pike, painted green, and a part of the iron-work of the cutter, burned at Port Famine during our absence; also some relics of the boat in which Mr. Ainsworth was drowned, which last they had doubtless found thrown up on the beach. Upon our inquiring how they became possessed of the iron-work, they pointed towards Port Famine; and I have no doubt they were concerned in the fire; but as we could not explain to them the mischief they had occasioned, it was thought better not to notice the affair, and the articles were returned to them. They could have had no idea of our being the owners of the boat, or they would have concealed all that belonged to her.

They conducted themselves very quietly during their stay on board, with the exception of one, who tried to pick my pocket of a handkerchief; the offender was ordered out of the vessel, and there was no further attempt to pilfer. They wished to go below; but this was not permitted, because the odour of their oily persons was scarcely tolerable, even in the open air. As to food, tallow-candles, biscuit, beef, plumb-pudding, were / equally liked, and swallowed most vora-

---

tions of which it is figured at plate 31: I rather think, however, that it is Molina's *Trochilus galeritus*. – (Molina, i.275.).

ciously. One of them was discovered taking the tallow out of the end of the deep sea lead and eating it, although mixed with sand and dirt.

Before sunset their canoes were despatched on shore to prepare the wigwams, during which operation three of the men remained on board; and as soon as the preparations were made they called for a canoe and went on shore. We obtained several spears, baskets, necklaces, bows and arrows from them in barter; but they seemed to have very few skins. Perhaps those they possessed were hidden in the bushes, because they had no wish to part with them.

One woman was covered with a guanaco mantle; another merely wore a seal-skin over her back and shoulders, which, while she crouched in the canoe, was sufficient to cover her person. One had a black stripe down the nose, but she was the only female among them who was so painted.

Next morning the Indians visited us with a fresh assortment of bows and arrows, in the manufacture of which they had evidently passed the night, for every one was quite new; the bows were of green wood, and the arrows not even pointed. They found, however, a ready sale. One of the party was a man who had been turned out of our vessel the preceding evening, for picking my pocket; but he was daubed over with a whitish pigment to deceive us, and would probably have escaped detection, but for the unusual ugliness of his person, which was not so easily disguised. He was much disconcerted by our recognition; and our refusal to barter with him made him angry and sullen.

The women had daubed their faces all over with bright red ochre; to add to their beauty, no doubt.

We sailed out of the port by the northern passage, and standing across the Strait, anchored in San Nicolas Bay. Mr. Graves went to Bougainville Harbour, to communicate with the Adeona, and take letters from me to Lieutenant Wickham. He brought back an account of all being well at Port Famine, and of the Beagle having sailed on the 17th.

When we left Port Famine my intention was to examine / the Magdalen Channel; but, upon leaving San Nicolas Bay (1st April), the weather was so favourable for our proceeding to the westward, that I changed my mind and steered round Cape Froward in order to get to Port Gallant, whence, with a westerly wind, we might more easily survey the coast in returning. An easterly breeze carried us near Cape Holland, into Wood's Bay, where we anchored, and obtained a bearing of Mount Sarmiento, which, being clear of clouds, was a conspicuous, and even splendid object; for the sun's setting rays, shining upon the projecting snowy ridges on its western side, gave it the appearance of a mass of streaky gold. It had been in sight the whole day, as well as the preceding evening, when its bearings were taken from the islet in San Nicolas Bay.

The next day was so calm that we only reached an anchorage in Bradley Cove, on the west side of Bell Bay, of which a plan was made; an extensive set of bearings was also taken on the west point of the bay, evidently that called by Sarmiento Tinquichisgua.* The conspicuous mountain at the back of the bay, on its south-eastern side, is particularly noticed by him, and, according to his opinion, is the 'Campana de Roldan' of Magalhaens.† Between Bradley Cove and Point Tinquichisgua are two coves, over which a high double-peaked mountain forms a conspicuous object upon rounding Cape Froward; and they were named in compliment to Mr. Pond, the late Astronomer Royal.[5]

While at Point Tinquichisgua we were discovered by some natives to the westward, who immediately got into their canoes, and paddled towards us; but, as we had no arms in the boat, I did not think it prudent to await their arrival; and therefore, after taking the requisite angles, embarked and returned to the Adelaide, examining the inlets under Mount Pond on our way. Nothing more was seen of the Indians until the following morning, when, as we sailed out of the bay, they made their appearance, but we did not communicate / with them. They were as vociferous as usual, and pointed to the shore, inviting us to land. One of them, who stood up in the canoe while we passed, was ornamented about the hair and body with white feathers.

This part of the Strait teems with whales, seals, and porpoises. While we were in Bradley Cove, a remarkable appearance of the water spouted by whales was observed; it hung in the air like a bright silvery mist, and was visible to the naked eye, at the distance of four miles, for one minute and thirty-five seconds before it disappeared.

A glance at the chart of this part of the Strait will show the difference of geological structure in the opposite coasts. The north shore, from Cape Froward to Port Gallant, forms a straight line, with scarcely a projection or bight; but on the opposite side there is a succession of inlets, surrounded by precipitous mountains, which are separated by ravines. The northern shore is of slate; but the other is principally of green-stone, and its mountains, instead of running up into sharp peaks, and narrow serrated ridges, are generally round-topped. The vegetation on both sides is almost equally abundant, but the trees on the south shore are much smaller. The smooth-leaved beech (*Fagus betuloides*) and Winter's-bark are the principal trees; but here and there a small tree was observed, like a cypress, which does not grow to the eastward, excepting on the sides of Mount Tarn, where it only reaches the height of three or four feet.

The scenery of this part of the Strait, instead of being as Cordova describes it, 'horrible,' is at this season exceedingly striking and picturesque.[7] The highest mountains certainly are bare of vegetation; but their sharp peaks and snow-covered summits afford a pleasing contrast to the lower hills, thickly clothed with

* Sarmiento, p. 213.

† Este monte es el que llaman las Relaciones antiguas la Campana de Roldan. – Sarmiento.[6]

trees quite to the water's side, which is bordered by masses of bare rock, studded with ferns and moss, and backed by the rich dark-green foliage of the berberis and arbutus shrubs, with here and there a beech-tree, just beginning to assume its autumnal tints.

In working into the narrow entrance of Port Gallant, the / schooner grounded upon a bank that extends off the mouth of the river; but the water being perfectly smooth, no damage was caused. As a secure cove, Port Gallant is the best in the Strait of Magalhaens; from the stillness of its waters, it is a perfect wet dock, and from its position it is invaluable. There are many coves as safe and convenient when once entered; but the prevailing steepness of the shores, as well as the great depth of water, are obstacles of serious importance. Here, however, is an exception: the bottom is even and the depth moderate; besides, Fortescue Bay, close by, is an excellent roadstead or stopping-place, to await an opportunity of entering.

For repairing a ship, Port Famine is more convenient, on account of the quantity and size of well-seasoned timber lying about the beach, and also from the open character of the country. At Port Gallant the trees are much stunted, and unfit for present use, while the shore, as is the case around almost every cove to the westward of Cape Froward, is covered with shrubs and brushwood, quite to the high-water mark; so that there is no possibility of walking easily to any distance from the sea-side. A shingle, or sandy beach, twenty or thirty yards in length, occasionally intervenes, but is scarcely preferable to a vessel's deck, for a walk. /

CHAPTER IX.

Detention in Port San Antonio – Humming-birds in snow showers – Fuegians – Geological remarks – Canoes – Carving – Birds – Fish – Shag Narrows – Glaciers – Avalanches – Natives – Climate – Winter setting in – Adelaide loses a boat – Floods – Lighting – Scurvy – Adelaide's survey – Bougainville Harbour – Indians cross the Strait, and visit Port Famine – Sealing vessels sail – Scurvy increases – Adelaide sent for guanaco meat – Return of the Beagle – Captain Stokes very ill – Adelaide brings meat from the Patagonians – Death of Captain Stokes.

OUR stay at this port was prolonged beyond my intention by thick snowy weather and hard gales, which cut off our communication with the shore; for notwithstanding we were in so sheltered a place, and the vessel had three anchors down, we did not consider her quite secure against the violent squalls. We had been fortunate in procuring observation, and took advantage of our detention to lay down the operation of the preceding days on paper. Muscles were found in great abundance on the mud flats. There are three varieties, one of which has a bitter, disagreeable taste, but the others are exceedingly good and wholesome. One of the latter is of large size (*Mytílus Magellanicus* of the Ency. Meth.) The other is of a more globose form than the bitter sort, and has a very obtuse hinge and margin. The bitter kind contains pearls, which are valueless, because small, and of a bad colour.

At first there were plenty of sea-birds* in the cove, which took refuge at the head of the bay; till after two days, they deserted us altogether. There appeared to be an abundance of fish; but as we had not provided ourselves with a seine, and they / would not take bait, we were confined for refreshments principally to shell-fish.

No traces of quadrupeds, excepting an Indian dog, were noticed. Here Wallis's people saw a large cloven-footed animal, which they described to be as 'big as

* Here we obtained a second species of the Steamer-duck, which is described in the Proceedings of the Zoological Society of London, as *'Micropterus Patachonicus,* Nob.' It differs from the *M. brahypterus* not only in colour but in size, being a smaller bird, and having the power of raising its body, in flight, out of the water. We called it the 'Flying Steamer.'

a jack-ass.' It was probably a deer, one or two of which had occasionally appeared at Port Famine.(*e*)

It has been mentioned that we found many humming-birds at Port San Antonio, which we attributed to the sheltered situation of the place, and the luxuriant growth of fuchsias and other plants, upon the sweets of whose flowers they feed. Here, however, one of the same species was seen sporting about in a most exposed place and during the falling of a snow shower, a proof of the hardy character of this little bird, which, if it does migrate upon the approach of winter to a warmer clime, lingers, at least, as long as it possibly can. This was the middle of April, the winter had, in fact, already commenced, and all the mountains around us were clothed with snow, while the ground was also coated with the same dazzling covering. Mr. Graves intended to ascend the Mountain de la Cruz; but a heavy fall of snow prevented the attempt, and we lost the opportunity of obtaining a round of angles from that elevation, which would have materially assisted our operations. We should also have obtained a bird's-eye view of the Barbara Channel and the Sounds on the opposite side of the Strait, whose extent and nature we did not know; for Cordova's notice of San Simon's Bay, and a deep inlet which exists to the westward of it, is very unsatisfactory.

There were no signs of a recent visit from the Fuegians, though at the entrance of the cove we found three or four wigwams in good repair; whence it seems probable, that the place is one of their frequent haunts. When the Beagle came here last year, some station staves were left standing; but, before her return, every one had been removed; and when Captain Stokes went down the Barbara Channel, to the relief / of the Saxe Cobourg's crew, those staves were seen in the possession of the Indians.

A fine morning (11th) induced us to leave this quiet anchorage, to examine the openings of the south shore; and in the afternoon, the anchor was dropped in a convenient place, on the west side of the western inlet, named by us Warrington Cove. While crossing the bay from Point Elvira, the north extremity of Cayetano Island, several 'smokes' were observed on the low land, at the bottom of the inlet; and after we anchored two canoes visited us, containing six men, four women, and two or three children. They approached very cautiously, and could not be induced to come alongside. At last the men landed, and invited us to communicate with them. I therefore went on shore with two or three officers, and remained with them half an hour, during which they gradually lost the distrust they had at first evinced; but each man still carried a number of pebbles in the corner of his wrapper, ready to repel any attack we might make upon them; from the knowledge we have since obtained of their character, I think it probable that they had lately committed some act of aggression on a sealing-vessel,

(*e*) Or the animal called by Molina 'Huemul.' – R.F.

and were afraid of retaliation. Our conduct tended to assure them of our friendship; and, shortly after we left the shore, they came alongside in their canoes, and were very familiar, eagerly bartering their necklaces and baskets. In their way to us they had probably landed their more valuable goods, such as otter and sealskins, as well as their weapons and dogs, without which they never go far.

The natives of this part are considered by the sealers to be the most mischievously inclined of any in the Strait, or Tierra del Fuego. The appearance of our visitors was certainly against them; but they did not commit themselves during our two or three day's communication, by any act which could make us complain, or cause suspicion of their honesty and friendship. We, however, kept too good a look-out, to enable them to take advantage of our seeming good-nature.

Among bushes behind the high beach were three wigwams, but the Indians had no intention of remaining with us for the / night. They went away, to our great satisfaction, at an early hour, and returned to the bottom of the sound, where a large party of their countrymen was assembled. Their departure enabled us to look round, in the vicinity of our anchorage, and examine its productions, which differed in no way from those of other parts of the cost. Its geological structure is, however, different: the rocks are greenstone, or granite, without slate. Mount Maxwell, rising immediately over the cove, is the termination of a rocky mountain range, whose summits are crowned with snow. The verdant sides of the hill, interspersed at intervals with large masses of bare rock, produced, from a distance, rather a pleasing effect; but, upon examination, the verdure was found to consist principally of moss, or a stunted vegetation, covering a soft and swampy soil. The upper portions of the mount are so precipitous as not to be easily reached; and, indeed, many parts rise with a perpendicular ascent for more than a hundred feet. On the south side of Mount Maxwell is Smyth Inlet, which contains anchorage on the north shore, particularly one in Earle Cove; but in the centre the water is deep, and on that account, it is not an inviting place for a ship During Mr. Graves's absence in Smyth Harbour, I examined the coast as far as Cape Edgeworth, where I obtained an extensive set of bearings. The afternoon was particularly favourable for the purpose, the snow-capped mountains of the north shore were perfectly distinct; and among them was a very high one, shaped like a Highland target, the peak of the mountain answering to the central spike of the shield. We never afterwards saw it, nor could I, on this occasion, fix its position better, than by estimating its distance. The rock is chiefly greenstone, accompanied by considerable masses of granite. A little islet, off Dighton Cove, is composed of granite, of a lamelliform structure. Mr. Graves brought me a specimen of lamelliform granite attached to a mass of greenstone.

The Indians visited us every day, their number being generally from twelve to sixteen, of which five or six only were men, the rest were women, and children of all ages. One of the latter could not have been more than three weeks old;

yet the / mother, apparently about sixteen years of age, was always occupied in the laborious employment of paddling the canoes. The child was secured in the mother's lap, with its head on her bosom, by a mantle, which was drawn tightly round both mother and child. Their canoes were similar to those of the eastern parts of the Strait, about ten feet long, holding four or five grown and weapons: they are formed of bark, and kept in shape by wooden cross supports secured to the gunwale, which is lined by a long, slender pole. They are divided into three compartments, the foremost occupying about one-third of the length, contains the spears, placed ready for immediate use; in the second are the grown persons, with the fire-place between them, the men sitting between the fire-place and the spears, to be ready to use them upon the approach of seals or porpoises; on the opposite side of the fire-place are seated the women who paddle the canoe, in which the men sometimes assist, when great expedition is necessary. Behind the women, in the third division, are the elder children and the dogs, the younger children being generally stowed away in the women's laps, for the sake of mutual warmth. The fire is made upon a layer of clay, several inches thick, at the bottom of the canoe; and above the fire, across the gunwales, are laid several pieces of half-burnt wood, for fuel.

During our communications with these visitors they conducted themselves peaceably, and made no attempt to pilfer, although there was some little roguery displayed by them in barter. One of the men having parted with all his disposable property, tendered one of his daughters, a fine girl of fourteen or fifteen years of age, for some mere trifle, and, being refused, became very pressing and importunate to close the bargain for the price that was jestingly offered; nor was it without difficulty that he was convinced we were not in earnest. They were as poor as the rest of their countrymen, very badly clothed, and possessing few skins to barter. Two of them exchanged their otter skin mantles for cotton shirts, which they continued to wear without complaining of cold. /

As their visits lasted all day they always brought their food, consisting of the blubber of seals and porpoises. The method used by them in cutting it up is nearly similar to that adopted by the Esquimaux Indians, as described by Sir Edward Parry in his second voyage,[1] and also resembles the process of the natives of King George's Sound, which I have described in the account of my survey of Australia (vol. ii. p. 140): a piece of blubber being held in the left hand, a corner of it is taken between the teeth, and it is then cut by a knife, held underhanded, into strips backward and forward, without passing the instrument entirely through: so that when the operation is finished the piece draws out into a long band, about an inch thick, formed by the connected strips. The whole affair from first to last is most offensive to the sight; and the countenance of the carver is beyond description, for his eyes being directed to the blubber, squint shockingly, and give his ugly face a hideous appearance. The strip of blubber is next divided among the party, each of whom proceeds to extract its oily juices by drawing it through his teeth and sucking it, after which it is warmed in the fire to facilitate its division into small pieces, which are swallowed or bolted without mastication. Morsels of this dainty food were given not only to the elder children, but even to infants at the breast.

On the 14th, while preparing to weigh, the Indians came on board and helped to heave in the cable, but without rendering us much real assistance. When the sails were loosed, the women in the canoes began to chatter and scream for fear we should carry off their friends, and their alarm was no sooner given than the deck was cleared of our visitors, who seemed to be quite as much frightened for their safety as the women were. In a few minutes afterwards we were proceeding to the southward, and first tried to anchor in a bay on the south side of Smyth Harbour, but finding the depth too great, I sent Lieut. Graves to sound behind an islet where there were indications of a place of shelter, but he returned unsuccessful. During his absence I went to a very narrow passage, which he had discovered, leading to a large channel or sound; but finding it / intricate, I deferred trying to enter with the vessel until a more favourable opportunity should offer, and we returned to the place south of Warrington Cove, called Dighton Bay, where we anchored off a sandy beach in twenty fathoms, and secured the vessel by laying the kedge on the shore. This sandy beach was the first we had found in the eastern part of the Strait. The sand is quartzose, of a white colour, and being a novelty, rendered the place interesting. A stream, supplied by the ravines of Mount Maxwell, runs over the beach into the sea, and from it an abundant supply of excellent water may be obtained without difficulty.

We observed no quadrupeds; but, of the feathered tribe, we found woodpeckers, kingfishers, and woodcocks, and in the sheltered nooks several humming-birds were darting about the flowery underwood of berberis, fuchsia, and arbutus. In the tide-way, at the narrow passage, the sea teemed with fish; over which hovered corvorants and other sea-fowl, preying upon the small fry that were trying to elude their voracious enemies, the porpoises and seals, thousands of which were seen sporting about as we proceeded on our way. Whales were also numerous in the vicinity, probably because of an abundance of the small red shrimp, which constitutes their principal food.

I went again to examine the passage, and the tide being against us, we were obliged to pull close to the western shore to benefit by the partial eddies, otherwise we could not have proceeded until the turn of the tide.

These narrows, named 'Shag' Narrows, from the quantity of birds there so called by seamen, are not a hundred yards wide. The south end is fronted by an island, from whose summit, about four hundred feet high, I hoped to obtain a good view southward, and after passing the narrows we landed and reached the summit. While looking around at the view, and preparing the theodolite, a woodcock started up from the long grass and walked away so leisurely, that Mr. Tarn nearly succeeded in striking it with a stick. This bird afforded us a name for the station, which we found to be at the northern side of a large basin, ten miles wide, and six long, terminated at / its south end by a channel leading to the open sea, but crowded with islands and rocks. A deep inlet or chasm in the land, at the N.W. corner of this basin, was filled with masses of floating ice, broken from an enormous glacier.

After obtaining all the bearings and embarking, we pulled three miles to the westward, and took a round of angles at Point Cairncross, the south-west point of Field's Bay, and again another set at the south head of Icy Sound, near Dinner Cove, where we found a very convenient anchorage for small vessels. Through Icy Sound we found some difficulty in penetrating, as the channel was much obstructed by ice.

Three miles within this sound the rocky shore became more precipitous, and at two miles farther, where the width across was not more than one hundred and fifty yards, the rocks rise perpendicularly on each side to the height of seven or eight hundred feet. Beyond this remarkable part the channel opens out to a basin about half a mile in diameter, bounded by a sloping glacier, from which immense masses of ice broke off frequently, and falling with a noise like the discharge of a ship's broadside, threw up the foaming water with terrific violence.

As we entered the basin, we were startled by a sudden roar, occasioned by the fall of one of these avalanches, followed by echoes which reverberated round the basin and among the mountains. We remained for half an hour afterwards waiting for another fall, but were not gratified. Several were heard at a distance, probably high up the sides of the glacier. The examination of Icy Sound occupied us until dark, when we returned to the schooner.

During our absence, Indians had again visited the Adelaide, the greater number of whom were strangers. We had also seen a party in a canoe close to Mount Woodcock, who were striking seal, and too intent upon their object to pay much attention to any thing else.

On the 16th, the term of our absence having expired, we left Dighton Bay on our return: at night we anchored in St. Nicholas Bay, and the day after arrived at Port Famine. / Natives had discovered and visited the ship while I was away, but Lieut. Wickham did not encourage them to remain; and two or three attempts to pilfer being detected, they were treated with very little ceremony; so finding their company was not desired, they went across the Strait to Lomas Bay, where for several days afterwards the smoke of their fires was seen. They were the same Indians whom we had met at Port San Antonio.

That these Indians should be received so coolly, may seem to have been impolitic on our side, when it is considered that our smaller vessels and boats might be met with, and their crews ill-treated by way of retaliation. It was, however, time that they should know our superiority; for, of late, several very treacherous attacks had been made by them on sealing vessels, and this party was the most forward and insolent we had seen. One of them was teazing several of the men to box, an accomplishment he had probably learnt from the crews of sealing vessels; among others, he fixed upon the serjeant of marines, who very unceremoniously pushed him over the side, and made him return to his canoe, which he resented by pushing off from the ship's side, and throwing a stone at the serjeant, who was standing at the gangway. As it missed him, and did no harm, no notice was taken of his mischief. We afterwards heard that the same party had visited Bougainville Harbour, where the Adeona was at anchor; but as Mr. Low neither gave them encouragement to remain, nor permitted them to go on board his brig, they very soon went away.

The difference between the climates of the western and eastern portions of the Strait was very striking. To the westward the country, being principally clothed with evergreens, such as the smooth-leaved beech, and Winter's-bark, with an underwood of arbutus and berberis, seems to possess a constant verdure, nor until the snow covers all, does it assume any thing like the appearance of winter. To the eastward, evergreens are less common, their place being occupied by the beech (*Fagus Antarctica*), whose leaves fall very early. Snow had also begun to cover the lower grounds, giving signs of winter. April terminated / with finer weather than we had experienced for some weeks, but May set in with north-easterly winds and much rain, succeeded by a heavy fall of snow.

> 'Tristis byems montes niveo velamine vestit.'

As yet the thermometer had not been very low. On one or two occasions it had fallen during the night to 28°, but generally it ranged between 45° and 33°.

The Adelaide was again despatched on the 30th April, to carry on an examination of the openings on each side of Cayetano Island; but she returned on the 21st of May, with the disagreeable intelligence of having had her only serviceable boat stolen by the Indians. This was a serious loss, not only on account of so much time being thrown away, but also because we had no other boat to substitute for her. To prevent delay, I sent to Mr. Low, at Bougainville Harbour, requesting that he would sell one of his boats; but he was himself so badly off, from similar losses, that he could only assist us by lending one for a few weeks, and as it was the only boat he possessed, it could not be spared to go far from his vessel. I, therefore, despatched Mr. Graves, in the Adelaide, to Bougainville Harbour, to employ himself in examining the coast thence to Cape Froward, and in the mean time began to build a whale-boat, to be ready for the Adelaide's use as soon as winter had passed over; for, from Mr. Graves's report of the state of the climate to the westward, very little could be done during the winter months.

The following is Lieut. Graves's account of the loss of his boat:– Upon leaving Port Famine he proceeded at once to Port Gallant, and surveyed Cordes Bay; after which he crossed the Strait to St. Simon's Bay, and anchored in Millar Cove, on its western side, immediately to the north of Port Langara, from which it is only separated by a narrow neck of land. The Adelaide remained there at anchor while Mr. Graves visited the different parts of the bay. Her presence had attracted a large party of Indians, who, occupying several wigwams near the entrance of the cove, paid daily visits to / our people, and were apparently very familiar and well-disposed.

But they had cast a longing eye on the whale-boat, which, when equipped for service, contained many things very useful to them, and they laid a plan to carry her off, which succeeded. One evening she was prepared for going away at an early hour the following day, and, to save time, every thing that might be required was placed in her, and she was made fast for the night. Two or three Indians were then on board, and observing what was done, laid their plan, and at sunset took their leave as usual. The night was pitchy dark, and at nine o'clock the boat was missed from alongside. The alarm was given, and instant search made at the wigwams of the Indians, who had all decamped, without leaving the least trace of themselves or the boat. The 'painter,' or rope by which she had been fastened to the vessel, had been cut through with some sharp instrument, most probably a knife, which our people had sharpened for them on the grindstone that very day.

Every possible search was made next morning, but without success; the boat that was left was one which could not be used with any advantage, and Mr. Graves returned to Port Famine. Vexatious as the accident was, I could not blame him for what had occurred, for no one had suspicions of such conduct from the Indians, who, on all other occasions, had kept at a distance from us after night-fall. The boat was properly secured alongside, and the night was so cold that no person would have thought the Indians would expose themselves to such a temperature (28°); for they must have swum alongside to cut her adrift, and then must have towed her away very gradually, to prevent the theft being discovered, for there were two persons walking the deck at the time.

Mr. Tarn, who accompanied Mr. Graves on this occasion, brought me a very fine sea-eagle (*Polyborus Novæ Zealandiæ*), and some other birds, and a specimen from a shrub which we had not before observed, a species of *Desfontanea*.

In order to prevent a similar loss in future, the Adelaide / was forthwith fitted with cranks outside, for hoisting up her boats when in harbour.

Winter advanced rapidly; the ground was constantly covered with snow, from one to two feet deep, and every night more fell. In the early part of June we had a gale of wind from the N.W., which flooded the low ground upon which our tents stood; but fortunately the large tent had been accidentally placed on a higher part, and escaped. This flood filled, and, of course, spoiled the water in all the ponds about the tents; and we had afterwards to procure our supplies from a considerable distance.

On the 8th of June much lightning was observed to the northward, and repeated rumbling noises were heard, which continued for long periods; one lasted distinctly for the space of twenty minutes. At first, they were thought to be eruptions of some distant volcano; but, from the frequent lightning, they were probably echoes of thunder, reverberating through the deep ravines that intersect the rocky ridges of the Cordillera, from which we were distant at least one hundred and fifty miles.

A succession of bad weather followed, during which the barometer fluctuated rapidly. On the 14th, the mercury fell to 28. 17. inches, after which it gradually rose, with fine settled weather, until it reached 30. 50., when bad weather again set in. The people at the tents experienced another inundation. Had the water risen six inches more, it would have carried every thing away; and as the wind was blowing dead upon the shore, while a heavy surf was beating upon the beach, we could have rendered them very little assistance from the ship.

The severity of the weather brought a most disagreeable accompaniment. Scurvy appeared, and increased; while the accidental death of a seaman, occasioned by falling down a hatchway, followed by the decease of two others, and also of Mr. Low, of the Adeona, whose body was brought to me for burial, tended to create a despondency amongst the crew that I could in no way check. The monotony of their occupations, the chilling and gloomy appearance of the country, and the severity of the climate, all tended to increase the number of the / sick, as well as the unfavourable symptoms of their disease. The Beagle's term of absence was, however, drawing to a close, and I caused a rumour to be spread, that upon her appearance we should quit Port Famine. To give a semblance of reality to this report, the topmasts were ordered to be fidded, and the ship otherwise prepared for sea, which had a manifest effect upon the scorbutic, of whom several were in a bad stage of that horrid disease, and many others were just attacked. We found ourselves now, too, thrown on our own resources for fresh food: scarcely a fish was taken with the hook, and the seine, although frequently shot, never caught anything. Of birds, only a few hawks and small finches were procured, which were all reserved for the sick, the greater number of whom lived on shore, at the tents, where they might walk about, and amuse themselves as they pleased.

The Adelaide returned from Bougainville Harbour on the 18th of June, having succeeded in the object for which she was sent. The extremity of Cape Froward, a bluff head, over which is a round-topped hill (precisely the French 'Morre') is what Sarmiento called the Morro de Santa Agueda. Any name given by this excellent old navigator is too classical and valuable to be omitted; therefore, while the extremity itself may retain the modern appellation of Cape Froward, the mountain by which it is formed may still be allowed to keep his distinction. Behind it, the land rises to a higher ridge, the edge of which is remarkably serrated, and probably of a slaty character.

The specimens procured from the Cape were clay-slate, much intermixed with iron pyrites, and crossed by small veins of white quartz. Of the anchorages examined by Mr. Graves, Bougainville Harbour, better known to sealers by the name of Jack's Cove, or Harbour, is the most sheltered.

It is surrounded on all sides by high precipitous hills, thickly clothed with trees. The depth is moderate, and the water so beautifully clear, that the anchors, and even shells and stones, were distinctly seen upon the bottom. It was here that Bougainville procured wood for the use of the settlement / at the Falkland Islands. Captain Stokes says of this place: 'After seeing the abundant supplies of timber which Freshwater Bay and Port Famine afford, I had shared in the surprise which Byron expresses, that any one should have come so far up the Strait to get it; but on examining the spot, I found that a happier selection could not have been made. It is a little cove, just round the eastern point of the Bay of San Nicolas, about a hundred yards wide and three times as long. Here, moored to the shore, a ship may lie in eight fathoms, perfectly sheltered from any wind, the water as smooth as in a wet-dock. Shapely trees, of all dimensions, are growing within a few yards of the shore; and the wood, when felled, may be hoisted on board from the beach, by tackles from the yard-arms. Here, too, with very little trouble, a supply of water may be got from the many streams that make their way through the underwood which skirts the cove. As we pulled up this sequestered nook, the unusual sound of our oars and voices put to flight multitudes of birds, and the surface of the water was broken by the jumping of fine fish. Some very eatable geese were shot. Our stay was too short to admit of hauling the seine; but my boat's crew contrived to half-fill the boat with excellent muscles and limpets, which are found here in great plenty.'

The geological character of the coast between Cape St. Isidro and San Nicolas Bay is clay-slate; near the beach, however, this rock is not visible, since it is there covered with a kind of breccia of rounded pebbles, in an indurated sandy rock, of green colour. The pebbles are principally of slate; but some were found to be of granite and other quartzose rock, perhaps green-stone.

One of the headlands, called by M. Bougainville Cape Remarquable,[2] was examined by Mr. Graves for fossil shells, of which the French navigator speaks. Half the rock was beaten to pieces, without detecting anything like organic remains. Living shells were in the greatest abundance about the base of the Cape, but that is the case every where. The species generally found are limpets and muscles, but with little variety and no novelty. /

On the 21st of June, after a heavy north-east gale, we had an unusually fine day. The hills at the bottom of Magdalen Channel were more distinct than we had ever noticed them, and Mount Sarmiento was particularly clear; indeed its outline was so sharply defined, that the distance did not appear to be more than ten miles. This extraordinary transparency of the air was at first considered a presage of wet weather; yet the clear and sharp appearance of the distant land was unlike that which usually precedes a fall of rain. The long series of rainy weather we had experienced made us look for a good result from such an unusual atmosphere, and we were not deceived. The following day our hopes were still further confirmed by seeing three Indian canoes, coming across the Strait, towards us, from Lomas Bay, which they would not have attempted had they not been sure of its continuing fine; for their canoes are ill adapted to encounter the short cross sea found during bad weather in mid-channel of the Strait.

Although the presence of the natives did not in general please me, because it naturally put a stop to all work; yet, on this occasion it was agreeable, as it tended in some measure to enliven the monotonous manner in which we passed our days.

Upon reaching the bay, the Indians did not approach the ship, but paddled into the coves under Point Santa Anna, where our boat was employed watering. Mr. Graves went to them, to prevent mischief, and found they were the same party who had before visited us. When our boats returned, they paddled over to the wigwam at the head of the port, about a quarter of a mile beyond our tents, and began to repair it, and by sunset were housed and sheltered for the night. We had, however, so lately experienced their treacherous disposition, that no confidence was placed in appearances. Sentinels were posted at the tents, to give the alarm, should any of them approach; and at eight o'clock a volley of musketry was fired, by way of intimidation, and to impress them with the idea that we kept a watch upon their movements, and were prepared.

While the wigwams were repairing, a few of the Indians visited our tents; but were not allowed to pass within a rope / that, by my orders, was stretched around our property, a restriction which they did not attempt to evade. At sunset all were told to go away, and they immediately, as well as cheerfully, complied.

The next morning, and indeed throughout the whole day, the neighbourhood of the wigwams exhibited the appearance of a fair. I visited them, and found that they had not only repaired an old wigwam, but erected another. Both together contained the whole party, consisting of twenty-six individuals, among whom were an old man, and two old women. They had brought over a collection of baskets, bows and arrows, stone heads of knives, &c. to sell to our people, who had always shown eagerness to possess these curiosities. The knife-heads were made generally of pitch-stone; but the greater number were of broken glass bottles, which they had collected when they visited us last year. A few strings of beads purchased all their riches; after which they sold their dogs, and Mr. Graves procured one of them for a knife and a string of beads. It was a remarkably fine animal, and showed great reluctance to be handled by our people, several of whom were bitten in their attempts to take him to the boat.

At night one of the canoes was despatched to collect shellfish, probably sea-eggs, from the reef of Rocky Bay. The following morning all their goods were embarked, and then they paddled their canoes to the beach, near the tents, where some of their men landed. They had nothing to offer in exchange for several things which tempted them, and were beginning to grow troublesome. One of them, the individual who threw a stone at the sergeant, persisting to pass the boundary that was marked upon the ground, which no one of them had before presumed to do, was pushed back by the sentinel; upon which he ran to his canoe and took out several spears, doubtless intending to try to force a passage; but the appearance of two or three muskets brought him to his senses, and the spears were returned to the canoe; after which he became familiar, and apparently friendly. This affair, however, was soon followed by their departure, which gave me much satisfaction. / They went southward, landing for the night in Voces Bay, and the following day went to the Adeona, in Bougainville Harbour, where they remained some days.

The day after the Indians left us, a boat came from the Adeona, to acquaint us that, in a day or two, she and her companions, the Uxbridge and Mercury, intended to leave the Strait for the Falkland Islands; upon which I prepared letters for England, and a report of my proceedings for the Secretary of the Admiralty. The ships passed by on the 30th, and took my letters.

This last month (June) set in with snow or rain, which continued until the 11th, when the weather assumed a very threatening appearance. On the 14th the barometer fell to 29. 27, and the wind blew a hard gale from N.E.; but in the afternoon it veered round to S.W., and the mercury rose rapidly. A gale from S.W. followed, and then to the end of the month we had a series of moderate weather, but much snow. The mean temperature for June was 32°, 97 the range being between 19°, 2 and 48°, 7.

July commenced with an unusually low temperature and a high barometer; the former, on the 4th was 12°, 2, and the latter, at the same time, at 30,5 inches, having risen since the 14th of June 1.82 of an inch. After this we had a few mild and fine days, but paid dearly for them; a northerly gale set in, bringing with it unwholesome damp weather, in which the temperature rose to between 35° and 42°, and melted much of the snow that had covered the ground, quite to the water's edge, during the last two months. Our sick-list, particularly of cases of scurvy, increased so much, during this damp, trying weather, that I determined upon sending the Adelaide to the northward, to procure a supply of fresh meat from the Patagonians; and, at the same time, to survey that part of the Strait lying between Cape Negro and the Second Narrow.

Lieutenants Graves and Wickham, and Mr. Tarn, went upon this service, the latter being most anxious to procure some change of diet for the sick under his care, for some of whom he was much alarmed. The appearance and severity of this disease, / although every precaution had been used, and subsequent attention paid to their diet, are not easy to account for: fresh provisions, bread baked on board, pickles, cranberries, large quantities of wild celery, preserved meats and soups, had been abundantly supplied; the decks were kept well-aired, dry, and warm, but all to no purpose; these precautions, perhaps, checked the disease for a time; but did not prevent it, as had been fully expected.[3]

The Adelaide sailed on the 16th of July, with every prospect of fine weather. The same evening, an American sealing schooner anchored near us, on her way to Staten Land. She had entered the Straits by Cutler and Smyth Channels, and in forty-eight hours arrived at Port Famine. After obtaining some trifling assistance from our forge, she sailed.

On the 25th, three new cases of scurvy appeared, one being the assistant-surgeon, which increased our sick-list to fourteen. Feeling the necessity of doing something, I ordered the hands to be turned up, 'Prepare ship for sea!' No sooner had the words escaped the boatswain's lips, than all was life, energy, and delight. The preliminary preparations were made, and every one looked forward with pleasure to the change, except myself. I had hoped to pass the twelve months at Port Famine, with the intention of completing a meteorological journal, for which this place afforded peculiar advantages. My plan was, on the Beagle's return, to despatch her and the schooner along the West Coast, and join them in the Adventure at Chilóe.

As our departure was now supposed to depend on the Beagle's arrival, every eye was on the stretch to watch for her, and every morning some one of our party ascended the heights, to look out. On the 27th she was seen, beating up from the southward; but as the wind was contrary, she did not anchor in the bay until the evening. Her return was greeted with three most hearty cheers; but on passing under our stern, Lieutenant Skyring informed me that Captain Stokes was confined to his cabin by illness, and could not wait on me. I therefore went to the Beagle, and found Captain Stokes looking very ill, and in low spirits. He expressed himself much distressed by the hardships / the officers and crew under him had suffered; and I was alarmed at the desponding tone of his conversation. He told me that the Beagle had been up the western coast as high as Cape Tres Montes, in latitude 47°, had surveyed the Gulf of Peñas and other portions of the coast, particularly Port Henry, at Cape Three Points, the entrance of the Gulf of Trinidad, and Port Santa Barbara, at the north end of Campana Island.

During the survey of the Gulf of Peñas they had experienced very severe weather, both stormy and wet, during which the Beagle's crew were incessantly employed, and had consequently suffered greatly. Captain Stokes seemed not to have spared himself. He appeared much gratified by my visiting him, and before we parted he was for a time restored to his usual energy, detailing the circumstances of the voyage, and conversing upon the plan of our future operations with considerable animation.

The return of the Beagle cheered our ship's company, and on the 30th the Adelaide came back, with a large quantity of guanaco meat, which had been procured from the Patagonian Indians at Peckett's Harbour.

When the Adelaide anchored there, about thirty natives appeared on the shore. Mr. Tarn landed, and communicated our wants, saying that he would give tobacco and knives for as much guanaco meat as they could procure; with them was the Fuegian, who seemed to be a leading man, and to have become one of the most active of the party. He was the principal spokesman, and upon commencing the hunt he pointed to the snow upon the ground, and called it 'bueno' (good), because it would show the traces of the animals, and the direction they had taken. Mr. Wickham thus described to me the manner in which they hunted: Two men ascended a hill, placed themselves one at each end of its summit, and stood motionless for some time, on the look-out. As soon as guanacoes were seen, their position and movements were communicated, by signs, to the men in the valley, who were thus enabled to approach their game unawares. The guanacoes are taken with the bolas, which entangle their legs and throw them down. As soon as they are killed, / they are skinned and cut up. The first night seven hundred pounds of meat were brought, and two thousand and forty-six pounds were obtained in a few days.

This ample change of diet inspired me with the hope that our sick, at least those affected by scurvy, would recover, and that after another large supply, which we now knew how to obtain, we might be enabled to prosecute our voyage as was first intended. All hands were therefore allowed fresh meat for a week, and the residue was placed at the disposal of the surgeon, for the use of the sick, but all ineffectually; the list still increased, and Lieutenant Wickham, with a violent cold, and Mr. Rowlett, with scurvy, were added to it. The assistant-surgeon's became the worst case of scurvy on board; and our people, finding that the preparations for quitting the place were not going on, began to despond again. Captain Stokes was anxious to prepare his vessel for another cruize, being very averse to giving up our plans and returning to Monte Video, since he thought the crews, from utter disgust at the privations and hardships they had endured, would not be persuaded to go on another voyage; but that if they were to go to Chilóe or Valparaiso, to refresh, they might recover their strength and spirits, and be willing to renew the survey; which, however, he himself seemed to dread, for he never mentioned the subject without a shudder. He was evidently much excited, and suspicions arose in my mind that all was not quite right with him. I endeavoured to prevail on him to give his people a longer rest, but he was the more anxious to make preparations. On the 31st July he sent an application for provisions, and in the evening I received a note from him, which was written in his former usual flow of spirits. The officers, however, knew more of the diseased state of his mind than I did; and it was owing to a hint given to me, that I desired Mr. Tarn to communicate with Mr. Bynoe, and report to me whether Captain Stokes's health was sufficiently restored to enable him to commence another cruize. This was on the 1st of August. The provisions had been sent; in compliance with his application, and the surgeons were on board the Adventure, / considering upon their report, which was, as I afterwards found, very unfavourable, when a boat came from the Beagle, with the dreadful intelligence that Captain Stokes, in a momentary fit of despondency, had shot himself.

The surgeons instantly repaired on board, and finding him alive, had recourse to every means in their power, but without hope of saving his life. During the delirium that ensued, and lasted four days, his mind wandered to many of the circumstances, and hair-breadth escapes, of the Beagle's cruize. The following three days he recovered so much as to be able to see me frequently; and hopes were entertained by himself, but by no one else, that he would recover. He then became gradually worse, and after lingering in most intense pain, expired on the morning of the 12th.

Thus shockingly and prematurely perished an active, intelligent, and most energetic officer, in the prime of life.[4] The severe hardships of the cruize, the dreadful weather experienced, and the dangerous situations in which they were so constantly exposed – caused, as I was afterwards informed, such intense anxiety in his excitable mind, that it became at times so disordered, as to cause the greatest apprehension for the consequences. On the return of the Beagle he got better; and the officers were so sanguine in hoping for his complete restoration to health, on account of his progressive recovery, that nothing which had transpired was communicated to me until after his decease.

His remains were interred at our burial-ground, with the honours due to his rank, and a tablet was subsequently erected to his memory. /

CHAPTER X.

Account of the Beagle's cruize – Borja Bay – Cape Quod – Stuart Bay – Cape Notch – Remarks on weather, and errors of Chart – Evan-gelists – Santa Lucia – Madre de Dios – Gulf of Trinidad – Port Henry – Puma's track – Humming-birds – Very bad weather – Campana Island – Dangers – Gale – Wet – Sick – Santa Barbara – Wager's beam – Wigwams – Guaianeco Islands – Cape Tres Montes – St. Paul's – Port Otway – Hoppner Sound – Cape Raper.

THE following account of the Beagle's cruize is drawn up from Captain Stokes's unfinished journal, and from detached memoranda, which were found amongst his papers.

It will be recollected that, on my departure from Port Famine, in the Adelaide, in the month of March, to survey portions of the southern side of the Strait, I left instructions with Captain Stokes to proceed in the execution of his orders as soon as the Beagle was ready. The details of those orders it is unnecessary to repeat here, as they were performed to my entire satisfaction; it will be merely requisite, as briefly as possible, to follow him through a most arduous and distressing service. It is the sequel that embitters the record.

'On the 18th of March, I sailed from Port Famine, and next day reached Port Gallant.

'On the 23d, we anchored in the little cove called Borja Bay, which, though very confined, and rather difficult of access, suited our purpose extremely well. (See Sailing Directions). While there we measured the height of one of the principal hills in the neighbourhood, and found it 1,800 feet.

'Bad weather detained us until the 26th, when we passed Cape Quod, and reached Stuart Bay. Many places were left unexamined, because my object was to hasten westward before the year was farther advanced.

'(27th.) We left Stuart Bay, and continued our progress to the westward, with westerly winds, thick weather, and rain. / The shores of the Straits were seldom visible to us, from a thick mist with which they were clouded: it is, however, a bold coast on each side, otherwise the Strait would be utterly unnavigable in such weather. Near Cape Notch the mountains spire up into peaks of great height, singularly serrated, and connected by barren ridges. About their bases there are generally some green patches of jungle; but, upon the whole, nothing can be more sterile and repulsive than the view. This afternoon we passed Playa Parda, and in the evening anchored in Marian Cove.

'In the course of the next day the wind freshened to a strong and squally gale from the W.N.W., with much rain; the weather was so thick that we could scarcely make out the coast. In this kind of weather, the lower parts of the shore are screened from view by mist, and the upper ones are seen looming through it in lofty masses, in a manner which would lead a stranger to believe that the ship was completely environed with islands.

'In the evening we anchored in the little cove called Half Port Bay, and next morning resumed our daily struggle against wind, tide, and weather.

'We crossed the mouth of a deep sound on the north shore,** where no tide or current was remarked: the delineation of the coast about this point is particularly defective in the old charts; fortunately, however, for the navigator, he has here to deal with shores where the omission of a whole island, or even the addition of a few that do not exist, is of less consequence to his safety than the exact limit of one sand-bank in other parts of the world. This night we anchored in Upright Bay, which, though affording excellent shelter from the prevailing winds, is bad with a southerly one; as, from the steepness of the bottom requiring a vessel to anchor close to the shore, sufficient scope is not left for veering cable.

'Sheltered by the high land under which we were anchored, / with the exception of occasional gusts down the ravines and sounds, we had the wind light at W.S.W.; but the rapid travelling of the scud over-head showed that the usual weather prevailed. We weighed early next morning (30th), and by noon had reached so far to the westward that the easternmost of the round islands in Cape Tamar Bight bore north about two miles. By nightfall we were off Cape Cortado; but the weather seeming settled and the wind drawing to the southward, I resolved to keep under weigh, and try to get out to sea that night. Circumstances favoured us; the weather was fine, the moon remained unclouded, and the wind held at S.S.W. An hour after midnight Cape Pillar bore W.S.W., distant about two miles, and thence we shaped our course for the Evangelists, which we passed at the distance of a mile.

'The Evangelists, as they are called by the early Spanish voyagers, or as they were afterwards named (1670) by Sir John Narborough, the Isles of Direction,† are a group of four rocky islets, and some detached rocks and breakers, occupy-

* Afterwards examined by Capt. Fitz Roy. It was called Xaultegua by Sarmiento, who very correctly describes it. – (Sarmiento's Voyage, p. 208.)

† Because they formed a capital leading mark for the Strait of Magalhaens.

ing altogether a space of three miles; they are exceedingly rugged and barren, and suited only to afford a resting-place for seals and oceanic birds. From the heavy sea prevalent there, and the raging surf that generally breaks around, landing on them can be rarely practicable; yet sealers effect it. The mate of a sealing vessel told me that he had landed on the largest in a whale-boat, and killed several thousand seals. The Evangelists are of sufficient height to be seen in clear weather from a ship's deck, at the distance of six or seven leagues, but the superior elevation of the coast on both sides will usually render it visible, before these islands can be observed.

'Immediately on rounding the Evangelists a cape was distinguished, appearing to terminate the northern coast line, which we made out to be 'Cape Isabel' of the Spanish charts. It is a steep, rocky promontory of great height, having at its base some detached columnar masses of rock, and at its summit a peak, and a serrated ridge; off it is a steep-sided island, / which proved to be that (Beagle Island) of which Lieut. Skyring and I took the bearing last year, when we were on the summit of Cape Victory.

'Northward of Cape Victory the land forms a deep bight, of which Cape Santa Lucia is the north-eastern headland. The coast in the interval is exceedingly rugged and mountainous. Cape Santa Lucia may be distinguished by a portion of flat table-land, about one-third of the altitude of the mountain from which it proceeds, and terminating at its outer face with a perpendicular precipice.

'The coast between Capes Isabel and Santa Lucia is dangerous to approach nearer than ten miles, for there are within that distance many sunken rocks, on which the sea only occasionally breaks. Some of these breakers were seen to seaward of us, as we proceeded along the coast, at the distance of five or six miles. When off Santa Lucia, whales were very numerous around us.

'The general aspect of this portion of the coast is similar to that of the most dreary parts of the Magalhaenic regions: bare, rugged, rocky, and mountainous, intersected by inlets, and bordered by islets, rocks, and breakers.

'The information we possessed respecting the prevalent winds on this coast was very scanty; yet, since all we could procure represented them as prevailing from the northward and north-westward, I considered it advisable to take advantage of the present southerly wind to proceed to the northern part of the coast assigned for our survey, instead of stopping to explore the bight between Cape Isabel and Cape Santa Lucia.

'From the bearings at sunset,* we ran along the land with bright moonlight, sounding every hour; and at daylight were about ten miles from the Island of Madre de Dios.

* Beagle Island N. 71°. E., Cape Isabel N. 32°.E., a remarkable mountain in the bight between Cape Santa Lucia and Cape Isabel N. 11°. W, Cape Santa Lucia N. 33°. W.; distance off shore three leagues; and soundings fifty fathoms, sandy bottom.

'We closed the land and proceeded to the northward, keeping at a distance of about three miles off shore, sounding / between twenty-eight and thirty-three fathoms, sandy bottom. The weather was clear and fine, and we were enabled to make observations, and take the bearings and angles, necessary for laying down the coast satisfactorily.

'At noon we were in latitude 50° 12' south, and in the meridian of Cape Tres Puntas, between which and a cape bearing from us N. 13° E. (magnetic), distant eight miles, there was evidently an inlet: this cape is marked on the chart as Cape William. The character of the land is the same with that which we had hitherto passed, bare, rugged, rocky mountains, with peaks, and sharply serrated ridges. From daylight to noon we had run twenty-one miles along the coast; in that interval only one inlet was seen, which was in the latitude of 50° 27' south, agreeing well with the 'West Channel' of the Spanish chart. It was four miles wide at its mouth, and appeared to follow a winding course to the eastward. The land of Cape William; at dusk we were abreast of Cape William, and two leagues off shore, where we lay-to till daylight, as I wished to examine the inlet between it and Cape Tres Puntas, which subsequently proved to be Sarmiento's Gulf of Trinidad. The old navigator thus describes its discovery:

' 'At daylight, 17th of March, 1579, in the name of the most holy Trinity, we saw land, bearing E.S.E., ten leagues distant, towards which we steered to explore it. At mid-day, being near the land, we observed the latitude 49 1/2°, but Hernando Alonzo made it 49° 9'. In approaching the shore we saw a great bay and gulf, which trended deeply into the land towards some snowy mountains. To the south there was a high mountain, with three peaks, wherefore Pedro Sarmiento named the bay 'Golfo de la Sanctisima Trenidad.' The highest land of the three peaks was named 'Cabo de Tres Puntas ó montes.' This island is bare of vegetation, and at the water-side is low and rugged, and lined with breakers; on the summit are many white, grey, and black-coloured portions of ground, or rock. Six leagues to the north of Cape Tres Puntas is the opposite side of the gulf, where it forms a large high mountain, backed / to the north by low land, and fronted by many islands. This high mountain, which appears to be an island from the offing, was called 'Cabo Primero.' '*

'The following night was clear, and the wind moderate from S.E., but in the course of next morning it shifted to N.E., with squalls, rain, and thick weather; we worked into the inlet notwithstanding, and by noon had reached three miles within its S.W. head-land, Cape William, and were abreast of a bay, into which I sent a boat to look for anchorage. On her return we stood into it, and anchored in the excellent harbour, afterwards named Port Henry, where we remained from

* (Sarmiento, p. 65.)

the 2d to the 5th of April, employed in making a correct survey of the harbour and its adjacencies, and determining the latitude and longitude.*

'The inner harbour, distinguished in the plan by the name of 'Aid Basin,' is perfectly land-locked, and sufficiently spacious to contain a numerous squadron of the largest ships in twenty fathoms water, over a mud bottom, and as completely sheltered from the effects of wind and sea as in wet-docks. At the south-west side of the basin is a fresh-water lake, which discharges itself by a small stream, whence casks might be conveniently filled by means of canvas hoses, and the shores around have wood for fuel in abundance; but, from the lofty surrounding mountains, some rising almost perpendicularly to an elevation of two thousand feet, the thick clouds with which this basin was generally overhung, and the dense exhalations that arose from it during the rare intervals of sunshine, together with the exceeding prevalence of heavy rain on this coast, this place must be disagreeable and unhealthy. Such objections do not apply to the outer harbour, for while its shores afford shelter, they do not obstruct a free circulation of air. It is sufficiently large to afford convenient and secure anchorage for five or six frigates.

'We hauled the seine with very poor success, as a few smelt only were taken; we had no better luck with our fishing-lines; / but the trial might have been more profitable at another season, judging from the number of seals we saw on the rocks off the Port, which live principally upon fish. Muscles, limpets, and sea-eggs abound here, and are good and wholesome of their kind. Birds are few in number, and of the species most common in these regions. No quadruped of any kind was seen; but the purser told me that he had observed, near the sandy beach, traces of a four-footed animal, resembling those of a tiger: he followed them to a cavern, and thence to the jungle. He also said that he had seen several humming-birds.

'With the exception of wild celery and the arbutus berry, I know not of any useful vegetable production that this place affords, unless the 'Winter's-bark tree' may be mentioned. Some coarse grass, fit perhaps for animals, may be there procured. The only signs of inhabitants were some wigwams on the western point, which seemed to have been long forsaken: in their construction they were precisely similar to those erected by the migratory tribes in the Straits of Magalhaens; and the shells of muscles, limpets, and sea-eggs, within and about them, showed that the former tenants of these hovels drew, like the Magalhaenic tribes, a principal part of their subsistence from shell-fish.

'Around the harbour are granite mountains, perfectly bare at their summits and north-western sides, but the lower parts are thickly covered in sheltered places and ravines, partly with trees, and partly with brushwood: among the

* The description of Port Henry is given in our Sailing Directions.

trees growing here we observed, as usual, two kinds of beech, a tree like the cypress, but of small size, and the Winter's-bark. The underwood is composed of all the various shrubs we had met with in the Straits of Magalhaens; and this brushwood is so thickly spread over the lower parts of the shores of the harbour, that it is only by crawling over it that the distance of a few yards from the rocks can be gained; and being generally of insufficient strength to support a man's weight, it frequently gives way beneath him, and he is so completely buried, as to make it difficult for him to extricate himself.

'Scarcely any of the trees attain a size to render them fit / for any thing but firewood; of those we felled there was scarcely one that was not more or less rotten at the heart, a defect probably caused by the extreme humidity of the climate.

'During our stay, the master, accompanied by our boatswain's mate, (*f*) an experienced sealer, went to take seal on the rocks, and returned in a few hours with some of the inferior sort, called 'hair seal,' which were numerous; but the surf was in most places too heavy to allow them to land without much risk. The fry of the young seals we thought extremely good, not exceeded even by the finest lamb's fry.

'On the morning of the 5th we worked to the westward, to clear the land on each side of the inlet; and at sunset, Cape Tres Puntas bore N.b.W. 1/2 W., distant two leagues. The northerly breeze, which we had worked with since leaving Port. Henry, increased rapidly to a hard gale, and by 8 P.M. we were reduced to the close-reefed main-topsail and reefed foresail. The gale continued with unabated violence during the 6th, 7th, and 8th, from the north, N.W., and S.W., with a confused mountainous sea. Our decks were constantly flooded, and we could rarely show more than the close-reefed main-topsail and reefed foresail. Only two accidents occurred: the little boat which we carried astern was washed away by a heavy sea that broke over us, while hoisting her in-board; and the marine barometer was broken by the violent motion of the vessel. At noon, on the 8th, Cape Corso bore from us, by account, S.E. (true), distant fifty-five miles. I had tried to gain a wide offing to get a less turbulent sea, and because not even an outline of the sea-coast of Campana Island was drawn in the chart. We had not, during these three days, a glimpse of the sun or of a star, for it blew a constant gale, accompanied by squalls, thick weather, and rain. According to the time of year, the season of winter had not arrived, but the weather seemed to say it was already come –

> Sullen and sad, with all it's rising train
> Of vapours, clouds, and storms.[1] /

(*f*) Thomas Sorrell, now boatswain of the Beagle (1837). He was boatswain of the Saxe Cobourg, when wrecked in Fury Harbour. – R. F.

'The wind abated at daylight on the 9th, and drew to the southward, and thence to the S.E. (the fair weather quarter of this coast). We bore up to make the land, and at about 10 A.M. the 'loom' of it was seen from the mast-head. At noon, high mountains were visible from the deck; our latitude, by observation, was 48° 51', and our longitude, by chronometer, 00° 27' west of Port Henry. No soundings were obtained with one hundred and ten fathoms of line. Hence we steered east (magnetic) towards a remarkable mountain, which, from our being nearly in the parallel of it at noon, has been marked in the chart as Parallel Peak. The coast we were upon was that of the Island 'Campana,' and, in its general appearance, did not differ from that of Madre de Dios. It was late before we got very close to the land; but, for a couple of leagues to the northward, and about a league to the southward of the parallel of our latitude at noon, we could distinguish rocks and breakers skirting the coast to a distance of two leagues from the shore.

'At dusk we hauled off for the night; but instead of being able to resume the examination of the coast next morning, we had to encounter another gale of wind from the N.W., which before noon, reduced us to close-reefed main-topsail and reefed foresail. This gale suddenly subsided in the western quarter, which was singular; for those we have experienced generally commenced at north, thence drew round to the westward, from which point to S.W. they blew with the greatest fury, and hauling to the southward, usually abated to the eastward of south.

'During the afternoon, we again made the land near Parallel Peak, but could not close it. Next morning (11th), with fine weather, and a fresh breeze at S.W. b. W., we once more saw the land about Parallel Peak; and when distant from the shore about eight miles, steered N.b.E. along the coast. At noon our latitude was 48° 47'.

'Throughout our run along the coast this day, we skirted a number of rocky islets, rocks, and breakers, lying off shore at the distance of three or four miles. Some of the islets were elevated several feet above the surface of the sea; others were / a-wash, and there were breakers that showed themselves only occasionally. Along this line the surf beat very heavily, and, outside, a long rolling sea prevailed, in which the ship was very uneasy.

This line of dangers is not altogether continuous; for there is an opening about two miles wide, abreast of Parallel Peak, to the southward of which is a bight, where possibly a harbour may exist; but, considering the prevalence of heavy westerly gales and thick weather, if there be one, few vessels would venture to run for it; and this line must, I should think, be considered as a barrier that they ought not pass. As seal are found on the rocks, vessels engaged in that trade might not, perhaps, be deterred by these dangers, but every other would give all this extent of coast a wide berth. We ran past the breakers at the distance of about a mile, having rocky soundings, from thirty to twenty-three fathoms.

The termination of the coast line northward was a high, rugged island, with a small peak at the north end. The extremity of the main land was rather a high bluff cape, whence the coast extends southward, with craggy, mountainous peaks and ridges, as far as Parallel Peak. At sunset, the N.W. end of Campana bore north (magnetic), distant three leagues, and from the mast-head I could see very distinctly the belt of rocks and breakers extending uninterruptedly to the northward, as far as the end of Campana.

We hauled off for the night, and had light variable airs, or calms, until 2 A.M. of the 12th, when a breeze from the northward sprung up, and freshened so rapidly, that by noon we were again reduced to a close-reefed main-topsail and foresail. The gale was accompanied, as usual, by incessant rain and thick weather, and a heavy confused sea kept our decks always flooded.

The effect of this wet and miserable weather, of which we had had so much since leaving Port Famine, was too manifest by the state of the sick list, on which were now many patients with catarrhal, pulmonary, and rheumatic complaints. The gale continued undiminished until the morning of the 13th, / when, having moderated, we bore up and steered N.E. to close the land. At noon a good meridional altitude gave our latitude 48° 30' south, and about the same time we saw the land bearing N.E.b.E., which we soon made out to be Parallel Peak. After allowing amply for heave of sea, and lee-way, we were considerably southward of our reckoning, which indicates a southerly current; but under such circumstances of wind and weather its exact direction, or strength, could not be ascertained.

We proceeded along the land, taking angles and bearings for the survey, and at sunset the N.W. end of Campana bore from us north (magnetic), distant five leagues. Being now off the N.W. end of the island of Campana, which forms the south-western headland of the Gulf of Peñas, I considered that, before I proceeded to examine its inlets, I ought to look for the Harbour of Santa Barbara, which has been placed on the old charts in this neighbourhood. Accordingly we lay-to during the night, and at 4 A.M. bore up to close the land; at daylight the extremes of it were seen indistinctly through a very cloudy and hazy atmosphere, from N. 39° E. to S. 53° E. About noon the weather cleared off, and we got the meridian altitude of the sun, which gave our latitude 48° 09' south.** We directed the course for the Dundee Rock, and when abreast of it, steered N.E.

* The N.W. end of the Island de la Campana bearing N. 71°. 40'. E. Two distant hummocky islands (answering pretty well in position with the Guaianeco Islands of the Spanish charts) N. 53°. 30'.E., and N. 55°. 48'. E., and a remarkable rock, the 'Dundee' of Bulkeley and Cummings, about forty-five feet high, rising like a tower from the sea, distant off shore five miles, bearing east of us, distant one mile. /

(compass) for an opening in the low part of the coast ahead, backed by very high mountains, which we found was the entrance of Port Santa Barbara. The coast to the southward was lined with rocky islets, rocks, and breakers, extending a league to seaward, and there were others to the northward. We were in a channel half a mile wide, through which we continued our course, sounding from fifteen to eleven fathoms, and in the evening anchored near the entrance of the harbour.

As our present situation was completely exposed to westerly winds, I went to examine a deep bight in the southern shore, which proved to be a good harbour, perfectly sheltered from all winds, with a depth of three and a half fathoms over a fine sandy bottom. In the afternoon we weighed anchor and warped into a berth in the inner harbour, where we moored in three fathoms. I found lying, just above high-water mark, half buried in sand, the beam of a large vessel.** We immediately conjectured that it had formed part of the ill-fated Wager, one of Lord Anson's squadron (of whose loss the tale is so well told in the narratives of Byron and Bulkeley): the dimensions seemed to correspond with her size, and the conjecture was strengthened by the circumstance that one of the knees that attached it to the ship's side had been cut, which occurred in her case, when her decks were scuttled to get at the provisions; all the bolts were much corroded; but the wood, with the exception of the outside being worm-eaten, was perfectly sound. Our carpenter pronounced it to be English oak.

The land about this harbour is similar to that about Port Henry. Its shores are rocky, with some patches of sandy beach, but every where covered with trees, or an impervious jungle, composed of dwarfish trees and shrubs. The land, in most places, rises abruptly from the shore to mountains, some of which attain an altitude of more than two thousand feet, and are quite bare at their summits and on their sides, except in sheltered ravines, where a thick growth of trees is found. These mountains, or at least their bases, where we could break off specimens, were of basalt, with large masses of quartz imbedded in it; but on some parts of the shores the rocks were of very coarse granite.

As in the vicinity of Port Henry, the thickness of the jungle prevented our going far inland; the greatest distance was gained by Lieut. Skyring, who, with his wonted zeal to prosecute the survey, ascended some of the mountains for the / purpose of obtaining bearings of remote points: he remarked to me, 'that many miles were passed over in ascending even moderate heights; the land was very high and very irregular; the mountains seemed not to lie in any uniform direction, and the longest chain that was observed did not exceed five miles. The flat land between the heights was never two miles in extent: the ground was always

* Length twenty feet five inches and a half, sided twelve inches, and moulded eight inches and a half.

swampy, and generally there were small lakes receiving the drainage of mountain-streams. Indeed the whole country appeared broken and unconnected.'

Some of the mountains were ascertained to be 2,500 feet high, but the general height was about 2,000 feet. A large island, on the northern side of the harbour, is an excellent watering-place, at which casks may be conveniently filled in the boat. It is also an object of great natural beauty: the hill, which forms its western side, rises to seven or eight hundred feet, almost perpendicularly, and when viewed from its base in a boat, seems stupendous: it is clothed with trees, among which the light-green leaves of the Winter's-bark tree, and the red flowers of the Fuchsia, unite their tints with the darker foliage of other trees. This perpendicular part extends to the northward till it is met by the body of the mountain, which is arched into a spacious cavern, fifty yards wide and a hundred feet high, whose sides are clothed with a rich growth of shrubs; and before it a cascade descends down the steep face of the mountain.

On the shore we found two Indian wigwams and the remains of a third; but they had evidently been long deserted, for the grass had grown up both around and within them to the height of more than a foot. These wigwams were exactly similar to those in the Strait of Magalhaens: one was larger than any I had met with, being eighteen feet in diameter. The only land birds I saw were two owls, which passed by us after dusk with a screeching noise.

On the patches of sandy beach, in the inner harbour, we hauled the seine, but unsuccessfully; we expected to find fish plentiful here, from seeing many seals on the rocks outside, and from finding the water quite red with the spawn of cray-fish. / Muscles and limpets were pretty abundant, and the shells (*Concholepas Peruviana*) used by the Magalhaenic tribes as drinking cups, were found adhering to the rocks in great numbers.

Nothing could be worse than the weather we had during nine days' stay here; the wind, in whatever quarter it stood, brought thick heavy clouds, which precipitated themselves in torrents, or in drizzling rain. We were well sheltered from the regular winds; but many troublesome eddies were caused by the surrounding heights, while the passing clouds showed that strong and squally north-west winds were prevalent.

On the morning of the 24th, we put to sea with a southerly breeze. The extent of coast from the eastern part of Port Santa Barbara to the outer of the Guaianeco Islands presents several inlets running deep into the land; but it is completely bound by rocks and rocky islets, which, with its being generally a leeshore, renders it extremely unsafe to approach. Observing an opening between some islets, of which we had taken the bearing at noon, we stood in to see whether it afforded anchorage; and approaching the extremity of the larger island, proceeded along it at the distance of only half a mile, when, after running two miles through a labyrinth of rocks and kelp, we were compelled to haul out, and in doing so scarcely weathered, by a ship's length, the outer islet. Deeming it useless to expend further time in the examination of this dangerous portion of the gulf, we proceeded towards Cape Tres Montes, its northwestern headland.

At sunset Cape Tres Montes bore N. 25° W., distant eighteen miles. In this point of view the cape makes very high and bold; to the eastward of it, land was seen uninterruptedly as far as the eye could reach. We stood in shore next morning, and were then at a loss to know, precisely, which was the cape. The highest mountain was the southern projection, and has been marked on the chart as Cape Tres Montes: but none of the heights, from any point in which we saw them, ever appeared as 'three mounts.' The land, though mountainous, seemed more wooded, and had a less / rugged outline than that we had been hitherto coasting, since leaving the Strait. We steered along the western coast of the land near Cape Tres Montes, and at noon, being three miles from the shore, observed, in latitude 46° 5. south, the cape, bearing N. 80° E. (mag.), distant seven miles. The northernmost cape in sight N. 26° W., distant ten miles, soundings ninety-seven fathoms. Shortly afterwards another cape opened at N. 37° W. (mag.).

The parallel of forty-seven degrees, the limit assigned for our survey, being already passed, I did not venture to follow the coast further, although we were strongly tempted to do so by seeing it trend so differently from what is delineated on the old charts. An indentation in the coast presenting itself between mountainous projections on each side of low land (of which the northernmost was the cape set at noon), we hauled in to look for an anchorage; but it proved to be a mere unsheltered bight, at the bottom of which was a furious surf. We then stood to the southward, along the land of Cape Tres Montes, with the view of examining the north side of the Gulf of Peñas.

The following morning was fine: Cape Tres Montes bore N.E., distant about three leagues. We lay off and on during the day, while the master went in the whale-boat, to examine a sandy bay (of which Cape Tres Montes was the easternmost point) for anchorage: he returned about sunset, and reported that it did afford anchorage; but was quite unsheltered from wind, and exposed to a great swell. The boat's crew had fallen in with a number of seals, and the quantity of young seal's fry they brought on board afforded a welcome regale to their messmates and themselves.

At daylight (27th) we were four leagues from Cape Tres Montes, bearing N. 68° W. (magnetic) a remarkable peak, marked in the chart the 'Sugar Loaf,' N. 19° E., distant twenty-four miles, and our soundings were sixty-eight fathoms. This peak resembled in appearance, the Sugar Loaf at Rio de Janeiro: it rises from a cluster of high and thickly-wooded islands, forming apparently the eastern shore of an inlet, of which / the land of Cape Tres Montes is the western head. Further to the N.E. stands a lofty and remarkable mountain, marked in our chart as 'the Dome of Saint Paul's.' It is seen above the adjacent high land. The height of the Sugar Loaf is 1,836 feet, and that of the Dome of Saint Paul's, 2,284 feet.

During the day we worked up towards the land, eastward of Cape Tres Montes, and at night succeeded in anchoring in a sandy bay, nine miles from the Cape, where our depth of water was twelve fathoms, at the distance of a cable and a half off shore. We lay at this anchorage until noon the following day, while Lieut. Skyring landed on some low rocks detached from the shore, where he was able to take some advantageous angles; and on his return we weighed and worked up the gulf, between the eastern land of Cape Tres Montes, and high, well wooded islands. The shores of the main land, as well as of the islands, are bold, and the channel between them has no dangers: the land is in all parts luxuriantly wooded. About a mile and a half to the northward of the sandy beach which we had left, lies another, more extensive; and a mile further, a considerable opening in the main land, about half a mile wide, presented itself, having at its mouth two small thickly-wooded islands, for which we steered, to ascertain whether there was a harbour. The water was deep at its mouth, from thirty-eight to thirty-four fathoms; but the comparative lowness of the shores at its S.W. end, and the appearance of two sandy beaches, induced us to expect a moderate depth within. As we advanced, a long white streak was observed on the water, and was reported from the mast-head as a shoal; but it was soon ascertained to be foam brought down by the tide, and we had the satisfaction of anchoring in sixteen fathoms over a sandy bottom, in a very excellent port, which we named Port Otway, as a tribute of respect to the Commander-in-chief of the South American Station, Rear Admiral Sir Robert Waller Otway, K.C.B.'

A deficiency here occurs in Captain Stokes's journal, which the Beagle's log barely remedies. From the 30th of April to / the 9th of May there was a succession of stormy weather, accompanied by almost incessant and heavy rain, which prevented the ship being moved; but proved, in one respect, advantageous, by affording a very seasonable cessation from work to the fatigued crew, and obliging Captain Stokes to take some little rest, which he so much required; but regretted allowing himself, and submitted to most reluctantly. He continues his journal on the 9th of May, stating that, 'Among the advantages which this admirable port presents to shipping, a capital one seems to be the rich growth of stout and shapely timber, with which its shores, even down to the margin of the sea, are closely furnished, and from which a frigate of the largest size might obtain spars large enough to replace a topmast, topsail-yard, or even a lower-yard. In order to try what would be the quality of the timber, if, in case of emergency, it were used in an unseasoned state, I sent the carpenter and his crew to cut two spars for a topgallant-mast and yard. Those they brought on board were of beech-wood; the larger being thirteen inches in diameter, and thirty feet in length.

On the 10th, the weather having improved, the Beagle was moved to the head of the inlet, to an anchorage in Hoppner Sound, and on the 11th I went with Lieut. Skyring to examine the opening, off which we were anchored.

On each side of it we found coves, so perfectly sheltered, and with such inexhaustible supplies of fresh water and fuel, that we lamented their not being in a part of the world where such advantages could benefit navigation. The depth of water in mid-channel was generally forty fathoms; in the bights, or coves, it varied from sixteen to twenty-five fathoms, with always a sandy bottom. We saw a great many hair seals, shoals of pie-bald porpoises, and birds of the usual kinds in considerable numbers. On several points of the shores were parts of the skeletons of whales; but we no where saw a four-footed animal, or the slightest trace of a human habitation. The unusual fineness of the morning, the smoothness of the water, and the proximity of the adjacent lofty mountains, clothed almost to their summits by the fullest foliage, with every / leaf at rest, combined with the stillness around to give the scene a singular air of undisturbed repose. We reached the extremity of the inlet, which we found was about six miles from its mouth; and thinking that it was the inner shore of an isthmus, of no great width, curiosity prompted us to endeavour to see its outer shore: so we secured the boat, and accompanied by five of the boat's crew, with hatchets and knives to cut their way, and mark the trees to guide us on our way back, we plunged into the forest, which was scarcely previous on account of its entangled growth, and the obstructions presented by trunks and branches of fallen trees.

Our only guide was an occasional glimpse, from the top of a tree, of the ranges of mountains, by which we steered our course. However, two hours of this sort of work were rewarded by finding ourselves in sight of the great South Sea. It would be vain to attempt describing adequately the contrast to the late quiet scene exhibited by the view we had on emerging from this dark wood. The inlet where we left our boat resembled a calm and sequestered mountain lake, without a ripple on its waters: the shore on which we now stood was that of a horrid rock-bound coast, lashed by the awful surf of a boundless ocean, impelled by almost unceasing west winds.

Our view of the coast was limited on each side by rocky mountainous promontories: off the northernmost, which I called Cape Raper, were rocks and breakers, extending nearly a mile to seaward. Having taken the few bearings our situation enabled us to obtain, we retraced our steps to the boat, and by aid of the marks we had left on the trees, reached her in an hour and forty-three minutes.

Some of the beech-trees of this wood were fifteen feet in circumference; but I noticed none differing in their kind from those already observed about Port Otway. A few wrens were the only living creatures we saw; not even an insect was found in our walk. In the beds of some of the streams intersecting the woods was a singularly sparkling sand, which had so much the appearance of gold, that some of our party carried / a bag-full on board to be tested. The shining substance proved to be, as I had supposed, the micaceous particles of disintegrated granite. It was not our good fortune to discover streams similar to those sung of by the poet,

> 'Whose foam is amber, and whose gravel gold.'[2] /

CHAPTER XI.

Leave Port Otway – San Quintin's Sound – Gulf of Peñas – Kelly Harbour – St. Xavier Island – Death of Serjeant Lindsey – Port Xavier – Ygnacio Bay – Channel's mouth – Bad weather – Perilous situation – Lose the yawl – Sick list – Return to Port Otway – Thence to Port Famine – Gregory Bay – Natives – Guanaco meat – Skunk – Condors – Brazilians – Juanico – Captain Foster – Changes of officers.

THE Beagle returned to Port Otway the following day, and in an interval of better weather obtained the observations necessary for ascertaining the latitude and longitude of the port, and for rating the chronometers.

Captain Stokes's journal continues on the 19th of May: 'We left Port Otway, and as soon as we had cleared its entrance, steered E.N.E. across the gulf; leaving to the northward all that cluster of islands, distinguished in the chart as the 'Marine Islands,' and went to within a mile from the eastern shore. Thence we ran four miles and a half parallel with the direction of coast E.S.E. (mag.), at the mean distance of a mile off shore. The aspect of the eastern and western portions of this gulf is very different, and the comparison is much to the disadvantage of the eastern. Ranges of bare, rugged, rocky mountains now presented themselves, and where wood was seen, it was always stunted and distorted. A long swell rolled in upon the shore, and every thing seemed to indicate a stormy and inclement coast. There are a few bays and coves, in which is anchorage depth, with a pretty good bottom of dark coarse sand: but rock-weed in large patches, seen in some of them, denoted foul ground; and they are all more or less exposed, and extremely unsafe. As night advanced, the weather became rainy and thick; so having reached a bight which seemed less insecure than others that we passed, I hauled in, and at about seven P.M., guided only by the gradual decrease of our soundings, from / fourteen to eight fathoms, and the noise of the surf, came to an anchor.

Next morning (20th) we found that we had anchored in a small bay, at about half a mile from a shingle beach, on which a furious surf was breaking so heavily as to prevent our landing any where. We were completely exposed to S.W. winds, with a heavy rolling sea; and the surf on all points cuts off communication with the shore. A breeze from the S.W. would have rendered it difficult to get out, and would have exposed us to imminent hazard. It is called on the chart Bad Bay. We left it eagerly, and proceeded to trace the coast to the E.S.E., until we were nearly abreast of a moderately high and thickly-wooded island, called Purcell Island. We passed to the northward of Purcell Island, leaving on the left a rock only a few feet above the surface of the sea, which lies about midway between that island and the main land. As we advanced to the eastward, a large and very remarkable field of ice was seen lying on the low part of the coast, which, at a distance, we took for a dense fog hanging over it, as nothing of the kind was observable in any other part. When nearly abreast of San Xavier Island, a deep sound was observed to the left, or north, which we concluded was the San Quintin Sound of the Spanish chart: it seemed to be about five miles in breadth, and following a westerly direction. We kept sight of the Sugar Loaf, and other points we had fixed, until more could be established, which enabled us to chart the coast as we went along. My next object was to trace the Sound of San Quintin to its termination, and at nightfall we succeeded in getting an anchorage at the entrance.

On the 21st we proceeded up the sound, passing to the northward of Dead Tree Island. Our soundings, until abreast of it, were from sixteen to ten fathoms, on a mud bottom; it then shoaled to four fathoms, and after running about three miles in that depth, we came to an anchor at the distance of a mile from the north shore of the sound, in four fathoms.

'Exceedingly bad weather detained us at this anchorage. From the time of our arrival, on the evening of the 21st, / until midnight of the 22d, it rained in torrents, without the intermission of a single minute, the wind being strong and squally at W., W.N.W., and N.W.

When the weather improved, on the 23d, we weighed, and made sail along the northern side of the sound, for the purpose of ascertaining whether it admitted of a passage to the northward. We kept within a mile of the shore, sounding from eight to fifteen fathoms, generally on a sandy bottom; and a run of seven miles brought us within three miles of the bottom of the inlet, the depth of water being four fathoms, on sand. The termination of this sound is continuous low land, with patches of sandy beach, over which, in the distance, among mountains of great height, we were again able to make out and take the bearing of that remarkable one, named the 'Dome of St. Paul's.' The shores of this inlet are thickly wooded; the land near them is, for the most part, low, but rises into mountains, or rather hills, from twelve to fifteen hundred feet in height, from which many streams of water descend. As soon as a ship has passed Dead Tree Island, she becomes landlocked; and as in all parts of the sound there is anchorage depth, with a muddy or sandy bottom, the advantages offered to shipping would be of great consequence in parts of the world more frequented than the Gulf of Peñas.

Whales were numerous, and seals were seen in this inlet, now called the Gulf of San Estevan.

Hence we went to Kelly Harbour, at the north-eastern side of the Gulf of Peñas, four miles N.E. of Xavier Island. The land around it is rocky and mountainous, but by no means bare of wood. Near the entrance it is low, as compared with the adjacent land; but in the interior are lofty snow-capped mountains.

A large field of ice, lying on the low land near Kelly Harbour, was remarkable. There was none on the low grounds at the other (southern) side of the port, though it was almost the winter solstice at the time of our visit.

Another day and night of incessant rain. In the morning of the 25th we had some showers of hail, and at daylight found / that a crust of ice, about the thickness of a dollar, had been formed in all parts of the harbour. The water at our anchorage being fresh at half-tide, was, no doubt, in favour of this rapid congelation. Lieutenant Skyring having completed the examination of the harbour, we left it and steered between St. Xavier Island and the mainland, through a fine bold channel, nearly four miles wide, with a depth of more than thirty fathoms. The land on both sides is closely wooded, and rises into high mountains. About dusk we stood into Port Xavier, a little bight, with a sandy beach, on the eastern side of the island; and, at a distance of two cables' length from the beach, anchored for the night in seventeen fathoms.

(26th). This sandy beach extended about half a mile between the points of the bay, and, at fifty yards form the water, was bounded by thick woodland, which rose with a rapid ascent to the height of a thousand feet. The trees were like those in the neighbourhood of Port Otway, and were stout and well-grown. A tree, large enough for a frigate's topmast, might be selected close to the shore. The Winter's-bark tree attains here a greater size than I had before seen. One, which was felled by our wood-cutters, measured eighty-seven feet in length, and was three feet five inches in circumference. All the trees were in full foliage and verdure, though the season corresponded to the latter part of November in our northern latitudes. At the south end of the sandy beach was a stream of fresh water, several yards in width, and various waterfalls descended from the mountains. The shore to the southward was composed of fragments of granite, lying at the base of a lead-coloured clay cliff, at least three hundred feet in height. In this cliff the mountain-torrents had formed deep chasms, and strewed the beach with its débris, and with uprooted timber. The only living creatures seen were steamer-ducks, king-fishers, and turkey-buzzards.

While on shore, I received a melancholy message, announcing the death of Serjeant Lindsey, of the Royal Marines. During the last few days he had suffered from inflammation of the bowels, which brought his existence to a close. /

The following day (27th) a grave was dug, and we discharged the last sad duties to our departed shipmate. A wooden cross was erected at the head of his grave, on which was an inscription to his memory: we also named the south point of the bay after him. About noon we left Port Xavier, and coasted the island, at the mean distance of a mile, examining it for anchorages, until, after a run of eight miles, we reached its south point. For the first four or five miles of that distance, the coast of the island consisted of a high steep cliff, having at its base a narrow beach, composed of various-sized masses of rock. In the interior there were heights, rising twelve or fourteen hundred feet, wooded nearly to the summits, with many streams of water descending from them; but for the remainder of the distance the coast was low, and the wood stunted and scanty. All along the shore rolled a heavy surf, that would have rendered any attempt to land exceedingly hazardous; there was no place fit for anchorage, except a small bight, near the extreme south point, into which we stood, and with some difficulty succeeded in anchoring at a cable's length from the shore. The bay proved to be that called by the Spanish missionary voyagers 'Ygnacio Bay.' Over the south point, – a narrow tongue of land, about five hundred yards across, with rocks and breakers stretching off shore, to the distance of two miles, – we took bearings and angles to various fixed points in the northern part of the gulf. The latitude, chronometric differences of longitude, and magnetic variation, were determined on shore at this southern point.

Our observations being completed, we left this anchorage; and as it is little likely to be visited again, it will be enough to say that it is exceedingly dangerous. Nothing would have induced me to enter it, but the duty of examining the coast for anchorage, and the danger of remaining under sail close to an unexplored shore.

Under an impression that the island of St. Xavier* was the / scene of the Wager's wreck, I wished to examine its western side; but a strong N.N.E. wind did not permit my doing so, without risking the loss of more time than could be spared for an object of mere curiosity. I steered, therefore, to the southeastward for an inlet, which proved to be the Channel's Mouth of the Spanish charts, and reached it, after running seventeen miles from the south end of Xavier Island. We got no soundings with ninety fathoms of line, when at its entrance; but making no doubt that we should get anchorage within, we left, at the distance of half a mile, the islets of the northern point; passed between two others distant apart only one-fifth of a mile, and shortly after anchored in twenty fathoms, sheltered by an island to the westward, but with rocky islets around us in all directions, except the S.E., some of which were less than a cable's length from us.† Here we were detained until the 10th of June by the worst weather I ever experienced:

* Xavier's Island is certainly the Montrose Island of Byron's Narrative. The Wager was lost, as will be seen, more to the southward, on the Guaianeco Islands. This group was afterwards called Hazard Isles.

† This group was afterwards called Hazard Isles.

we rode with three anchors down and the topmasts struck; and though we lay within a couple of hundred yards of the islands and rocks, and less than half a mile from the shores of the inlet, such a furious surf broke on them all, that it was but rarely a boat could land, even in the least exposed situations the inlet afforded. The evening of our arrival was fine, and we put up the observatory tent, on the island to the westward of us; but the weather was so bad, during the next day, that we could effect no landing to remove it, although we anticipated the result that followed, namely, its being washed away.

'In the short intervals of the horrible weather that prevailed, boats were sent to the northern shore of the inlet, for the purpose of procuring water and fuel; but though they sometimes succeeded, by dint of great perseverance, in landing through a raging surf, it was but seldom they could embark the small casks (barecas) which had been filled, or the wood they had cut.

'Upon this shore the master observed remains of some Indian wigwams, that seemed to have been long forsaken, and / he described them to be exactly like those we had hitherto met to the southward.

'This was the northernmost point at which we noticed traces of human beings.

Finding the boats' crews suffer much from their unavoidable exposure during continually wet weather, I ordered some canvas to be given to each man for a frock and trowsers, to be painted at the first opportunity, as a protection against rain and spray.

Nothing could be more dreary than the scene around us. The lofty, bleak, and barren heights that surround the inhospitable shores of this inlet, were covered, even low down their sides, with dense clouds, upon which the fierce squalls that assailed us beat, without causing any change: they seemed as immovable as the mountains where they rested.

Around us, and some of them distant no more than two-thirds of a cable's length, were rocky islets, lashed by a tremendous surf; and, as if to complete the dreariness and utter desolation of the scene, even birds seemed to shun its neighbourhood. The weather was that in which (as Thompson emphatically says) 'the soul of man dies in him.'[1]

In the course of our service since we left England, we have often been compelled to take up anchorages, exposed to great risk and danger. But the Beagle's present situation I deemed by far the most perilous to which she had been exposed: her three anchors were down in twenty-three fathoms of water, on a bad bottom of sand, with patches of rock. The squalls were terrifically violent, and astern of her, distant only half a cable's length, were rocks and low rocky islets, upon which a furious surf raged.

I might use Bulkeley's words in describing the weather in this neighbourhood, and nearly at this season: 'Showers of rain and hail, which beat with such violence against a man's face, that he can hardly withstand it.'[2]

On the 10th, the wind being moderate, and the weather better, preparations were made to quit this horrid place. We put to sea, with a moderate breeze from N. b. W., which / increased rapidly to a strong gale; and scarcely were we fairly freed from the channel, than we found ourselves in a heavy confused sea. Anxious to clear the entrance, I had not waited to hoist in the yawl, with which we had weighed one of our anchors, expecting to find smooth water as we went out; but the sea we met made it unsafe to tow her, and while hauling up to hoist her in, she was so badly stove by blows received from the violent motion of the ship, that we were obliged to cut her adrift. This was a heavy loss. She was a beautiful boat, twenty-eight feet in length, – pulled and sailed well, and was roomy, light, and buoyant; her loss was second only to that of the ship.

We endeavoured to clear the Guaianeco Islands, by carrying a heavy press of sail, but soon after midnight were obliged to furl the reefed mainsail. Before daylight the wind shifted suddenly to W.b.N., taking us aback by a violent squall, with much vivid lightning and heavy rain. Our admirable little vessel paid off without sustaining any damage; but for a minute her situation was critical. At daylight, the land of Cape Tres Montes bore W. 1/2 N. (magnetic), distant four leagues. The violence of the gale we had just had put it out of our power to clear the gulf; and, from the state to which we were reduced by the loss of our yawl, both gigs being in bad condition, and our cutter so much stove as to be useless, I considered that it would not be justifiable to attempt proceeding in a lone ship to an unknown and most stormy coast, without a single efficient boat; so I resolved to hasten to Port Otway, and put the boats into an effective state. We had baffling winds all day; but in the evening succeeded in reaching the harbour, and anchoring nearly in our old berth. On the 13th and 14th, we had a continued hard gale, with the usual accompaniment of heavy rain. The carpenters were, however, kept constantly at work to render the cutter effective. On the 15th, the state of the sick list caused me to require from the surgeon, his opinion as to the 'necessity of a temporary cessation of surveying operations.' Mr. Bynoe's reply stated 'that in consequence of great exposure to a long-continued succession of / incessant and heavy rain, accompanied by strong gales, the health of the ship's company had been seriously affected, particularly with pulmonic complaints, catarrhal, and rheumatic affections; and that, as a recurrence of them would probably prove fatal in many instances, a temporary cessation would be of the greatest advantage to the crew, by affording an opportunity of recruiting their health.'

On receiving the above communication from the surgeon, I ordered the yards and topmasts to be struck, and the ship covered over with sails. Precaution was used to prevent the people from being subjected to frequent exposure, by not employing any of them in boats, except once a day in procuring muscles, and every thing was avoided that could in the least interfere with the recovery of their health: but this place is exceedingly ill adapted for the winter quarters of a ship's company, as the woods that surround it, down even to the water's edge, allow no space for exercise on shore, and there is neither game nor fish to be procured, except shell-fish; of which, fortunately (muscles and clams), we found an abundance, and they proved useful in removing symptoms of scurvy, besides affording a change of diet. The place being destitute of inhabitants, is without that source of recreation, which intercourse with any people, however uncivilized, would afford a ship's company after a laborious and disagreeable cruise in these dreary solitudes. Every port along this coast is alike ill suited for a winter's residence, and it was only our peculiar situation that induced me to determine on making a short stay at this place.'

Here poor Captain Stokes's remarks and notes end. Those who have been exposed to one of such trials as his, upon an unknown lee shore, during the worst description of weather, will understand and appreciate some of those feelings which wrought too powerfully upon his excitable mind.

The Beagle remained quiet until the 29th of June, when the surgeon reported 'the crew sufficiently healthy to perform their duties without any material injury to their constitutions.' / Leaving Port Otway, she steered along the coast with, strange to say, easterly winds and fine weather, which enabled Lieut. Skyring to add much to the survey of the coast of Madre de Dios. Captain Stokes now began to show symptoms of a malady, that had evidently been brought on by the dreadful state of anxiety he had gone through during the survey of the Gulf of Penas. He shut himself up in his cabin, becoming quite listless, and inattentive to what was going on; and after entering the Strait of Magalhaens, on his return to Port Famine, he delayed at several places without any apparent reason; conduct quite opposite to what his would naturally have been, had he then been of sound mind. At last, want of provisions obliged him to hasten to Port Famine; and the day on which he arrived every article of food was expended.

The fatal event, which had cast an additional gloom over every one, decided our quitting the Strait. Both ships were immediately prepared, and we sailed on the 16th August; but previously, I appointed Lieutenant Skyring to act as commander of the Beagle; Mr. Flinn to be master of the Adventure; and Mr. Millar, second master of the Adventure, to act as master of the Beagle. The day we sailed, Mr. Flinn was taken ill; and, Lieutenant Wickham being on the sick list, I was the only commissioned officer able to keep the deck. As the wind was from the N.W., we were obliged to beat to windward all night, and the next morning were off Sandy Point; but it blew so very strong from the westward, and the weather was so thick from snow-squalls, which passed in rapid succession, that we bore up, and anchored in Freshwater-Bay, where the ships were detained by northerly winds until the 21st, when we proceeded; the wind, however, again opposing, we anchored about half a mile from the shore, in a bight, seven miles southward of Sandy Point. The following day we were underweigh early, and reached Gregory Bay. When off Elizabeth Island, I despatched the Beagle to Pecket's Harbour to recall the Adelaide, in which Lieutenant Graves had been sent to procure guanaco meat. The Beagle worked through, between Elizabeth Island and Cape Negro, and was seen by / us at anchor off Pecket's Harbour before we entered the Second Narrow.

Upon our anchoring under Cape Gregory, two or three Patagonians were seen on the beach, and before half an hour had elapsed others joined them. By sunset several toldos, or tents, were erected, and a large party had arrived. When the Adelaide first went to Pecket's Harbour, Mr. Tarn told the Indians that the Adventure would be at Gregory Bay in twenty-five days, and, accidentally, we arrived punctually to the time. The Patagonians must have been on their way to meet us, for they could not have travelled from Pecket's Harbour in the short space of time that we were in sight. To their great mortification, however, we held no communication with them that evening, and the next day the weather was so bad we could not even lower a boat. At noon the wind blew harder than I had ever witnessed; but since we were on good holding-ground, and the water was smooth, no danger was anticipated.

As the snow-squalls cleared off, we looked towards the Patagonians, with the full expectation of seeing their huts blown down:– to our astonishment, they had withstood the storm, although placed in a very exposed situation. We counted twelve or fourteen of them, and judging by our former experience of the number belonging to each, there must have been, at least, one hundred and fifty persons collected. During the gale they kept close; and it was only now and then that a solitary individual was observed to go from one toldo to another.

The weather having moderated, the Beagle and Adelaide joined us on the following day. They rode the gale out, without accident, off the entrance of Pecket's Harbour. The next morning being fine, we prepared to proceed; but previous to weighing I landed, and communicated with our old acquaintances. Maria was with them, and, if possible, dirtier, and more avaricious than ever. We collected the guanaco meat they had brought for us; distributed a few parting presents, and then returned on board.

The Adelaide brought sixteen hundred pounds of meat, which, with what was first obtained, amounted to four thousand / pounds weight; and cost altogether ten pounds of tobacco, forty biscuits, and six pocket-knives. At first a biscuit was considered equivalent to forty or fifty pounds of meat; but as the demand increased, the price rose four or five hundred percent. With the Patagonians were two of Mr. Low's crew, who had left him. They were Portuguese, in a miserable state, and appeared to be thoroughly ashamed of being the companions of such a dirty set: they could not speak English, and could give us very little information. They had not then assumed the Indian garb, although, from the state of their clothes, they would very soon be obliged to adopt it.

At Pecket's Harbour a few words of the native language were collected, which are very different from those given by Falkner, in his description of the Patagonian natives: he says himself, that the language of the northern Indians differs materially from that of the 'Yacana Cunnees.'[3]

During Lieutenant Graves's communication with the natives, at Pecket's Harbour, he obtained some interesting information respecting these Indians, which will be given in a subsequent part of the work.

The Adelaide brought me a few very gratifying additions to my zoological collection, among which was the Zorillo, or Skunk, of the Pampas; differing in no way whatever from the species found about the River Plata, in such numbers as to impregnate the air with their disagreeable odour for many miles around.

I have frequently found the scent of this offensive little animal distinctly perceptible when I was on board the Adventure, lying at anchor about two miles from Monte Video, with the wind blowing from the land.*

The yagouare moves very slowly, and cannot run. It produces two young ones, which are placed at the bottom of its burrow. The unconquered Indians of the Pampas make mantles with the furs of the fox, cavia, or other animals, and border them with the skins of the yagouaré, which are very soft and fine, and would be fit for being employed by the furrier were it not for the disagreeable odour which they impart to every thing they touch. The Indians eat the flesh of this animal, which they irritate until its only means of defence is unavailing, and it can be captured without offensive consequences. /

A very large condor was shot by one of the Adelaide's party, which measured, in length, four feet three inches and a half, and nine feet two inches between the extremities of the wings. It was presented to the British Museum. Many exaggerated accounts of this bird have been given by old voyagers; but the largest dimensions stated, of whose accuracy there exists no doubt, are those of one that was preserved in the Leverian Museum,[4] which measured thirteen feet one inch, from wing to wing. This, however, must have been an old bird; for the one

* D'Azara, in his Essai sur I'Historie Naturelle des Quadrupèdes de Paraguay, gives the following account of this animal, which he calls Yagouaré. It burrows in the ground, eats insects, eggs, and birds, when it can surprise them, and moves about the plains and fields both by day and night in search of food; brushing the ground with its body, and carrying its tail horizontally. It regards not the presence of man or beast; unless an attempt be made to injure or take it, when it gathers up its body, bristles up the hairs of its tail, erecting it vertically; and in this position awaits the approach of its enemy, at whom it ejects its urine, which produces so unbearable a smell, that neither man, dog, nor tiger, will attempt to touch the animal.

we killed is larger than the usual size of specimens which have been obtained. Molina states, in his account of this bird, vol. i. p. 298, that the largest he ever saw measured fourteen feet and some inches (Spanish measure), from the tip of one wing to that of the other. M. Humboldt also gives a detailed description.

'It is with the condor,' says this celebrated voyager, 'as with the Patagonian, and many other objects of natural history; the more they are examined, the more they diminish in size.'[5] They inhabit the highest mountains of the Andes, and only descend to the plains when pressed by hunger. Frequently, in troops, they attack cattle, deer, guanacoes, and even the puma, and always succeed in killing them; but their principal food is carrion, of which, in a country so abundantly stocked with quadrupeds, there is probably no want.

Our departure from the Strait was attended with beautiful weather; the moon was full, and the wind fair and moderate. / Cape Virgins was passed soon after sunset, and we proceeded on our course with rapidity.

The timely supply of guanaco meat had certainly checked the scurvy, for we had no new cases added to the number of the sick, now amounting to twenty. The Beagle was not so sickly; but, during the last cruise, upwards of forty cases, principally pulmonic, had occurred, and several were not yet recovered. On the passage, a man fell overboard from the Beagle, at night, and was drowned.

In latitude 45° S. we were delayed three days, by northerly winds and damp foggy weather, after which a fresh S.W. gale carried us into the River Plata. Having obtained good chronometer sight in the afternoon, we steered on through the night, intending to pass to the westward of the Archimedes Shoal; which would have been rather a rash step, had we not been well assured of the correctness of our chronometrical reckoning. At this time Brazil and Buenos Ayres were at war,[6] and some of the blockading squadron of the former were generally to be met with in the mouth of the river; but we saw none, until half-past two in the morning, when several vessels were observed at anchor to leeward, and we were soon close to a squadron of brigs and schooners, whose number was evident by a confusion of lights, rockets, and musketry, on board every vessel. I bore down to pass within hail of the nearest, which proved to be the Commodore's, the Maranao of eighteen guns;[7] and on approaching, explained who and what we were; but they were so confused, I could not even make myself understood. The breeze, at the time, had fallen so light, that, fearing to get foul of the brig, the ship was hove up in the wind, and the anchor ordered to be let go. Unluckily a stopper was foul, and before another bower could drop, the Brazilians had fired several muskets into us, happily without doing any mischief; and threatened us, if we did not immediately anchor, with a broadside, which, in their utter confusion, I am astonished they did not fire. Having anchored, and I am astonished they did not fire. Having anchored, and lowered the topsails, I sent a boat to inform the Brazilian who we were, and to request, that in consequence of the

A. Earle. T. A. Prior.

MONTE VIDEO.--CUSTOM HOUSE.

number of / / our sick (we had only ten serviceable men on deck), we might not be detained, as even a few hours might prove of serious consequence; but all I could urge was unavailing, and we were detained until daylight with trifling excuses. We were so situated, that unless the brig veered her cable, or dropped out of our way, we could not move without getting foul of her, else I should have proceeded without permission. After daylight, the brig gave us room, by tripping her anchor; and upon an officer coming on board to release us, I told him my opinion of the affair, and said I should report the captain's conduct to his admiral. This report was afterwards made, in a very spirited manner, by Captain Henry Dundas, of H.M.S. Sapphire; but the admiral defended the conduct of his officer by saying that he had merely acted, 'magna componere parvis,' as an English blockading squadron would have done in a similar case.[8]

Whether the act was borne out, or not, by the law or custom of blockade, it was very uncivil; and one for which, after the explanation given, and the proofs offered, there could not be the slightest occasion. Owning to this detention, we did not reach the anchorage at Monte Video until too late in the day to procure refreshments for the sick. We found, to our sorrow, that fresh provisions were so extremely scarce, owing to the war, that none could be procured for our ships' companies; and had it not been for the kindness of Senor Juanico, a wellknown and, highly esteemed resident at Monte Video,[9] who supplied us plentifully with bitter (Seville) oranges, we might have been much distressed. The free use, however, of this fruit alone caused a rapid change in the health of those affected by scurvy, and in less than a week every man was at his duty.

A few days after our arrival, through the intervention of the British minister, a peace was concluded between the belligerents, in which Buenos Ayres gained all it had contended for, and Brazil gave up what she had so imperiously demanded.[10]

I was extremely gratified by meeting, at this port, the late Captain Henry Foster, in H.M.S. Chanticleer, on his pendulum voyage.[11] He was established at an observatory on a small island, called Rat, or Rabbit Island, whither I lost no / time in proceeding, and found him deeply engaged in that series of observations which has reflected so much honour upon his memory.

Before he sailed, I made an arrangement to meet the Chanticleer, either at Staten Land or Cape Horn, for the purpose of supplying her with provisions, to enable him to proceed thence to the Cape of Good Hope, without returning to Monte Video.

On the 13th of October, we sailed for Rio de Janeiro to procure some stores, which had been sent from England for our use, and to be caulked and refitted. The Beagle remained at Monte Video, to prepare for our next cruise. Before we were ready to leave Rio de Janeiro, the Commander-in-chief, Sir Robert Otway, arrived from Bahia, in his flag-ship, the Ganges. Sir Robert acquainted me, that he considered it necessary for the Beagle to be hove down and repaired; – that he intended to supersede Lieutenant Skyring; and had sent the requisite orders to Monte Video. When the Beagle arrived, Lieutenant Robert Fitz Roy, flag lieutenant of the Ganges, was appointed as commander; Mr. J. Kempe, mate, as lieutenant; and Mr. M. Murray, second master of the Ganges, as master.

Although this arrangement was undoubtedly the prerogative of the Commander-in-chief, and I had no reason to complain of the selection he had made to fill the vacancies, yet it seemed hard that Lieutenant Skyring, who had in every way so well earned his promotion, should be deprived of an appointment to which he very naturally considered himself entitled.

The conduct of Lieutenant Skyring, throughout the whole of his service in the Beagle, – especially during the survey of the Gulf of Penas, and the melancholy illness of his captain, – deserved the highest praise and consideration; but he was obliged to return to his former station as assistant surveyor: and, to his honour be it said, with an equanimity and good-will, which showed his thorough zeal for the service.

Captain FitzRoy was considered qualified to command the Beagle: and although I could not but feel much for the bitterness of Lieutenant Skyring's disappointment, I had no other cause for dissatisfaction. / /

A. Earle. S. Bull.

CORCOVADO MOUNTAIN, RIO DE JANEIRO.

CHAPTER XII.

Adventure sails from Rio de Janeiro to the River Plata – Gorriti – Madonado – Extraordinary Pampero – Beagle's losses – Ganges arrives – Another pampero – Go up the river for water – Gale, and consequent detention – Sail from Monte Video – Part from our consorts – Port Desire – Tower Rock – Skeletons – Sea Bear Bay – Fire – Guanacoes – Port Desire Inlet – Indian graves – Vessels separate – Captain Foster Chanticleer – Cape Horn – Kater's Peak – Sail from St. Martin Cove – Tribute to Captain Foster – Valparaiso – Santiago – Pinto – Heights – Chilöe – Aldunate.

THE Adventure sailed from Rio de Janeiro on the 27th of December 1828, leaving the Beagle to complete her repairs, and follow to the River Plata. The day before our arrival at Maldonado, we were overtaken by the Commander-in-chief, in H.M.S. Ganges, and entered the river in company. The Ganges proceeded to Monte Video; but we went into Maldonado Bay, where I had determined to wait for the Beagle.

Since our last visit to this place, the Island of Gorriti had been occupied by Brazilian troops, who, before going away, set fire to the buildings, and destroyed all the wood-work. As one object of my stay was to obtain observations for the latitude and longitude, I erected our portable observatory, and set up an azimuth altitude instrument.

On the 30th of January, after some intensely hot and sultry weather, we experienced a very severe 'Pampero.' It was preceded by the barometer falling to 29.50, and by a strong N.W. wind, which suddenly veered round to S.W., when the pampero burst upon us. Our ship and boats fortunately escaped any bad effects from the violence of the squall, which was so strong as to lay the former, at anchor, upon her broadside; but on shore our tent was blown down, and a boat that had been lately built, and fresh painted, on the Island Gorriti, was completely destroyed. The part above the thwarts, was torn away from the bottom of the boat, and carried, by the violence of the wind, for two hundred yards along the beach. A boat, also, / on the opposite shore, was blown to atoms. When the squall commenced, one of our boats was coming off from the island; the officer being quite unconscious of the approaching hurricane, and as she was overloaded with people, I felt very uneasy until after the squall cleared away, when I observed her beached on the opposite shore, many yards above high water mark, to which position she had been driven by the force of the wind. The violence of this pampero, during the twenty minutes it lasted, was terrific. Old inhabitants of Maldonado declared, that they had experienced nothing like it for the last twenty years. The spray was carried up by whirlwinds, threatening complete destruction to every thing that opposed them. In less than half an hour it had diminished to a strong S.W. gale, which lasted during the night.

Just before the pampero commenced, L'Arethuse, French frigate, was observed over the point of land under all sail; but not being seen after the squall cleared off, we were much alarmed for her safety. At daylight, however, the next morning, she was seen at anchor under Lobos Island, and near her was our consort, the Beagle, of whose approach we had known nothing; bnt she appeared to be lying quietly, with topmasts struck, under the lee of the island. L'Arethuse slipped her cable in the afternoon, and ran out to sea.

On the 1st of February the wind moderated, and enabled the Beagle to join us, when we found that she had been nearly capsized by the pampero; and had suffered a considerable loss of sails and masts, besides injury to her boats. Both topmasts, and jib-boom, with all the small spars, were carried away; and her jib and topsails, although furled, were blown to pieces. The vessel was on her beam ends for some time; but letting go both anchors brought her head to wind and righted her, which prevented the necessity of cutting away the lower masts. To add to their misfortune, two men were blown overboard, from aloft, and drowned.

These severe losses caused considerable detention; but, fortunately, the Ganges arrived, and rendered every assistance in repairing and replacing the Beagle's damages. /

On the night of the 2d of February we experienced another very severe pampero, during which one of the Beagle's boats, hauled up on shore, was blown to atoms. The barometer had previously fallen to 29.39.

On the 9th of February, we went to Monte Video, and on the 17th ran up the north side of the river for water; but did not find it fresh until we were within four miles of Cape 'Jesus Maria.' The wind was against our return, so that we had to beat down the river, in doing which the Adelaide grounded, but without receiving any injury. We anchored twice in our passage out, and, at the second anchorage,* experienced a very heavy westerly gale. In attempting to weigh at its commencement, our windlass was so much injured, that we were obliged to ride the gale out, which we did by veering to one hundred and ten fathoms of chain cable; and the Beagle, to one hundred and fifty fathoms. Owing to a short heavy

* From which the Mount (at Monte Video) bore N. 11°. W., distant eight leagues.

sea, in which the Adventure frequently pitched her bowsprit and stern alternately under water, her jolly-boat was washed away. This loss we could ill afford, as we were already three boats short of our establishment, and wants; and as the Adelaide had suffered severely, by losing her topmast and jib-boom, and carrying away the head of her bowsprit, we were obliged to return, very reluctantly, after the gale had subsided, to Monte Video; whence we finally sailed on the 1st of March. On the 5th a S.S.E. gale separated us from our consorts, our course, therefore, was directed for the first rendezvous, at Port Desire.

When off Cape Blanco, the high land of Espinosa, in the interior, was clearly distinguished at a distance of sixty miles, and might probably be seen twenty miles further; so that its height must be, at least, four thousand feet. This range is of irregular form, and has several peaked summits, so very different from the general features of this coast, where the heights are either flat-topped, or of an undulating outline, that I suppose the rock to be of a character unlike that of the porphyry hills common hereabouts. /

On anchoring off Port Desire (14th), we found that the Beagle had arrived, but had not met the Adelaide. The following afternoon I landed to examine the Tower Rock, a very conspicuous object, on the south side of this harbour, having the appearance of an enormous dead tree with its branches lopped off. On our way to it we passed over an undulating plain, composed of a sandy light soil, lying on a rocky basis, which in many parts protruded. The soil was so poor, as only to produce a few tufts of grass, and here and there a straggling bush of Berberis, or Piccoli, a dwarf woody shrub, which is much esteemed as firewood by the sealers who frequent the coast. Sir John Narborough, in describing this place, says, 'The soil is gravelly and sandy, with tufts of dry seared grass growing on it;' again: 'from the tops of the hills I could see a great way into the land, which is all hills and downs, like Cornwall, toilsome travelling to those who were not used to it.'[1]

The Tower Rock is evidently the remains of what was once probably a considerable rocky mass, which has either been partially destroyed by some convulsion, or, more probably, has been gradually worn away by the effect of weather. Like all the débris around, it is of a fine-grained red porphyritic claystone, much decomposed, but very hard, and difficult to break.*

It stands erect at the summit of a mound or heap of broken stones, of all sizes, some being very large blocks, from ten to twenty, or thirty tons weight. It is about forty feet high, and twelve in diameter, having its upper portion cleft, as it were, for about one-third down the middle, which gives it a resemblance to the forked branch of an immense tree. It is covered with moss and lichen, and, from its peculiar shape and prominent situation, presents a very remarkable object.

* Specimens of this rock are deposited in the Geological Society's Museum, Nos. 3 and 3 – 1.

Near it we observed traces of an Indian visit, among which was a horse's skull. From the sterility of the soil and absence of fresh water, it is probable that it is but little frequented by them. Port Desire is celebrated as being the place where / Schouten, the Dutch navigator, is said to have found skeletons measuring eleven or twelve feet in length![2]

Captain Fitz Roy informed me that he had not seen the Adelaide since we separated. The Beagle had lost another boat in the gale; the eleventh we have lost in the expedition since leaving England. As the Adelaide did not make her appearance, I determined upon proceeding in the Adventure to Sea Bear Bay, a few miles to the southward of Port Desire, to await her arrival with the Beagle. While standing into the bay, we were amused by a chase of a novel description: a guanaco was observed following a fox, which had much difficulty in keeping his pursuer at a distance. As the guanaco is not carnivorous, it may have been in playfulness: Reynard, however, by his speed, and anxiety to escape, did not seem to think it an amusement. How the chase terminated we did not see, for they disappeared in a valley.

While the ship was being moored, I landed to examine some wells near the outer point, which have been said to afford some tuns of good water. I found them to be deep holes in the solid rock, within the wash of a heavy surf, and large enough to contain two hundred gallons of water; but in one only was the water fresh, the sea having broken into the others, and, of course, spoiled their contents. They receive the rain from the ravines, and are much depended upon by sealing vessels which frequent this coast.

Sea Bear Bay was discovered in the voyage of the Nodales, in the year 1618; they describe the place, but give it, as it deserves, a very poor character. 'The port,' they say, 'for a short stay, is not bad, since it affords a good depth of water and a clear bottom; but otherwise it possesses nothing to make it worth a ship visiting it, for there is neither wood nor water, which are what ships most require.' Nodales called the bay 'Sea Lion,' from the multitude of sea-lions (*Phoca jubata*) found on Penguin Island. Why it has been changed to Sea Bear Bay I cannot determine.

In one of Mr. Tarn's excursions into the country, he observed a sail in the offing, which he thought was a whale-boat; and / supposing it might be in distress, if not one of the Adelaide's, kindled a fire to attract attention. As the grass was very dry, it blazed furiously, and spread rapidly around, yet without exciting fear that it could do us any injury; but the next morning flames being observed on the crest of the hills, behind the valley in which our tent had been erected, a boat was sent to save it, and remove the instruments. Our men had just left the ship, when, fanned by a land breeze which rose with the sun, the flames flew on with rapidity, descended the valley, and before the boat reached the shore, had consumed every vestige of the tent, and several articles of minor consequence. The sextant and artificial horizon, lying on the ground, escaped destruction, and the dipping-needle had fortunately been taken on board. Before the fire burned itself out, the whole country for fifteen or twenty miles around was completely over-run, so that all hope of procuring guanacoes was destroyed. Previous to the fire, Mr. Tarn had shot one; but being young, the carcase only weighed one hundred pounds, and was scarcely worth the trouble of sending fifteen miles for; however, as an amusement to the people, I sent a party to bring it on board, and it proved sufficient to furnish the ship's company with a fresh meal.

We had seen several herds within four miles of the ship before the conflagration; but the country was so very level and open, that these shy animals were always warned of the approach of our people by their vigilant scouts. So watchful and attentive is the look-out at his post, that he never drops his head even to feed, and it is only with the greatest cunning and care a man can get near the herd. The best way is, to lie concealed near the water holes, and await their coming to drink. A small stream of fresh water trickled over the beach into the bay, fringed by a patch of grass which the fire had spared, at which having once observed a guanaco drinking, we set a watch; but whether the animals were aware of it or not, none came until the morning we sailed, when a small herd walked down to the place quite unconcernedly, having no doubt first ascertained that there was no danger. /

The little vessel Mr. Tarn saw was an American sealer, which anchored in the bay next morning.

Besides the guanacoes, and fox, above-mentioned, we saw no quadrupeds, although two or three sorts of cavia and the puma are common in this neighbourhood. Of birds, nothing interesting was seen, except a plover (*Totanus fuscus?*), oyster-catcher (*Hæmatopus niger, rostro rubro, pedibus albis*), and one of the night bitterns, very much resembling the young of the European bird;* but these three species had previously been found at Port Famine. Several lizards were taken, and preserved.

This extremely sterile and barren country is very unfavourable for animals of any kind. The soil is like that already described about Port Desire. The rock is of the same character as at Port St. Elena and Port Desire: red porphyritie claystone.†

On the 23d of March, a week having passed since we came to Port Desire, my anxiety for the Adelaide's safety was much increased; especially as both wind and weather had been favourable for her approach to this rendezvous. I therefore despatched Lieut. Wickham overland to Port Desire to order the Beagle to join us, and proceed with us to the other points of rendezvous, Port San Julian and

* See Zoological Journal, vol. iv. p. 92.

† Nos. 1 and 2 in the Geological Society's Museum. A new species of Solen (*Solen Scalprum*, nob., Zool. Journ. V. 335. No. 5.) was found on the beach; and the camerated nidns of *Buccinum muriciforme*, nob., Zool. Journal, 1.c. No. 62.

Cape Fairweather. Lieut Wickham reached Port Desire after a fatiguing walk, and early next morning the Beagle was beating into Sea Bear Bay against a very strong wind which increased, and detained us. I seized this opportunity of completing our consort's provisions to five months. Captain Fitz Roy informed me that he had taken advantage of his stay at Port Desire, to ascend the inlet to the head. It extended for thirty miles, and the water was salt to its very extremity; but, from the height of the old banks on each side, it appeared likely that at times there may be considerable freshes. At the head of the river he lighted a fire, / which spread, and soon joined that which Mr. Tarn had made. Their union probably burned many square leagues of country.

On the 27th, we were still detained by a southerly gale. Captain Fitz Roy accompanied me in search of Indian graves, which are described to be on the summits of the hills. We found the remains of two, one of which had been recently disturbed, but the other had been opened a considerable time.

No vestiges of bones were left. It is said that the corpse is extended in an east and west direction, on the top of the highest pinnacle of the hill, and then covered over with large stones until secure from beasts of prey. Decomposition takes place, or the flesh is consumed by small animals or insects, without the bones being removed, so that complete skeletons are formed. According to Falkner,[3] the bones are collected at a certain period, and removed to some general cemetery, where the skeletons are set up, and tricked out with all the finery the Indians can collect. The avidity they evince for beads and other ornamental trifles is, perhaps, caused by this desire of adorning the remains of their ancestors.

The next morning we left Sea Bear Bay and proceeded to San Julian, off which we anchored for a few hours, while Captain Fitz Roy entered the port to look for the Adelaide, or for some vestige of Lieutenant Graves's visit. Finding nothing in the port, nor any tracks upon the shore, we went on towards Cape Fairweather, and in our way met the Adelaide. After parting from us during the gale in which all her sails were split, she went to Port Desire, where she arrived first, and, not seeing us, proceeded to the two other places of rendezvous, and had been lying at anchor eight days off Cape Fairweather. Finding we were not there, she was returning to Port San Julian, when we met her.

The weather being calm, so good an opportunity of supplying the Adelaide with provisions was not lost, and she was completed to six months.

On the 1st of April we were off Cape Virgins, and parted from the Beagle and Adelaide; Captain Fitz Roy having previously received orders from me to proceed through the Strait of Magalhaens, and despatch the Adelaide to survey the Magdalen / and Barbara Channels, while he was to survey part of the south shore of the Strait and the Jerome Channel, and then proceed, in company with the Adelaide, to Chilóe.

The Adventure then proceeded along the coast of Tierra del Fuego towards Staten Land, for the purpose of communicating with the Chanticleer, or obtaining some intelligence of her. The appointed rendezvous was New Year's Harbour, and the day on which I had promised to be there was past.

It was so foggy that no part of the coast of Tierra del Fuego could be seen; but as any detention might cause Captain Foster inconvenience, I did not wait for fair weather, but went at once to the place appointed.

When crossing Strait le Maire, we were very nearly drifted through by the tide, which, however, changed just in time to admit of our keeping on the north side of Staten Land.

With a strong squally breeze we entered New Year's Harbour, and seeing nothing of the Chanticleer, should have sailed without further investigation, had we not observed a cleared white space on one of the islands, which being near the place where I had requested Captain Foster to leave a document, I concluded was intended to attract our attention. The anchor was therefore dropped in twenty-five fathoms (the island bearing from N. to N.W. 1/4 W.), nearly in the spot where Captain Cook anchored, and a boat was sent to the white mark, near which a flag-staff was observed, at whose foot was a tin canister, containing a letter from Captain Foster, which informed me of his having been obliged, in consequence of a longer detention here than he had anticipated, to alter his arrangements, and requesting me to meet him at St. Martin's Cove, near Cape Horn, about this day. We therefore lost no time in getting under weigh, but in doing so, broke an anchor. We passed round Cape St. John, and with a fair wind made rapid progress to the westward. At noon, the next day, being seventy-five miles from Cape Horn, bearing W. by S., the high mountains on the S.E. end of Tierra del Fuego came in sight, among which the 'Sugar Loaf' (*g*) was a conspicuous object. / By an angular measurement of its altitude, and the distance given by the chart, its height must be nearly five thousand feet, and the average height of its neighbouring mountains full three thousand.

A south-west gale now set in, and delayed our reaching Cape Horn until the 16th, when we anchored off the entrance of St. Martin's Cove and found the Chanticleer moored within. A boat soon after came with the welcome information of all being well on board her. We were not able to warp into the Cove until next day, and in doing so found much difficulty, owing to the violence of the squalls, which repeatedly obliged us to slacken the hawsers quickly, else we should have carried them away.

The Adventure was moored in seventeen fathoms, about a cable's length within the low green point on the south side: and the Chanticleer lay in ten fathoms near the head of the Cove. The summit of Cape Horn being in a line

(*g*) Campana, or Bell Mountain. – R. F.

with the south point of entrance, we were quite land-locked, and perfectly sheltered from all winds, excepting the williwaws, or furious gusts from off the high land, which sometimes suddenly struck the ship, and threw her on her broadside; but being as momentary in duration as they were sudden in approach, we found them more disagreeable than dangerous.

During our stay here I made a partial survey of the Bay of St. Francis, which has since been completed by Captain Fitz Roy. St. Joachim's Cove, to the southward of St. Martin's Cove, is more exposed than the latter, but is of easier depth. These coves are separated from each other by a steep and precipitous mass of hills of greenstone, which in many parts appear to be stratified, the dip being to the westward, at an angle of 40°. I landed at the point, and ascended the hill, which I found more difficult to do than I supposed, the whole surface being covered with stunted beech bushes, so thickly matted or interwoven together, that I was obliged to walk or crawl over their tops. Among them were occasionally seen the berberis ilicifolia and veronica, the latter of very small size. Another day, Lieutenant Kendall,[4] of the Chanticleer, accompanied me / to Weddel's Port Maxwell, which is evidently St. Bernard's Cove of D'Arquistade. (*h*) Port Maxwell is contained between Jerdan Island, Saddle Island, and a third island, forming a triangle. It has four entrances, the principal one being to the north of Jerdan Island, and affords tolerable anchorage in the centre, in nineteen and twenty fathoms, sand;* nearer the shores of the island the depth is more moderate, but the bottom is very rocky.

The summit of Saddle Island, which I ascended for bearings, is composed of large blocks of greenstone rock, on one of which the compass (Kater's Azimuth,[5] without a stand) was placed; but the needle was found to be so much influenced by the ferruginous nature of the rock, composed of quartz and feldspar, thickly studded with large crystals of hornblende, that the poles of the needle became exactly reversed. An experiment was then made, by taking bearings of a very distant object, at several stations around, about fifty yards from the magnetic rock, when the extreme difference of the results amounted to 127°. The block upon which the compass stood, in the first instance, is now conspicuously placed in the museum of the Geological Society.†

Saddle Island, like the others near it, is clothed with low stunted brushwood of beech, berberis, and arbutus, and the ground is covered with a species of chamitis, and other mountain plants. While Mr. Kendall and I were absent from the boat, the crew caught several kelp fish, which are very delicate and whole-

(*h*) I do not think the bay adjacent to Cape Horn is that which was named by D'Arquistade 'St. Francis,' and, if my supposition is correct, Port Maxwell is not the place which was called 'St. Bernard's Cove.' See Second volume. – R. F.

* According to Capt. Fitz Roy the best berth is in sixteen fathoms. (Sail. Directions.)

† Nos. 268 to 271, Geo. Soc. Museum.

some food. On the following day, while going with Mr. Kendall to Wollaston Island, we passed a great many whales, leaping and tumbling in the water. A blow from one / of them would have destroyed our boat, and I was glad to cross the Sound without getting within their reach. We returned by the west side of Jerdan Island, where there are bights which might afford shelter to a small vessel.

The Sound that separates Wollaston Island from the Bay of St. Francis, I named after Sir John Franklin, and the harbour to the east of the point on which we landed, after Lieutenant Kendall, who was one of Sir John Franklin's companions in his last journey to the north-west coast of America.

On the west point of Kendall Harbour, I observed a magnetic property in the rock, which is of the same character as that on Saddle Island. Weddel noticed the same at St. Martin's Cove; but I placed the compass in various parts of that cove, without observing any difference from the correct bearing. This was, perhaps, owing to the rock being much covered with soil; for, being of the same character with that of the places above-mentioned, it should cause a similar effect.

The next day S.W. gales and thick weather set in, and confined us almost to the ship. Taking advantage of a short interval of more moderate weather, I ascended the highest peak on the south side of the cove, immediately over the anchorage, taking two barometers, one of the Englefield construction, and the other a syphon barometer, on M. Gay Lussac's plan, made by Bunten, of Paris.[6] Mr. Harrison accompanied me, taking charge of one barometer, whilst I carried the other. My coxswain carried a theodolite. On landing, the barometers were set up at the edge of the water and read off, and at the same moment the barometer on board was read off. We then ascended, but the rise was so precipitously steep as to offer very great impediments; and had it not been for a water-course, in whose bed we climbed for the first part, the ascent, with delicate instruments, would have been almost impracticable. We had ascended but little way, when the unfortunate theodolite escaped from my coxswain, rolled down the ravine, and was much damaged. It was an excellent magnetic transit, and for that purpose was irremediably injured; but, as a theodolite, it was yet useful.[7] The first third of the ascent, from the comparative / facility offered by the water-course, was only impeded by loose stones, which frequently yielded to the foot, and rolled down the gully, to the great danger of those who followed. The banks of the ravine were saturated with water, and covered either with spongy moss, or matted with plants,* which afforded no assistance; had it not therefore been for straggling shrubs of arbutus, or veronica, and tufts of rushes, growing on the steeper parts, we should have had many a fall; and however unimportant we might think bruises and scratches, a broken barometer would have been a serious

* A species of Gunnera (*Dysemore integrifolia*, Banks and Solander), and the green-stemmed Cineraria (*Cin. leucanthema*. Banks and Solander).

accident, and much care was required to avoid it. We had to leave the bed of the torrent, when it became full of wood, and then our difficulty increased much; for in many places we had to scramble over the thickly-matted and interwoven branches of the stunted bushes of beech which frequently yielded to our weight, and entangled our legs so much, that it was no easy matter to extricate ourselves.

At the height of one thousand feet, vegetation became much more stunted; we found the plants and shrubs of very diminutive size, consisting principally of the deciduous-leaved beech, one plant of which, though not more than two inches high, occupied a space of four or five feet in diameter, its spreading branches insinuating themselves among wild cranberry, chamitis, donacia, arbutus, and escalonia, so closely matted together, as to form quite an elastic carpet. For the last two hundred feet, we walked over the bare rock, in which no other vegetation was observed than lichens. The summit of the peak is formed by a loose pile of green-stone rock, in which the hornblende appears in very varied forms, sometimes in large crystals, and again so small and disseminated, as to be scarcely visible; on the summit it is seen, in very long, narrow (? filiform) crystals, and the feldspar predominating, gives it a white appearance.*

The only living creatures we saw were a solitary hawk and / one insect, a species of Oniscus. Nothing, in fact, could be more desolate, and we had only the satisfaction of a good observation for the height, and an excellent bird's-eye view of the surrounding islands and channel, to repay us for the labour of the ascent. On reaching the top, the barometers were suspended under the lee of the rock, twelve feet below its summit, and I then proceeded to set up the theodolite, which I found more damaged than I had anticipated; but not so much as to deprive me of a very extensive round of angles, in which were contained bearings of the Ildefonso Islands. We were thus occupied about an hour and half, which afforded me an opportunity of obtaining two good readings of the barometer.

The view to the N.W. was very extensive, and bounded by long ranges of snow-clad mountains of great height; the atmosphere was remarkably clear, and every object unusually distinct. Bearings of the islands of Diego Ramirez would have been taken, but for the extreme force of the wind, which more than once blew me from the theodolite, and once actually threw me on the ground. The temperature was not below 38°; but, owing to the wind, the cold was intense, and the rapid evaporation produced the most painful sensations, particularly in our feet and legs, which were thoroughly wet when we reached the top.

Our descent was not effected in less than an hour and twenty minutes, owing to the difficulty of passing through the beech thickets; but we reached the base without injury to the barometers, which was being more fortunate than I

* Nos. 283 to 286, in Geol. Soc. Museum.

expected. They were again set up on the beach, and read; after which we returned on board, amply gratified and rewarded for our fatigue.

The height of the peak, which, from its vicinity to the station selected by Captain Foster for the pendulum experiments, could not receive a more appropriate name than Kater's Peak,[8] was found to be 1,742 feet above the high-water mark.*

The following is the result:

By Bunten's Syphon.		By Jones 509.
Ascent	1743.4	1749.3
Descent	1738	1739.1
Mean	1741	1744.2

Mean of the two instruments 1742.4 feet. /

The next day, after a beautifully clear and mild morning, with a fresh northerly breeze, the weather became cloudy, and the wind veered to the S.W. blowing excessively hard, with hail and rain. The gusts, or williwaws, rushed through the valley of the cove with inconceivable violence, heaving the ship over on her broadside every minute, so that we were obliged to have every thing lashed as if at sea. Fortunately, we had completed wood and water, and now only waited for observations, to rate the chronometers, for our run to Valparaiso, whither it was my intention to proceed. Days, however, passed without a glimpse of the stars, and the sun only appeared for a few minutes above the hills. Captain Foster had completed his observations, and embarked all his instruments, excepting the transit, which remained for taking the passages of stars; but the bad weather continued, with little intermission. On the 3d, the gale was most violent, and the williwaws became short hurricanes, in some of which the ship drifted and fouled her anchors. On the 10th, we had a dry and fair day, which permitted us to sight the anchors and moor again.

The fine weather was of only a few hours duration, when the gale again sprung up, and lasted, with little intermission, until the day of our departure (the 24th). From the 4th to the 22d the sky was so perpetually clouded, that the only transits obtained in that interval were, one of Antares, one of Regulus, and one of the limb of the moon, though Captain Foster even slept close to the telescope, in the greatest anxiety to obtain observations. On the night of the 22d four stars were / observed, by which the error of the clock was satisfactorily ascertained.

Captain Foster's pluviameter,[10] a cubic foot in size, placed on a stand two feet above the ground, at an elevation of forty-five feet above the sea, contained eight

* The changes of pressure, during the intervals of ascent and descent, were obtained by registering the ship's barometer, which was done by signal from the stations on shore, when the readings were taken. During the ascent the column fell 0.039 inches, and during the descent rose 0.041 inches. Corrections were made for the dew point, as observed by Daniell's hygrometer at the base and summit, and the calculations were made according to the formula in Daniell's Meteorological Essay.[9]

inches and a quarter of rain, after standing thirty days; therefore, with the quantity evaporated, at least twelve inches must have fallen. The day after the above was registered, the vessel only contained seven inches and a quarter; so that in twenty-four hours one inch had evaporated, by which an idea may be formed of the sort of weather we experienced, and of the humidity of the climate.

With respect to the geological features, I can only add, that all the islands on which I landed, and, I believe, all the others, are composed of green-stone of various characters. The lower portion, or base, being less decomposed, is a fine-grained green-coloured rock, in which the component parts are so blended as not to be distinguished from each other. It appears sometimes in strata, dipping at various angles, from 20° to 45° from the vertical; and is very similar to the rock which alternates with granite in the Straits of Magalhaens, at the entrance of the Barbara; and also to that about Pond Harbour, and Bell Bay. At a greater elevation the feldspar predominates, the hornblende is observed in distinct crystals,* and the rock contains a considerable quantity of iron, which is observed in the reddish tinge of its surface. I have before noticed the magnetic property of this rock, which was more or less according to the quantity of hornblende: the beach-stones are different sorts of green-stone.

The lower parts of the hills, around St. Martin's Cove, are thickly wooded with the smooth-leaved, evergreen beech, which I have before described. Its leaves were as fresh and vivid, when we sailed, as if it were the height of summer; but those of the deciduous-leaved beech had assumed their autumnal tint, and were falling fast. Neither species attained a greater size, in diameter, than six or eight inches. The Winter's-bark was / found in sheltered places, but not larger in dimensions than the beech.† Where no trees are produced, the ground is covered with tufts of chamitis and donacia, which, being of a bright-green colour, give the sides of the hills a lively and verdant appearance. Had the state of the weather permitted our boats to leave the neighbourhood of the cove, or had the woods afforded any addition to collections for natural history, our detention would have been more agreeable; but, with the exception of a few corvorants, divers, and 'steamers,' with now and then a solitary hawk, or a Patagonian 'warbler,' we saw no traces of animal life. No Indians came near us, having been frightened away by the Chanticleer; for when Captain Foster was absent at night, after attempting to land at Cape Horn, several rockets were fired off as signals, and

* This rock is very similar to the boulders and pebbles which we found on the beach at Point St. Mary (Freshwater Bay).

† The underwood is composed chiefly of *Arbutus rigida* – *Berberis parvifolia* and *ilicifolia* – (*sempervirens* of Banks and Solander). *Veronica* (*decussata?*) and, in moist places, *Cineraria leucanthema*, and *Dysemore integrifolia*; both of which are found in all the sheltered corners of Tierra del Fuego. No *Fuchsia* was seen, but Mr. Anderson gathered the sweet-scented *Callicene marginata*, and a species of *Escalonia*, on the hill sides.

a few Indians who were then in the cove were so much alarmed, that they went away next day, and never afterwards showed themselves, although I dare say we were very narrowly watched by them.

Having supplied the Chanticleer with the provisions she required, we prepared to leave St. Martin's Cove. On the 24th the Chanticleer sailed, and in two hours after we also left this dismal cove, in which we experienced a succession of very bad weather, an almost constant S.W. wind, and for the last month a scarcely ceasing fall of either rain, hail, or snow. The Chanticleer bore away round Cape Horn, and was soon out of sight.

This was my last meeting with Captain Foster, who, the night before we sailed, communicated to me a presentiment, which he could not shake off, that he should not survive the voyage. I cannot now resist indulging in the melancholy satisfaction of saying a few words to the memory of my late excellent / friend, and lamenting, with many others, the severe loss which science suffered in his death. He was a fellow of the Royal, and Astronomical Societies, and to the former had contributed, to use the words of His Royal Highness the Duke of Sussex, as President of the Royal Society, a most valuable and extensive series of observations upon the diurnal variation, diurnal intensity, and dip of the magnetic needle; and upon other subjects connected with the terrestrial magnetism and astronomical refraction, which formed an entire fourth part of the Philosophical Transactions for the year 1826. For these papers he received the Copley medal; and the Lords of the Admiralty acknowledged their sense of the honour which was thus conferred upon the profession to which he belonged, by immediately raising him to the rank of Commander, and by appointing him to the command of the Chanticleer, upon a voyage of discovery and observation in the South Seas. The address of the President of the Royal Astronomical Society, at the anniversary meeting,** also bears ample testimony to his active and useful services in the expedition, under Captain Parry, towards the North Pole; as well as to his ardent zeal, very great attention, and accuracy, in every thing which he undertook for the promotion of science; and concludes the notice of his death in the following words: 'In the premature death of this young and accomplished officer, the Society has to deplore the loss of a zealous and active votary to science; and his memory will be long held dear by those who were more intimately acquainted with him in the relations of private life.' Captain Foster was unfortunately drowned, near the close of his voyage, while descending the River Chagres in a canoe.[11]

No sooner had we cleared the land, than we found a strong westerly wind, and a heavy sea; so that if we had entertained any expectation of making a quiet passage to the westward, we should have been disappointed.

* Ann. Meeting, 30th Nov. 1832.

The land of Hermite Island, and its vicinity, has a most remarkable appearance when seen from the south. Its outline is a series of peaks, following each other in regular succession, / and resembling the worn teeth of an old saw. Mount Hyde is made sufficiently distinct by its rounded apex, and by being higher than any land near it. Kater's Peak also is remarkable in this view, from its conical form and very pointed summit, and from being situated at the eastern end of the island. The 'Horn' itself needs no description; it cannot easily be mistaken.*

Westerly winds carried us as far as 60° south latitude before we could make any westing, and then we had a slant from the eastward, followed by variable winds. Our run to Valparaiso was much like all other voyages in this climate; we had the usual quantity of foul and fair winds, with a share of tempestuous weather, and arrived at Valparaiso Bay on the 22d of June. While remaining here our chronometers were cleaned, and some of them repaired; and the ship was refitted and provisioned, with a full supply for the Beagle and Adelaide as well as herself.

At the latter end of July, Lieutenant Wickham accompanied me to Santiago, the capital of Chile, ninety miles from the port, for the purpose of waiting upon General Pinto, the Director;[12] and communicating to him the purpose of our voyage, to prevent exciting suspicion, or receiving any interruption on the part of the authorities of places we might visit, particularly Chilóe, where our stay might be viewed with distrust or apprehension; for rumour had already said that the English were about to take that island. Ridiculous as such a report was, I deemed it sufficiently important to induce me to explain to the Chilian Government our views and orders, which could be done better by personal explanation than by a correspondence.

We commenced our journey early on the 11th of July, travelling in a covered chaise, drawn by three horses, one in / the shafts, and the others outside, attached to the carriage by a single trace of hide; and preceded by a drove of horses, from which, at the end of every stage of twelve or fifteen miles, we selected a relay. The day was so very stormy, that we saw but little of the country. Immediately after leaving the Almendral, or suburbs of Valparaiso, we ascended twelve hundred feet, and then descended about four hundred feet to an extensive plain, reaching to the Cuesta de Zapato, the summit of which, at least the highest part of the road over it, we found by barometrical measurement to be 1,977 feet above the sea. In the interval we passed through the village of Casa Blanca, lying eight hundred and three feet above the sea. After passing the Cuesta de Zapato, between

* The Survey of this part now presents the navigator with the means of ascertaining his position, to a nicety, by angles taken with a sextant between Cape Horn summit and Jerdan's Peak, or Mount Hyde, and Kater's Peak; and if Jerdan's Peak and Mount Hyde be brought in a line, and an angle taken between them and Cape Horn summit, the operation will be still more simple.

it and the Cuesta de Prado, is another extensive valley, through which runs the River Poangui. At Curacavi, where we crossed the river, the height above the sea is six hundred and thirty-three feet;** and the road proceeds by a gentle ascent to the foot of the Cuesta de Prado, near which is the village of Bustamente, eight hundred and eight feet above the sea.

This 'cuesta' is passed by a very steep road, and is ascended by twenty-seven traverses, which carry one to a height of 2,100 feet above the plain, or 2,950 feet above the sea. When we reached the summit of this mountain the weather was so cloudy, that the Andes were almost concealed from view. Beneath us was the extensive plain of Maypo, with the city of Santiago in the distance, a view of considerable extent, and possessing very great interest; but from the state of the weather, its beauty would not have been seen to advantage, had not portions of the towering Andes, raised by optical deception to apparently twice their height, appeared at intervals among the clouds. On / a fine day, when the range of mountains is uncovered, the view is grand; but not so imposing as when their lower portions are concealed, and their summits partially exposed. This part of the Andes rises about 11,000 feet above the plain, and is covered half way down the sides with snow, the lower edge of which is regularly defined, and presents a change of colour so abrupt and horizontal as to appear unnatural, and therefore diminish the grandeur of the scene very much. But under whatever circumstances this view is seen from the Cuesta de Prado, it is magnificent, and produces an effect beyond description. The road descends down the eastern side of this Cuesta, to a plain about 1,100 feet below the summit. So much rain had fallen during the two preceding days, and last night, that our driver expressed some doubt whether we should be able to cross the Podaguel, a river which is frequently impassable from the strength of its current. The idea of spending a night at the miserable hovel we were leaving was enough to induce us to run a considerable risk, and we set off to make the attempt. The water was very deep, and the current sufficiently strong to render it a performance of some danger; but, this difficulty being passed, we soon reached the city of Santiago, and in the house of Mr. Caldcleugh,[14] enjoyed the hospitality and society of a warm-hearted friend.

I waited on the Director (Pinto), who received me with the greatest politeness. He entered into the particulars of our past voyage with much interest, assuring me that every facility should be afforded, and every assistance rendered, whenever it might be required; and in this assurance we never found ourselves deceived, for on all occasions the conduct of the executive authorities towards

* Miers, in his account of Chile,[13] gives a table of barometrical measurements of the heights of the land between Valparaiso and Mendoza, from which it appears that he has deduced the height of Curacavi to be 1,560 feet. As my determinations are the results of observations made on my way to and from Santiago, I have no doubt of their correctness, and think that the registered height of Miers's table should be 29.355 instead of 28.355.

us was marked in attention, and even kindness. I make this observation with the more pleasure, as it was very unusual in our communications with the authorities of those governments we had previously visited, to find the objects of our voyage considered in the least interesting.

Although the weather, during our visit to Santiago, was not there considered fine, we left the city and its neighbourhood with a strong impression of the salubrity of the climate, and the / mildness of its temperature, which even in the middle of winter, and at the height of nearly 2,000 feet above the sea, ranged no lower than 45° Fahrenheit, and during the day the maximum height of the thermometer never exceeded 62°.* (*k*)

We returned to Valparaiso on the 26th of July, and made preparations to sail; but were detained by a strong northerly gale for many days, in which we were enabled to render assistance to a large Indian trader that would otherwise have been wrecked. On the 10th of August, we sailed for Chilóe; and on our way were greatly delayed by southerly winds, which carried us in sight of the island of Juan Fernandez. We reached our destination on the 26th, and found the Beagle, to our great delight, arrived, and all well. Captain Fitz Roy came on board before we anchored, and gave me an outline of his proceedings, and those of the Adelaide, which had not / returned, but was daily expected, having been despatched to survey some interior channels on her way to Chiloe. Our anchorage was off Point Arenas, which is not only the best in the bay, but appeared to

* The following are the results of the barometrical determination of the height of various points on the road between Valparaiso and Santiago:–

	Feet above the sea.
Casa Blanca, ten leagues from Valparaiso	803
Highest point of the road over the Cuesta de Zapata	1,977
Inn at Curacavi	633
Plain near Bustamente	808
Summit of Cuesta de Prado (not certain to 200 feet)	2,949
Inn, or post-house, at the base of the east side of the Cuesta de Prado	1,804
Santiago, by mean of numerous observations	1,821

Miers makes the above places above the sea as follows:–

				Feet above the sea.	
Casa Blanca				745	
Summit of Cuesta de Zapata				1,850	
Curacavi				1,560	
Summit of Cuesta de Prado				2,543	
Post-house, Prado				1,773	
Santiago, mean of two observations				1,691	
Do.	by Malespina	2,463	Spanish	2,254	English.
Do.	Mercurio Chileno	1,693 ½		1,550	

(*k*). Sharp frosts sometimes occur. – R.F.

be well adapted to our wants. The Beagle had arrived early in July, and had sent to Valparaiso for stores with which to refit, and make preparations for another cruize to the south.

The harbour master, Mr. Williams, an Englishman, visited us soon after our anchoring, and by him I forwarded to the Yntendente (or governor), Don Jose Santiago Aldunate,[15] the letters brought for him from Chile.

In the afternoon I received his acknowledgments, and offers of all the assistance in his power to render. As it was probable that our stay would occupy some weeks, I established myself at a house in the town, obtained by his kindness; and there fixed my portable observatory, and set up an azimuth altitude instrument. /

CHAPTER XIII.

Beagle and Adelaide anchor in Possession Bay – Beagle passes the First Narrow – Fogs – Pecket Harbour – Adelaide arrives with Guanaco meat – Portuguese Seamen – Peculiar light – Party missing – Return – Proceed towards Port Famine – Fuegians – Lieut. Skyring – Adelaide sails to survey Magdalen and Barbara Channels – Views – Lyell Sound – Kempe Harbour – Cascade Bay – San Pedro Sound – Port Gallant – Diet – Rain – Awnings – Boat cruise – Warning – Jerome Channel – Blanket bags – Otway Water – Frequent rain – Difficulty in lighting fires.

THE following is an account of the Beagle's and Adelaide's operations, after separating from the Adventure, on the 1st of April, at the entrance of the Strait of Magalhaens.

Light northerly winds were favourable for their entering the Strait, and they reached Possession Bay the first night. The following day was foggy, and almost calm, until the afternoon, when both vessels weighed, and proceeded with the tide. At sunset the Adelaide anchored on the north shore; but the Beagle stood on, and entered the Narrow. After dark, when within it, with a rapid tide running, the wind fell light, and an anchor was let go, under the north shore, in eight fathoms; but the cable being accidentally checked too soon, snapped like a small rope, and the vessel was hustled out into deep water. As it would have been both useless and imprudent to let go another anchor, the Beagle was kept underweigh, and worked to the westward, aided by a very powerful tide, which speedily carried her through the Narrow, without accident, although the night was dark; and they had no guide but the chart and lead. At eleven o'clock she was anchored within the Narrow, in twelve fathoms, soon after which the tide turned, and ran with great strength; but the night was calm, as well as the next morning.

While waiting for wind, and the change of tide, several Patagonian Indians were observed on horseback hunting guanacoes. / A very large dead cod-fish was also seen, floating past, which was taken on board; on its skin were several parasites.*

* Probably they are the same as we observed on the fish taken by us off Cape Fairweather, and which, I believe, to be nearly allied to the one that is figured in Cuvier's Regne

With the evening tide the Beagle reached Gregory Bay; and the next day (April 4th) worked through the Second Narrow, and anchored in Pecket Harbour.

As soon as she arrived people were sent on shore to make a large fire, to show the natives where the ship was, and attract them to her. Next morning, the 5th, it had spread very much, and overrun several acres of ground, which showed either a very dry soil, or that there had not been much rain for some time. The ground was covered with cranberries; so much so, that it had quite a red tinge; they were very good. Plenty of wild celery was found, but no wood of any kind. Water was obtained in small quantities, from a spring about eighty yards from the beach, abreast of the anchorage: it may also be procured by sinking wells. Early on the 6th of April the Adelaide anchored near the Beagle. Captain Fitz Roy went on board, and found that Lieutenant Graves had seen the Indians in Gregory Bay; and had anchored there for the purpose of obtaining guanaco meat, of which he got about nine hundred pounds weight. Thick fogs had prevented his getting through the First Narrow until the 4th. At Gregory Bay, Lieutenant Graves took three Portuguese seamen on board, who claimed his protection, having been left by an English sealing vessel nearly a year before.[1] One of them asked to be again put ashore, and was landed on Quoin Hill to carry a message to the Indians, from whom he promised to bring a supply of meat in two days. The other two were entered on the books as supernumeraries, and employed in the Adelaide. Having given the Beagle two-thirds of the meat, the Adelaide weighed; and in two hours was out of sight, on her way to Port Famine. /

The following are extracts from Captain Fitz Roy's journal of this cruise of the Beagle.

Monday 7th April. Several of our people were employed in gathering cranberries, and preserving them for future use; they are anti-scorbutic, as well as the wild celery, much of which has been used with our guanaco soup.

Wednesday, 8th. I went to Oazy Harbour with Lieutenant Skyring, who surveyed the harbour while I examined the cove to the northward.

Oazy Harbour appears large, but the part where there is anchorage is very small, and a strong tide sets in and round it, by which a bank is thrown up, a short distance inside the entrance; there is very little wood, and some difficulty in obtaining fresh water, even in a small quantity. The anchorage outside might be more convenient for procuring guanaco meat from the Indians than Gregory Bay, but it is exposed to winds between W.S.W. and S.S.E.

At my return to the Beagle, I was much surprised to find that Lieutenant Kempe, Mr. Bynoe, and a boy, had not yet come back from a shooting excursion. A boat had been to the appointed place at sun-set, and had waited an hour with-

animal, Plate XV. figure 5, a species of *Lernæa*, or *Entomoda* of Lamarck, iii. 233. The species is new.

out seeing them. At seven, a light was seen on the top of Quoin Hill, and I sent a boat to the spot, with cautions about landing, being in doubt whether it was shown by them or by the Indians; but the boatswain, who went with her, could find no person, nor any light. He waited some time, and returned on board. (*l*) A similar light was again seen, more than once, during the dark and gloomy weather, with small misty rain falling, and a light breeze from the westward, which we had all night.

Thursday, 9th. No signs of our officers, nor any appearance of the Indians. Fearing that some accident had happened, I sent two boats away, with arms and provisions, to look for them all round the harbour, and the large lagoon which communicates / with it. Both boats were thoroughly cautioned about the Indians, for I had thoughts of their treachery. Just as the boats got out of sight, three people were observed on the ridge of a hill, about six miles distant; and, at the same time two other persons appeared, much nearer the ship, on the east side of the harbour. Which was our party, and who the others were, it was perplexing to say. Both disappeared again for about two hours, when our stragglers came over a hill, very near the ship. Upon their arrival on board, they were scarcely able to move: they had been on their legs, almost without food, and without shelter from the rain, since they left the ship. Their intention had been to walk round the harbour, which appeared an employment for two hours only; but at its head they found a lake, and beyond that lake a much larger one, joined to the first by a passage, which they could not cross. When they arrived at this passage, it was too late to return by the way they went, and their best chance seemed to be going on. After dark, they tried to make a fire, but the rain prevented them. It was too dark to see their way, and the cold rain obliged them to keep moving about, though in one place. When daylight came, they travelled on, and until they reached the ship at two o'clock, were constantly walking.

The other people seen by us must have been Indians; none were met by our wanderers, but several places were passed where fires had been made by them.

April 10th. Directly our boats returned, we weighed and made sail; but the wind soon failed, and the tide setting against us, obliged me to anchor.

April 11th. Made sail towards the passage between Elizabeth Island and Cape Negro, and anchored there to wait for the tide, which ran past us when at anchor, at the rate of three knots an hour. About Cape Negro the appearance of the land entirely changes. A low barren country gives way to hills covered with wood, increasing in height, and becoming more rocky and mountainous as you go southward.

(*l*) This was a remarkable instance of what I often observed afterwards in those regions, a kind of 'ignis fatuus,' which sometimes was stationary, like the light of a lanthorn, and at others suddenly flitting, like the flashes of pistols, at a distance. It was only seen upon the lower hills.[2] – R.F.

On the 13th, when working near the land, against a light southerly breeze, we saw a small canoe paddling along shore, / and some people walking on the beach. While the ship was standing off, I went to them, being the first savages I had ever met. In the canoe were an old woman, her daughter, and a child, and on shore were two Fuegian men with several dogs. Their figures reminded me of drawings of the Esquimaux, being rather below the middle size, wrapped in rough skins, with their hair banging down on all sides, like old thatch, and their skins of a reddish brown colour, smeared over with oil, and very dirty. Their features were bad, but peculiar; and, if physiognomy can be trusted, indicated cunning, indolence, passive fortitude, deficient intellect, and want of energy. I observed that the forehead was very small and ill-shaped; the nose was long, narrow between the eyes, and wide at the point; and the upper lip, long and protruding. They had small, retreating chins; bad teeth; high cheek-bones; small Chinese eyes, at an oblique angle with the nose; coarse hair; wide ill-formed mouths, and a laugh as if the upper lip were immoveable. The head was very small, especially at the top and back; there were very few bumps for a craniologist.[3] They asked earnestly for 'tabac, tabac', but seemed very timid. We bartered some biscuit and old knives for a few of their arrows, skins, spears, &c.

Their canoes, twenty-two feet long, and about three wide, were curiously made of the branches of trees, covered with pieces of beech-tree bark, sewed together with intestines of seals. A fire was burning in the middle, upon some earth, and all their property, consisting of a few skins and bone-headed lances, was stowed at the ends.

The young woman would not have been ill-looking, had she been well scrubbed, and all the yellow clay with which she was bedaubed, washed away. I think they use the clayey mixture for warmth rather than for show, as it stops the pores of the skin, preventing evaporation and keeping out the cold air. Their only clothing was a skin, thrown loosely about them; and their hair was much like a horse's mane, that has never been combed.

April 14th. Anchored in Port Famine. /

April 16th. Lieutenant Skyring went on board the Adelaide with Mr. Kirke, five seamen, and one of the Beagle's whale-boats. Mr. Bynoe, the assistant-surgeon, also went as a volunteer.

April 17th. The Adelaide sailed to survey the Magdalen and Barbara Channels; after which she was to rejoin the Beagle at Port Gallant. She soon got into a strong southerly wind, and could make no progress, as the current was against her; she therefore again stood into the bay, and anchored.

A sharply cold night made us remember we were far south, although the weather by day had been mild. I have said little about this anchorage, as it has already been described. The appearance of the surrounding country is striking and picturesque. Mount Tarn, with its patches of snow, rising from thick woods, and the high snow-covered mountains in the distance, with dark blue sea at their base, are very remarkable objects.

We sailed on the 19th with the Adelaide, which had been prevented from going sooner by strong and unfavourable winds: and about noon we parted from our consort, whose course was southerly, into the Magdalen Channel, while we went towards Lyell Sound.

I cannot help here remarking, that the scenery this day appeared to me magnificent. Many ranges of mountains, besides Mount Sarmiento, were distinctly visible, and the continual change occurring in the views of the land, as clouds passed over the sun, with such a variety of tints of every colour, from that of the dazzling snow to the deep darkness of the still water, made me wish earnestly to be enabled to give an idea of it upon paper; but a necessary look-out for the vessel, not having a commissioned officer with me who had been in the Strait before, kept my attention too much occupied to allow me to make more than a few hasty outlines. Under the high land the Beagle had but little wind, and night closed upon us before we could gain an anchorage in Lyell Sound, so we shortened sail after dark, and kept near mid channel until the morning. /

The night was one of the most beautiful I have ever seen; nearly calm, the sky clear of clouds, excepting a few large white masses, which at times passed over the bright full moon: whose light striking upon the snow-covered summits of the mountains by which we were surrounded, contrasted strongly with their dark gloomy bases, and gave an effect to the scene which I shall never forget.

At daylight, on the 20th, we were close to Lyell Sound, and stood along its west side, looking for an anchorage, until we found a very good harbour, about a mile inside Mazaredo Point.

I then went away, with two boats, to examine the Sound, leaving the master to sound and plan the inner harbour.

Kempe Harbour, within Lyell Sound, would hold six large ships in security; but, like most of the harbours hereabouts, access is difficult, on account of the squalls off the high land, which are so irregular, and often violent.

During the night of the 21st, it blew strong in squalls, and the chain-cable kept us awake by rattling very much over rocks; yet when the anchor was hove to the bows next morning, it appeared to have been well bedded in stiff clay. To these sounds we afterwards became familiarised.

Wednesday, 22d. Strong squalls from the south-eastward during the night, and in the morning; when, being anxious to reach Cascade Bay, I weighed, though the weather was thick, and the wind against us. The flaws were so variable, that we were two hours knocking the helm and sails about before we could clear the anchorage, and move half a mile in still water. I should recommend warping in and out of these harbours, in preference to making sail: as it is far easier, if a ship is provided with small hawsers and kedges:[4] and the hawsers can often be made fast to the rocks, or roots of trees.

The tide rises about four feet in Kempe Harbour; and there is a place where a vessel might be grounded or careened with perfect safety.

Mazaredo Peak (Bougainville's Sugar Loaf) is an excellent guide to Kempe Harbour; the valley-like appearance of / the land also shows its situation to a vessel in the Straits. What at first appears to be Lyell Sound is Kempe Harbour, the Sound lies more to the left.

After passing Mazaredo Point, the land is rugged and less woody; it is not very high, and has a peculiar, rounded appearance, like the tops of loaves of bread.

There was slate in Kempe Harbour, which seemed to me fit for roofing purposes.

In Cascade Bay we found the abundance of limpets and muscles usual on these shores, and of particularly good quality. The Indians live almost entirely upon them and sea-eggs, though birds, and occasionally a seal, add to their subsistence. Vegetation, both on shore and in the water, is most abundant. At every step one sinks knee-deep in moss, grass, fern, or low bushes. Trees seem to arrive but seldom at perfection; the climate is so moist that they rot while growing, before they attain any size. Moss grows every where; each bough is covered with it: and the water appears to be as favourable to the growth of kelp as the land is to that of plants. The large kind (*Fucus giganteus*) shoots up, from many fathoms depth, to the surface, with strong stalks and large leaves.

23d. A bad day; blowing strong, and at times raining. Mr. Murray, Mr. Stokes, and I, went with three boats to continue our work of exploring and sounding.

Saturday, 25th. We weighed and made sail; but the breeze failed, and flaws came against us. While laying out warps, and hanging by the stream-cable, a squall took the ship and drove her against the rocks, but without doing her any injury, for they were quite wall-sided.[5] The main-yard and spankerboom were among the trees. We again laid out warps, and had made some progress, when another strong squall obliged us to go back into our anchorage, to remain until the hail, snow, wind, and rain should cease.

26th. An unpromising and wet morning; but the heavy rain being over, we weighed, and in a few hours reached the western side of San Pedro Sound.

About a mile from the point we anchored in Murray Cove, / which affords good shelter from westerly winds, and is very easy of access, being a small roadstead rather than a harbour.

27th. We set out early with the boats, but the weather was too bad to do much; however, something was done, and at dusk we went ashore on a small island in the Sound. It rained very hard all the afternoon and during part of the night. We sheltered ourselves as well as we could with the boat's sails and tarpaulins; but during the night the wind shifted, and blew so hard, that it threw down our slight shelter, and made me very anxious about the ship; for I was doubtful of the security of the anchorage where she lay.

28th. This morning was very cold, it rained hard and blew strong; but when it cleared away for a short time, we set to work again, to explore what appeared to be a channel.

After a three hours' pull against wind, snow, and hail, my channel proved to be only one of the numerous inlets which encroach upon the Fuegian territory; and the boats returned to the Beagle, with the help of strong squalls from the S.W. I was not a little glad to see the ship in the place where I had left her. During the night another anchor had been let go; but she had not moved from her position. This anchorage is so easy of access, that I hope it will be of use to vessels passing through the Strait. There is room for one large sized ship to lie conveniently, or for two or three small craft.

The weather has not yet been so cold as I expected it would be: snow lies on the deck a short time, but the thermometer has not been lower than 31° (Fahrenheit).

29th. A rainy, blowing morning: Mr. Stokes and I set out in the boats; but it rained so much, that we could only make a fire to dry our clothes, and remove the numbness, caused by sitting a long time in the wet.

On the 3d of May, we anchored in Port Gallant: though perfectly secure, this is a dismal harbour in winter, being so surrounded by high mountains, that the sun is seldom visible. Until the 7th, in addition to our usual daily duties, we were occupied in preparing for an excursion, in boats, to the Jerome Channel. Salt provisions were entirely withheld from the crew / for three days, and instead of them, preserved meat, shell-fish, and a large pig, brought from Monte Video, were substituted. We found in this, as in almost every Fuegian harbour, abundance of muscles, limpets, and wild celery; some fish and some wild-fowl. Many of our party thought shags good eating, but only one person could be found daring enough to try whether old Sir John Narborough was quite warranted in saying that a fox was 'savoury food,' and that one repented of his experiment during a week's serious illness.[6]

'My reason for entirely stopping the use of salt-meat, for a few days, was the belief that, at least, two or three days' change of diet is necessary to cause any real alteration in the system; and that it is better to give fresh provisions for three days in succession, and salt-meat during the remainder of three weeks, than to give fresh-meat at three separate intervals in the same period.

'During the wet weather of these regions, we derived great benefit from awnings, painted for the purpose, while refitting at Rio de Janeiro and Maldonado: they kept the lower, and a great part of the upper deck quite dry, even in heavy rain.

'May 7th. Mr. Stokes and I set out with a cutter and whale-boat, to explore the Jerome Channel. We were well provided, with as much as the boats could stow, of what we thought likely to be useful during a month's cruise. Of water we took but little, trusting to the wetness of these regions for a supply. Each man had his clothes covered with canvas, or duck, well painted; and instead of a hat, every one had a 'south-wester' (like a coal-heaver's cap).

'Our provisions, being sufficient for twenty-eight days, made the boats rather deep; and I soon found the cutter pulled very heavily, and was obliged to take her in tow. All our party slept in the cutter the first night, the whale-boat being made fast astern. Towards midnight it blew fresh, and as the boats were anchored near the wash of the beach, they rolled a good deal; and soon afterwards, feeling the whale-boat hanging heavily on her rope, I hauled her up alongside, and found she was almost swamped; in a few minutes she / must have sunk with all her heavy cargo, to us invaluable. The plug had worked out by her rolling:– I seldom left her afloat at night after this warning. Having saved the boat, made me think less of all our things being wetted, and of some of the instruments being almost spoiled.

'At daylight, on the 8th, we pulled along shore, with the wind against us, and reached Point York before the tide made strongly; but that place we could not pass; and sooner than give up an inch of ground, let go our grapnels, in the middle of a race of tide, that tumbled in over both gunwales, and ran past us at the rate of five knots. At one P.M. it slackened, and we pulled on into Bachelor River, very glad to get so good a place to dry our clothes, and put the boats to rights. Three deserted wigwams gave us shelter; and while some made fires, others went to collect shell-fish, or shoot birds. Though the season was so far advanced, some shrubs were in flower, particularly one, which is very like a jessamine, and has a sweet smell. Cranberries and berberis-berries were plentiful: I should have liked to pass some days at this place, it was so very pretty; the whole shore was like a shrubbery. I cannot account for the exaggerated accounts of the Fuegian coasts given by some voyagers: it is true that the peaks of the mountains are covered with snow, and those sides exposed to the prevailing west winds are barren, and rugged; but every sheltered spot is covered with vegetation, and large trees seem to grow almost upon the bare rock. I was strongly reminded of some of the Greek islands in winter, when they also have a share of snow on their mountains.

May 9th. The tide carried our boats rapidly up the Jerome Channel, which, though narrow, is quite free from danger. The west shore is very high, and steep, and well covered with wood; the eastern is lower, and less woody.

'Having passed this channel, we entered the mysterious Indian Sound, with all that anxiety one feels about a place, of which nothing is known, and much is imagined. I hoped to find a large river; and the strong tide setting up the channel convinced me that there was a body of water inland, but of what nature / remained to be discovered. At dusk we put into a small creek, and secured the boats, hauling up the whale-boat on the sand. When too late to remove, we found the place of our bivouac so wet and swampy, that nearly two hours were occupied in trying to light a fire. Supper and merry songs were succeeded by heavy rain, which continued throughout that night and the next day without intermission.

'10th. Continual hard rain prevented our moving: the whale-boat's men were thoroughly drenched in their tent during the night; but made a better one in the morning. The cutter, having a tarpaulin cover, gave her crew a better lodging; and although a small and loaded boat, only twenty-four feet long, could not be expected to allow much room to a dozen sleepers, during such weather, with the help of our blanket bags, *(m)* we did very well.

'11th. During this night, also, it rained very hard. Early the next morning, however, it cleared a little, and we got under-weigh. When in the fair-way our hopes were much excited; for beyond a high island, like a sugar-loaf, appeared an opening without land. I tasted the water repeatedly, fancying it less salt, and that we were approaching a river.

'Less salt it might have been, from the number of waterfalls dashing down the mountains on each side of the channel, which is here about two miles wide, with a current, or rather stream of tide, running at the rate of two knots an hour.

'At noon, we reached the Sugar Loaf: it cost a struggle to get to the top with the instruments; but the view repaid me. For three points of the compass towards the north-east, I could see no land, except two islands; and the farthest extreme to the eastward, appeared to me distant, at least, thirty miles. No mountains or high land could be seen to the north or east; the country seemed there to change its character, and become / lower and less wooded. This was, indeed, an animating view: I stood considering what might be the boundary of this water, till I recollected, that the longer I thought about it, the longer I should be finding it out; so we pushed on with the boats, of course taking the necessary bearings and angles, until we reached the 'Point of Islets' in 'Otway Water.'

'On the 12th, our oars were going early.

The two islands, 'Englefield' and 'Vivian,' were the only land upon the horizon for six points of the compass. The southern coast trended away nearly east from Cape Charles, preserving the high mountainous character of the Fuegian shores, while that to the northward was low, though as yet well wooded.

I was nearly tempted to try whether Fanny Bay led towards the Gulf of Xaultegua; but fortunately did not, as I should have regretted the time so employed.

Point Hamond is thickly-wooded with evergreens, similar to those of the Strait; and with a species of pine, about thirty or forty feet in height.

(m) Each officer and man, when detached from the Beagle during a night, carried a blanket, or large poncho (sewed up, and with a drawing-string, like a large bag), in which they slept, and found much comfort and warmth. – R. F.

To the S.E. three remarkable promontories stand out in bold relief from the Fuegian shore; but beyond them the land sinks into the tame flatness of Patagonia.

The water on the west shore is not deep; from ten to thirty fathoms at a quarter of a mile off shore, but getting more shallow advancing northward. There is anchorage for a vessel after passing Indian Channel, the whole way along; and as the prevailing winds are off shore, it would generally be safe. In Indian Channel I only know of two anchorages, Cutter Bay and Bending Cove.

'Such constant rain fell during this evening, that it was not until after much trouble that we at last made fires. Carrying dry fuel in the boats we found indispensable, and I would recommend any person who passes a night on shore in this wet climate, with a boat, to carry a sheet of copper, or a piece of flat iron, in preference to any boat-stove, as a fire can be lighted upon it much more easily, and it does not take much stowage: the great difficulty about fires here is getting fuel to burn when the ground is wet, or when snow lies on it. /

13th. Raining so steadily all day, that it was useless to proceed: I could neither see my way, nor notice any thing but wind and rain.

14th. So mild was the weather, that I bathed this morning, and did not find the water colder than I have felt it in autumn on the English coast; its temperature, at a foot below the surface, averaged 42°; that of the air was 39°. From this place, Point Hamond, I saw seven points of the compass clear of land, my eye being twenty feet above the level of the sea. The water was quite salt, therefore we were certain of being in an unexpected inland sea, or large lagoon. Four miles from Point Hamond lie Englefield and Vivian Islands, rather low, but well wooded with evergreens. They are the only islands of any note in the Otway Water. The farthest point I could discern I called Cape Marvel, for much I wondered at the hitherto unsuspected extent of this inlet.

At noon we were off the north end of Englefield Island. Mr. Stokes and I observed the sun's meridian altitude satisfactorily from the boats, so smooth was the water. This quiet day was too fine, for it was hard work pulling from nine till five, without any help from sails. Towards evening a breeze sprung up in our favour, and with its assistance we ran along the land about ten miles. Taking advantage of the moonlight, I did not look out for a resting-place till past seven o'clock, when we had a great deal of trouble in landing; the coast having quite changed its character; and instead of deep water with a rocky shore, we found a flat shingly beach and shoal water, with very large stones scattered between high and low water marks, so numerously as to make it dangerous for a boat, especially at night. Upon landing, we found the ground quite changed into a fine light soil, with stunted bushes and trees; and so dry was the wood, that a fire was easily kindled, but not a drop of water could be got any where to cook our supper. A considerable rise and fall of tide was observed, much greater than near Indian Channel.

15th. No breakfast this morning, for want of water – a decided proof of the change of climate and country. North of / us the sky was clear; but to the southward, over the Strait, hung thick clouds. The trees were not evergreen, and at this time their leaves were withered and falling.

While pulling along shore, and passing a low projecting point, we saw the smoke of three fires, and approaching nearer; observed four canoes lying on the beach, near several wigwams. Their owners soon appeared, running along the shore, hallooing and jumping. The first who came near us reminded me of an old-fashioned sign of the 'Red Lion,'[7] for he was painted red all over, and looked more like a wild beast than a human being; another was covered with a bluish mixture; a third was quite black. Several had the lower half of the face blacked, and the oldest men and women were painted entirely black. There were about eight men, six or eight boys, and perhaps a dozen women and girls. Some had a skin over their shoulders, but others had no covering at all, except paint; they seemed apprehensive, and hid several skins and other things in the wood, as soon as they saw us approaching.

When they found we were peaceably disposed, and had tobacco and knives, they were eager to barter with us. How they have learned the use of tobacco is curious, but they are fond of it to excess. Guanaco, as well as seal and otter skins, are in their possession; therefore they probably barter with the Patagonians. They have also the skins and horns of a deer, which, as I understood them, inhabits their country. *(n)* They catch small animals with snares, made of whalebone, just like hare-snares. This tribe was very rich in Fuegian wealth, such as skins, arrows, lances, &c. They appeared to be of a race similar, but superior, to the Fuegians, being stronger, stouter, more lively, and more active. I persuaded one of their boys to have his face washed, and found his natural complexion was scarcely darker than that of a European. Their language sounds like that of the Fuegians, and the huts and weapons are precisely similar to their's. We asked them for water, and they pointed to a place about a mile further, making / signs to us that we must dig in the earth for it. We went there, and near a green-looking spot some good water was found. We then landed, and enjoyed our breakfast at one o'clock, being not a little thirsty.

The natives were still with us; they seemed inquisitive and cunning; and shewed great surprise at a sextant and artificial horizon, by which they sat down, attentively watching what was done. I put my watch to their ears; they were much astonished, and each came in his turn to hear it tick. I pointed to the watch and then to the sky; they shook their heads and suddenly looked so grave, that from their manner in this instance, and from what I could understand by their signs, I felt certain they had an idea of a Superior Being, although they have

(n) Like a roebuck; supposed to be the 'Huemul' mentioned by Molina. – R.F.

nothing like an image, and did not appear to us to have any form of worship. We could learn scarcely any words of their language, because of their trick of repeating whatever we said.

They saw how we lighted a fire, by means of a tinder-box, and took an opportunity to tread it out of sight. Our loss was not known until leaving the spot, when that material necessary was missed. It was evident they had stolen it; and while I was meditating a reprisal, one of our men by chance trod upon the missing box, which was artfully hid under the sand. After this discovery, they seemed rather inclined for a skirmish, all having clubs, while our men appeared to have no weapons. However, we parted without a quarrel.

The features of these people differed from those of the Fuegians whom I had previously seen, in being better formed, and having a less artful expression.

We pulled hence along a low shore until evening, when distant land began to show itself, stretching to the northward and eastward, and bounding this supposed inland sea. At dusk we discovered an opening, which appeared to be either a river or a channel, and I steered for its north bank, securing the boats for the night in a place we named Donkin Cove,[8] as a mark of respect to the preserver of meat, to whom we had been so often thankful. A little of this meat, mixed with wild fowl, and some wild celery, makes a wholesome and / agreeable mess. On boat service, meat preserved in tin is particularly useful, being already cooked, and therefore fit for dinner without the aid of fire.

We were surprised at the mildness of the weather. Indeed, the change of climate was as pleasant as it was sudden and unexpected.

16th. At daylight, we found ourselves in the entrance of what was thought a river. Under this impression, I hoped to penetrate into the interior of the country, and meet some new tribes of Patagonians. As soon as we could get underweigh, we pulled and sailed along a winding channel, on one side of which was a pleasant-looking, woody country, extending towards Tierra del Fuego; and on the other, a low, barren district, like Eastern Patagonia. The banks on both shores were from five to forty feet high, sloping, and covered with grass.

The current was in our favour, which with the saltness of the water, inclined me to think it a channel, and not the mouth of a river. In this opinion I was confirmed in a short time, by seeing surf breaking against some land beyond an opening, which showed that we were approaching a large body of water. Soon after, we reached the extreme west point of this small channel; and, to our surprise, saw an expanse, at least thirty miles across from east to west, and twenty from north to south. I thought it more at first, but probably was deceived. West and south of it I observed high snow-covered mountains; and the summit of one was remarkable, being like a castle with a high tower. Northward, the land was low; excepting a few ranges of down-like hills with large plains between them.

It happened to be a very clear day, and all that could be seen at any time was visible. In two places there seemed to me to be openings to the westward; in the southernmost I could see no land at all; the other was backed by distant mountains, but still had the appearance of an opening. After this I went to the top of a hill near me, about three hundred feet high, to gain a better view, yet so small an elevation made but little difference, and I rather though the opposite coast farther off than I had at first supposed. /

'Having sent the cutter back a short distance, to make a fire and land our things, I crossed the channel to a fine level plain, and measured a base line. In crossing, I found a most rapid tide, at least five or six knots at neap tides, and to pull against it was out of the question. It caused a considerable swell and race at the entrance, which is not a quarter of a mile wide, though it averages twelve fathoms in depth. On the plain was growing thick grass, like that in the vicinity of the river Plata. So rich and good was the grass and trefoil, that I saved a few seeds, hoping some day to see their produce in England. No tree was seen; the soil seemed dry, rich, and light. Skunks, and a small kind of cavy, had burrowed every where, which proves the climate to be of a different nature from that of the Strait. The bones and traces of guanacoes were numerous, and some horses' tracks were found; as also part of a dead guanaco, which appeared to have been a prey to wild beasts. Water was not so plentiful as to the southward; but quite sufficient for all useful purposes, many small brooks being noticed, besides springs in the sides of the low hills. We shot a swan *(o)* and some coots; the swans were so fat, or so tame, that they would not rise from the water.

17th. While on Whitestone Plain, a very heavy squall of wind and hail passed over from the S.W., so cuttingly cold, that it showed me one reason why these plains, swept by every wind from S.S.W. to N., are destitute of trees.

After dark, we returned to the cutter and partook of a large mess, made of the swan we had shot, the coots, some limpets, and preserved meat. The shortness of the days was becoming very inconvenient; from eight to four were the only hours of daylight; but some of the nights were so fine, that I got many sets of observations of the moon and stars. /

(o) Black-necked swan, noticed elsewhere by Captain King.[9] – R.F.

CHAPTER XIV

Place for a Settlement – Frost – Boats in danger – Narrow escape – Sudden change – Beagle Hills – Fuegian painting – Tides – Medicine – Water warmer than the air – Jerome Channel – Mr. Stokes returns to the Beagle – Cape Quod – Snowy Sound – Whale Sound – Choiseul Bay – Return to the Beagle – Adelaide returns – Plan of operations – Difficulties removed – Preparations – Wear and tear of clothing – Ascend the Mountain de la Cruz – Sail from Port Gallant – Tides – Borja Bay – Cape Quod – Gulf of Xaultegua – Frost and snow – Meet Adelaide – Part – Enter Pacific – Arrive at Chilóe.

'18th of May. VERY cold, raining heavily, and blowing strong from S.W. The tide turned this day (full moon), and set to the westward at 1.15. I only say 'turned,' because I could not distinguish the ebb from the flood, so little rise and fall was there. No sooner had the tide ceased to run in one direction, than it began to run as strongly in the other, for about six hours. For the last four nights I noticed, that soon after sunset the sky was suddenly overcast, a trifling shower fell, and afterwards the heavens became beautifully clear. The climate must be much like that of the east coast of Patagonia, as shrubs grow here like those I saw at Port Desire. While walking, the leaves and dry sticks crackled under foot, which is very different from what one observes about the Strait of Magalhaens, where everything is wet and spungy. I was inclined to think this place suitable for a settlement. There is water, wood, and good soil, fit for planting, besides pasture land; the climate is not bad; and probably the Patagonian Indians might be induced to trade in guanaco meat, as they now do at Gregory Bay; while any of their hostile incursions would be prevented by the channel.

'19th. Two natives, a man and a boy, came to our boats this morning; they seemed to have neither curiosity, nor fear, nor even a relish for tobacco. They took a piece of tinder, picked up a stone, and went away to some wigwams, at a little distance, where we soon afterwards saw a fire burning. /

'During this night and the preceding it froze sharply; but the sky was so clear, that I observed many sets of distances, on each side the moon.

'20th. We went eastward through the little channel. Every thing was frozen; and the boat's sails were useless until thawed. We left Donkin Cove directly after noon, and with a fresh and fair wind, steered towards Pecket Harbour. I may as well mention here my reasons for taking this course, instead of going farther westward.

'Considering our very limited time, and provisions, I wished to do first what was most useful; and to find a new passage, seemed to me the primary object. Having surveyed the narrow winding channel, and proved its navigability for vessels of any class; I thought it desirable to ascertain next the nature of the separation between Otway Water and the Strait of Magalhaens, between Laredo Bay and Pecket Harbour.

'A western passage might be sought by the Adelaide schooner, or by myself, at a future time. If we tried to cross the Skyring Water, our success would be very doubtful, for during the whole time we had been in the channel, the wind blew strong from S.W., raising so much sea, that it was with great difficulty I could sound outside the western entrance, even in a whale-boat.

'A fine breeze carried us rapidly eastward; but it freshened too fast, reef after reef was taken in, until at two o'clock we were obliged to lower the sail, and pull to windward; for as far as we could see, the shore continued unbroken, flat, and low, with a high surf breaking on it. To have attempted to land, would have been folly; and as the wind continued to increase, and a current setting to windward caused a very short awkward sea, I sent Mr. Stokes off in the cutter, under his small close-reefed sails, to hang to windward as long as he could carry sail, while I kept the whale-boat head to wind. At three o'clock, we were embayed, and about a mile from the shore. My boat was deeply laden, and as our clothes and bags got soaked, pulled more heavily. We threw a bag of fuel overboard, but kept everything else to the last. At sunset the sea / was higher, and the wind as strong as ever. I saw the cutter a little before, about three miles from us, standing to the eastward on a wind; but whether she would clear the shore I could not make out.

'After dark, finding we could not well be worse off as to risk, I bore up, and pulled with the sea rather abaft the beam, twisting the boat 'end on' to each wave as it came, hoping to get into smoother water to the westward. Night, and having hung on our oars five hours, made me think of beaching the boat to save the men; for in a sea so short and breaking, it was not likely she would live much longer. At any time in the afternoon, momentary neglect, allowing a wave to take her improperly, would have swamped us; and after dark it was worse. Shortly after bearing up, a heavy sea broke over my back, and half filled the boat: we were baling away, expecting its successor, and had little thoughts of the boat living, when – quite suddenly – the sea fell, and soon after the wind became moderate. So extraordinary was the change, that the men, by one impulse, lay on their oars, and looked about to see what had happened. Probably we had passed the place where a tide was setting against the wind. I immediately put the boat's head towards the cove we left in the morning, and with thankful gladness the men pulled fast ahead. In ten minutes the sea was smooth, and the breeze so moderate, as not to impede our progress. Our only anxiety was then about the cutter; for we could not tell how she had weathered the gale. I was sure she would have prospered if kept by the wind; but some accident, or change of purpose, was to be feared.

'About an hour after midnight, we landed in safety at Donkin Cove; so tired, and numbed by the cold, for it was freezing sharply, that we could hardly get out of the boat. The embers of our morning fire were still burning; so we put on some wood, and lay down round them. No men could have behaved better than that boat's crew: not a word was uttered by one of them; nor did an oar flag at any time, although they acknowledged, after landing, that they never expected to see the shore again. We resolved to start early to look for / the cutter, and fell asleep: but before daylight I was roused by some one, and to my joy, saw Mr. Stokes standing by me. He had just arrived with the cutter, having kept his wind till the sea fell; and since that time had been pulling towards this spot: with what thankful feelings all hands lay down to sleep may be easily supposed.

'21st. This morning I believe no one waked before ten o'clock. Drying our clothes, and putting the boats to rights, occupied most of the day. Our time was now so short, besides having almost expended our provisions, that I gave up the idea of crossing the Otway Water, and decided to return nearly the way we came, after taking a view from the higher ground.

'22d. A sharp frost, during the past night and this day, hardened the ground, and with four of my boat's crew, I walked to the Beagle Hills. Our way led through a scattered wood, the only one seen on the north side of the channel, and in which most of the trees appeared to have been burned. We gained the summit of the heights soon after noon, and were amply rewarded by an extensive view.

'Although not more than eight hundred feet above the sea, I could discern the Gregory Hills (so plainly as to make out their yellowish brown colour); Cape Bartholomew, Nassau Island; Cape Monmouth; the high peaks over Cape Forward; the range of mountains thence to the Jerome Channel, and from the Jerome, westward to all those about Cape Phillip, and Cape Parker; and the whole extent of the Otway and Skyring Waters; the latter being bounded to the N.W. by down-like hills, about six or eight hundred feet high. North of the Beagle Hills, a range of similar downs extended; and to the east was a succession of lagoons, completely intersecting the flat country towards Pecket Harbour.

'We left a memorial, cut in lead, at the foot of a post sunk in the ground; but the air was so cold, that the men, who wished to add their names, were unable to mark them on the lead. It was eight o'clock before we regained our bivouac, much fatigued by the day's work.

'23d. I went into a wigwam, where there was a woman / and two children. A rough likeness made of her did not please at all, because it was white: she took out her red paint, and put some on her own cheeks, as drawn on the paper, and then was quite satisfied, sitting as still as a mouse, while I made another sketch.

In return for the compliment paid to her countenance, she daubed my face, as well as my coxswain's, with the same red mixture.

'24th. A sharp frost during the night. We left Donkin Cove, as soon as I had taken observations for the chronometers. A fine breeze in our favour carried us rapidly along, and at dusk we were near Englefield Island. The last few nights have been so clear, that two or three of the men, and myself, have slept in the open air without any other covering than our blanket-bags, and clothes. My cloak has been frozen hard over me every morning; yet I never slept more soundly, nor was in better health.

'We had a good view of Mount Misery this day. It is about 3,000-feet in height; twice as high as the surrounding mountains, and quite bare, even of snow, on the summit. The night tides here rise more than those of the day at this season: the times of high water do not differ much on the opposite shores. About an hour after dusk we reached Englefield Island, having made a capital run, with a fresh and fair wind. Creeping in the dark, along shore, we at last found shelter for the boats, and formed a snug place amongst the bushes for our tent and fires. One of my boat's crew was ill this day; the first man that had been seriously so, although several had been slightly affected by the muscles and limpets; and one had fits. A draught of hot port wine and Winter's bark, certainly seemed to be an efficient medicine for the slighter complaints.

'25th. Blowing strong from the westward, with much rain. I forced a way, with much difficulty, among thick bushes, to the top of the island, and when I got there found, to my mortification, that by no possible contrivance could I see round, for I was encompassed by lofty trees of nearly equal height.

'26th. We crossed over to the east shore; the temperature of / the water, between Englefield Island and the nearest land, one foot beneath the surface, was 42°; the air at the same time being 38°. While the sea water preserves this temperature, it must tend much to moderate the severity of cold, one would naturally expect in this latitude, near so many snow-covered mountains. We arrived at the Point of Islets, soon after sunset, on the 27th.

'28th. Almost every night I observed that the wind subsided soon after sunset, the clouds passed away, and the first part of the night was very fine; but that, towards morning, wind and clouds generally succeeded. From Point of Islets, we sailed southward; and were again close to the mountains: from whose appearance at this spot, no one would suppose that any passage lay between them; so intricate and winding are the channels.

'I was sorry to leave the open country, behind me; but time pressed; and there was yet much to do with our loaded boats, which could not make very great progress in the short daylight afforded by this season. After passing Bennett Island the land became rugged, and mountainous on each side, covered, however, with wood and vegetation wherever it could grow; and we were again in the Magalhaenic regions.

'This day I examined as much of the west side of the channel, as time would allow, and reached Corona Creek at about eight o'clock. What I called the Sugar Loaf must be the Corona Island of Cordova's officers; for at some distance it looks somewhat like a crown. It is singular that they inserted (in their chart) an island near their Corona, which cannot be distinguished from the main-land, until one is within two miles of it; and as at that distance the Otway Water is plainly visible, must they not have seen the opening? Tired of their job, did they return without prosecuting the discovery, or was the weather too thick to see far? Their description of the Jerome Channel, leads to the supposition of a continual current setting through in one direction, instead of a regular ebb and flood; and the surest sign of a passage between places in Tierra del Fuego, is a current or stream. / Many large inlets and sounds look like channels; but on going a short distance into them, you find dead water.

'29th. We passed through Jerome Channel, and reached the bar, off Bachelor River, after dark; but the cutter got aground, and gave us some trouble to float her again. Afterwards one of the men was landed on the bar, and by his walking in the deepest water, and the whale-boat going next, we got into the little river at nine o'clock, not sorry to be in safety. There are tide races between the Jerome Channel, and Bachelor River, which are sometimes dangerous; but as the breeze was moderate, we passed them without difficulty.

'May 30th. Employed chiefly in stowing the cutter afresh, packing specimens, and preparing my boat to take what remained of our provisions. At two next morning, when the tide served, Mr. Stokes set out to return to the Beagle: and having both wind and tide in his favour arrived early at Port Gallant.

'The wind increased after daylight, and blew strong, with squalls. I waited a short time, but, having no hopes of its improving, left the river. My boat was much lumbered, having the chronometer-box, and more instruments than before; yet she pulled pretty well, even against the heavy squalls. After landing at the west side of the entrance to the Jerome Channel, to take bearings and angles, we pulled along shore to the westward, and at dark hauled the boat up in a small sheltered corner. After she was secured, we employed ourselves looking for limpets and muscles for supper, by the light of a lanthorn, as we had good appetites, and our provisions were scanty.

'June 1st. We pulled along shore against a strong and squally wind, and before evening nearly reached Cape Quod; but not being able to pass it, stopped in a cove on the east side.

'2d. At the oars again, early, having a fine clear morning, with the tide rather in our favour. By eleven, Cape Quod was astern of us; and a long view of the Strait presented itself. This part is very rugged and barren, and looks triste, indeed; still wherever a tree can take root it tries to grow. This night was passed on a small island at the west point of Snowy Sound. /

'3d. We began at daylight, and worked, from point to point, up the sound; thinking it a channel. Two good anchorages were found on the west side, but none on the east, except a trifling cove between the little island and the land, which would only shelter a small vessel. The night was passed on an island five miles within the sound. It rained hard for an hour before we landed, and all the night afterwards. Our rest was not the most satisfactory, as the ground was wet and swampy.

'Two of the boat's crew got into a hole under a tree thinking they should be warm; but in the middle of the night they complained of not being able to get up, and of being half frozen.

'4th. The rain ceased at times this morning, but the wind continued. After going to the top of an island, we pulled and sailed onwards, not having a doubt of soon getting into Whale Sound. At noon, the passage appeared suspiciously small; yet I could not doubt the fine large opening laid down in our old charts, and proceeded until the shore made a sudden turn, when, to my astonishment, I saw a high black cliff stopping farther progress. After a hearty growl, we turned back, and landed to look for a sleeping place. Not a spot could we find that was not wet like a sponge; but night was closing in, and obliged us to stay where we were. It was bitterly cold, all of us were wet through, the ground was a mere swamp, we could not get a fire to burn, and the frost was sharp.

'After daylight on the 5th, we succeeded in making a large fire, and spent two hours drying our clothes and warming ourselves. In order to lighten the boat, no one carried more clothes, since leaving the cutter, than those he wore, except one shirt. We hastened back towards Charles Island, passing some very remarkable glaciers, one of which looked like an enormous frozen river, covering the whole side of a mountain. Many portions were of a transparent blue colour, which, contrasted with the snowy whiteness of others, and with the dark shadows of bare rocky places, had a very striking effect. At noon, we passed out of the sound, and steered for Charles Island, with a / light breeze in our favour. Seeing a canoe coming across, we made towards it, and found a wretched-looking family, consisting of a man, his wife, and three children, with some small dogs, seemingly more miserable than their owners. A few woodenheaded spears were all the property they possessed, excepting the worn-out skins thrown over their shoulders. The man sold me a little dog for a bit of tobacco, and afterwards wanted to have him again, because his wife would not consent to the bargain. However, I kept the dog, and they began to abuse us in right earnest, the woman alternately crying and scolding, and the man apparently calling on the wind and water to destroy us. His gestures were very expressive and animated. I was surprised to see so much feeling for a wretched little half-starved puppy, and made them happy by returning him, without asking for the tobacco.

'El Morrion (*p*) (the helmet) was certainly an excellent name for the promontory we passed this day. It reminded me of the 'Castle of Otranto.'[1]

'We reached a small islet, at the west point of Charles Bay, and passed a good night on the top of a bare rock. So often had we slept in wet places, that a dry, though stony berth, was thought very comfortable. The boat's two sails, oars, and boat-hook, formed our tent.

'6th. We left the islet as soon after day-light as we could get breakfast, and take the required bearings and angles; went into Spot Cove, thence crossed to Charles Island, and to the narrow opening between it and the nearest land. Ulloa's memory can no longer be preserved here in an island, though it may in a peninsula.[2] This small channel is narrow, and has a, strong tide setting through it. There is anchorage all the way, though generally over a rocky bottom, and it is navigable for small vessels: its average width is a quarter of a mile, and its length about three miles. For a boat going westward through the Strait, it is far preferable to the regular channel. Two old Fuegians were living here, a man and a woman.

'When in Whale Sound, appearances were such that had I / not been to the bottom of Snowy Sound, I should have thought they joined. After going far enough, to see quite to the end, we returned, hauled the boat on a shingle beach, and secured her for the time. When a bit of shingle beach could be found, it was a prize; for on it we could always make a good tent, and have a dry bed, besides hauling the boat up easily. There is a greater rise and fall of tide here, than at the other side of Charles Island, being not less than seven or eight feet, at springs. During the night, a dog stole a small piece of pork, which we had reserved for our last dinner; and, until his track was discovered, there was no little distrust among our party.

'Whale Sound is a large and deep inlet, ending in a valley between mountains. On the south side, a vessel may anchor in one place, at the west side of Last Harbour; but there, though the harbour appears large, the anchorage is small, and close to the shore. We pulled and sailed along the south shore, landing occasionally to take bearings, until we reached Choiseul Bay, and in a cove, at its west side, we passed the night. This is a place no ship need approach: it is a large, deceiving bay, full of islets and patches of kelp, under which, probably, there are rocks, and between the islets the water is deep and unfit for anchorage. The temperature of the sea this day, in the middle of the sound, one foot below the surface, was 45°.

(*p*) Noticed previously by Captain Stokes. – R. F.

'8th. As it rained heavily, we remained under such shelter as we could obtain; and prepared for our return to the Beagle, by making use of the only razor we had. When the rain ceased, we left the cove and sailed across to Port Gallant, with a fresh breeze. The smoke of natives' fires was seen near the entrance of the Barbara Channel; and on Prince Island, where we stopped a few minutes, the first man seen had on an old pair of sailor's trowsers, which he had obtained from the Beagle tied round his legs in six places. The wigwam these people were living in was not half covered: both wind and rain passed through it. How they bear the cold is surprising, being without clothes: one minute sitting close to the fire, and the next perhaps up to the waist in water, getting muscles or sea-eggs. The women dive for sea-eggs, even in the middle of winter; / but the water is never very cold (42° to 44°). (*q*) In the afternoon we saw the Beagle's mast-heads, and soon afterwards arrived on board, and enjoyed the happiness of finding all hands well, and every thing ready for farther progress. Lieutenant Kempe had turned the few hours of light, each day afforded, to the best account. Those who have had the care of ships in remote places, will know my feelings at finding all as it should be, after a long absence, in a country little known. Not a man had been ill; and the weather had been very tolerable compared with what was expected. There was less snow on the mountains than when I left Port Gallant early in May. One thing only disappointed me, – the Adelaide had not arrived. It was past the time appointed for her, but she might have found much more to do than was expected, or might have been obliged to return by the Magdalen, instead of coming through the Barbara Channel.

'During my absence, two sealing vessels had been at Port Gallant, on their way through the Strait. From one (an American), which arrived on the 7th of May from Staten Land, information was received that the Adventure had not been there. The Chanticleer had remained some time, but had sailed for the Cape of Good Hope. The master of the American had a brother staying with a boat's crew in Staten Land, during the whole of April, who would probably have seen the Adventure, had she called. The other was Mr. Cutler's vessel, the Uxor, bound to the United States; he had been through a channel which leads from the Gulf of Trinidad to Cape Tamar, and spoke well of it; but could give no drawing, nor precise information; having passed through rapidly.[3]

'Lieutenant Kempe had been at the summit of the Mountain de la Cruz, and left a memorial. No rare animals had been seen, nor any new birds. Small fish were still caught with hook and line, but very few with the seine.

'I never was fully aware of the comfort of a bed until this / night. Not even a frost-bitten foot could prevent me from sleeping soundly for the first time during many nights.

(*q*) At the western entrance of the Strait the water is said to be generally a few degrees warmer than at the eastern. – R.F.

'9th. At one o'clock this day, I heard an exclamation of 'The schooner!' and soon saw her standing across from the Barbara Channel with a fair wind. Before she anchored in Port Gallant, I went on board, and, to my joy, found Lieutenants Skyring and Graves, and all their companions well, having thoroughly completed the work they had to perform, without loss, or even an accident. The difficulty of their task was increased by very bad weather; but they succeeded in tracing and surveying the Magdalen Channel to its junction with the sea, and thence returned by the Barbara Channel to Port Gallant; carrying on a regular chain of triangles, and connecting their work with points previously fixed in the Strait of Magalhaens. A multitude of small islands, and much bad weather, detained them longer than was expected.

'While Lieutenants Skyring and Graves, assisted by Mr. Kirke, were employed surveying, Mr. Bynoe collected geological and other specimens.*

'11th. We had nearly reached the shortest day; the sun did not rise above the hills until past eleven; it disappeared again before two (the land being less high towards the N.W.), and even in those three hours was seldom visible.

'12th. Finding that Lieutenant Skyring agreed with me in thinking that the channel from Cape Tamar to the Gulf of Trinidad might be surveyed by the Adelaide, in her way to San Carlos de Chilóe, I resolved to send him and Lieutenant Graves on that service, hoping that it would lead to the discovery of a passage into the Skyring Water, and give vessels another way of getting into or out of the Strait, should thick weather or adverse winds oppose them in the usual channel.

'In making this arrangement there was much to be considered. As I had received no orders from Captain King to employ the Adelaide in surveying, after her return from the Magdalen Channel; and as I had been desired to repair, with her, to San Carlos, in Chilóe, during which voyage Lieutenant / Skyring was to be on board his own vessel, the Beagle, it would be incurring considerable responsibility, to order a new piece of service to be undertaken, which might not be successful, and would require officers, men, a boat, provisions, and stores from the Beagle.

'I did not doubt that the measure would be approved by Captain King, because he had discussed the feasibility of such a plan with me, and had expressed a wish that it should be tried; but as I had not received any orders, I could not decide without anxiety.

'Another, though a minor difficulty, arose from sending Lieutenant Skyring in command of the Adelaide, over Lieutenant Graves, her proper commander, who had expected to take her to Chilóe, and was quite competent to undertake this or any other service in which she might be employed. Both these officers

* Geol. Soc. Museum, Nos. 176 to 205, and Zool. Mus.

excelled in their professional duties; but Skyring had been on the western coasts of Patagonia before, and was the senior.

'Much to the credit of Lieutenant Graves, he removed one weight, by volunteering to go any where I thought proper to direct, either alone or with Lieutenant Skyring, and the necessary orders were forthwith given. (See Appendix).[4] Mr. Kirke was again to form one of their party, as well as Mr. Bynoe, who exchanged temporarily with Mr. Park. The Beagle's whale-boat was also lent, with five able seamen to man her; and good care was taken that nothing the ship could give should be wanting in their outfit for a service which, at that time of year, must be severe and tedious.

'Anchors and cables, hawsers and kedges, were abundantly supplied, because in warping into unknown places, or anchoring hastily, many an anchor is unavoidably broken or lost.

'The boat's crew, who had been away with the Adelaide, and were going in her again, were supplied with extra clothing at the expense of Government, the wear and tear of their clothes having been far beyond what they could be expected to make good out of their pay.[5] /

'As an instance, I may mention, that a careful north countryman carried with him, when he left the Beagle, two new pair of shoes (besides those on his feet), and three pair of new stockings: but brought back only a ragged pair of stockings and the remains of one shoe. The others had been fairly worn out, or lost, in scrambling over rocks and ascending mountains.

'One height ascended by Lieutenant Skyring was so steep, that the men were obliged to pass the instruments from one to another, at a great risk of their own lives; and when they reached the summit, the wind was so strong, that a heavy theodolite and stand, firmly placed, was blown over; and even a Kater's compass could scarcely be used.[6]

'With good clothing and provisions, weather may be almost defied, and work may be done at the less unfavourable times; but without them, ill-humour and ill-health must inevitably appear in such a climate as this.

'14th, Sunday. I had the satisfaction of keeping this day in a proper manner, for the first time since we entered the Strait. So much had depended upon employing every minute of our time while the weather would allow, that there had been little distinction of days.

'17th. The morning being fine, with not much wind, though a sharp frost, I left the ship with Mr. Murray and four men, and landed in Fortescue Bay, intending to ascend the mountain 'De la Cruz,' if the snow and ice did not prevent me.

'On the beach, close to the water, I suspended the mountain barometer, and let it remain half an hour before we began the ascent, which, from the snow lying so deep, was troublesome; for at one step a hard rock received one's foot, and at the next, perhaps, a deep hole amongst broken trees. Sometimes we tumbled head foremost into soft snow, slightly covering rotten mossy boughs and swampy ground; and at others, slipped between the concealed trunks of trees, which, though much decayed, were hard enough to cause many a bruise. Each movement of our arms or legs shook down a shower of snow from the trees, among which we were forcing our way.

'At noon we gained the part that is clear of wood, but / so very steep and slippery was the summit, that we were obliged to go on our hands and knees, forcing them as deeply into the snow as possible, to avoid sliding down again. The highest point is not visible from Port Gallant.

'While I took angles with the theodolite, the seamen made a fire. It was well we carried some fuel and a tinder-box, with a sheet of copper, upon which to kindle it; for without a fire we should have been quite numbed. Standing in one place for two hours, after being much warmed by exertion, made us more sensible of the cold. The highest spot is but a few yards wide, and by barometrical measurement is 2,280 feet above the sea.** The height is, in truth, small; but as the mountain is so steep, and rises so abruptly from the sea, it appears considerable.

'When we had finished our observations with the barometer and theodolite, we deposited a Memorial, containing a list of the officers and crews of the Beagle and Adelaide – an account of the object of their voyage, how far it had succeeded, and where we were going – and a collection of coins, well-soldered up in a tin case – upon the bare rock; and made a great pile of stones over it.

'Having again examined the barometer, we began to descend; for the sun disappearing behind the distant mountains, warned us that it was time to return. We had enjoyed a magnificent view on all sides, and were reluctant to leave our station. In descending, we made rapid progress at first, sliding many yards together down the soft snow; but, by the time we reached the woody part, it was getting dark, and having foolishly tried to return by a straight line, instead of going round, we found steep cliffs, and ravines covered with rotten trees, which perplexed us exceedingly. Darkness, and the deep snow, much increased our dilemma; yet we could not resist laughing heartily at the ludicrous scrapes some of the party got into: one man was rather a-head, looking for a way to descend a steep place, when the snow slipped from under him, and down he went, about eighty feet, partly sliding, / partly falling, but quite against his consent. What he did by accident, we were obliged to do, because there was no alternative; so away we slid, one after another, like so many sledges upon Russian ice-hills, holding the instruments as we could, by one hand, while the other was employed to check or steady us. With a little more of this sort of work, and some struggling

* By angular measurement it was found to be 2,270 feet.

through the wood at the bottom, we reached the shore, where a boat was waiting for us, and at about eight arrived on board, in a half-wet, half-frozen condition.*

'19th. Every thing was brought on board, the ship unmoored, and all made ready for our departure next morning.

'20th. Sailed from Port Gallant, leaving the Adelaide to rate her chronometers, and rejoin us before leaving the Strait. In the evening we anchored in Elizabeth Bay, after a severe day's struggle against a strong and contrary wind, with much rain.

'21st. Blowing hard again this morning from the N.W., with a great deal of rain. Weighed and made sail under reefed courses and treble reefed topsails, but the wind and tide were more than a match for us, so we stood across into Whale Sound, and worked up under the lee of Carlos Island, finding the tide there rather in our favour. The 'williwaws' (I know no better name for the sudden gusts that come off the high land) gave us some trouble, occasionally laying us almost on our beam ends. At half past two I was induced to anchor under the lee of the south-east extremity of Carlos Island, and thought our day's work was repaid by a snug position close to a weather-shore, besides having made some little progress; but after dark the wind became more violent, and a williwaw drove us out into deep water. We set the storm sails, which, with the weather-tide, known to be then making strongly, I hoped would take her a-head sufficiently to clear Rupert Island (lying under our lee), and all hands then went to the capstan; but while heaving-in the cable, our bower anchor again caught the ground and brought us up. We veered away cable / directly, let go another anchor, and rode our the rest of the gale, which was extremely violent, without driving.

'The instant our anchor caught, I knew we must be on a ridge, of which Lieut. Skyring had spoken to me, lying between Rupert and Carlos Islands, across which the tide makes strongly, at the rate of about three knots. Rupert Island was still under our lee, distant less than half a mile.

'22d. Blowing hard and raining. At 9 A.M. it cleared and moderated, but so strong a tide set past us, to the south, that we could not attempt to weigh. It differs here from that in mid-channel by two hours, which may much assist a vessel if she manages so as to take eight hours tide in her favour.

'At eleven we unmoored, and got ready for moving at the turn of tide.† At one we weighed and made sail with a moderate wind from N.W., and by keeping close to Carlos Island, and making short boards, we had a weather-tide, while in the fairway of the Strait the stream was running to the S.E. We anchored in Bachelor's Bay (or York Roads), choosing an outside berth in order to have more

* The wristbands of our shirts, and all our outer clothes, were coated with ice, while our inner clothing was wet through.

† On heaving up the best bower, we found it had lost one fluke.[7]

room to weigh again and work with the morning tide. It blew hard in the night, but we rode securely, although the tide ran at least three knots where we were.

'23d. We started and worked to the westward, and at nine were abreast of Borja Bay; but by trying for too much, nearly lost all that we had gained, for in standing across from the bay, hoping to weather Cape Quod, the flood tide took us so strongly, that it cost three hours close working to get to an anchor even in Borja Bay. We had rain and sleet continually through the day, and it blew hard at night, but as plenty of chain was out, the topsails and courses were close reefed, and the top-gallant masts on deck, we were ready for anything.

'24th. Heavy squalls, with almost constant rain, prevented our moving westward, and similar weather continued throughout the day, becoming worse at night. Had we had plenty of provisions I should not have minded this delay, because we might have remained at anchor till it was over; but so much / had been said about the difficulty sometimes found in working through the Strait, that it concerned us greatly not to lose a chance of making progress. During this night the squalls were very heavy. The holding ground must have been excellent, for williwaws drove the ship from one side to the other as if she had been a chip upon the water.

'26th. Weighed this morning, weathered Cape Quod, and worked to the westward, the weather having cleared and become very fine. The part where most tide is felt was then past. Cape Quod projects so far south that the Strait is there extremely narrow, and though very deep, has a strong tide.

'27th. At daylight we found ourselves to windward of Marian's Cove. Looking eastward upon the land about Cape Quod, it has a very bleak and rugged appearance. The almost perennial west winds prevent vegetation from growing on the heights exposed to their action. Hence the desolate look of the western shores of Tierra del Fuego. We saw a sail beyond Cape Notch, and, just before we moored, close to the shore in Half-port Cove, we made her out to be the Adelaide.

'28th. A bad morning, snowy and blowing, but the wind being moderate between the squalls, I went in a whale boat to examine the Gulf of Xaultegua, and pulled along the south shore towards Cape Monday. Having gained some distance to windward, while the snow was so thick it was impossible to see the shore, we made sail across the Strait, and hit the place within a cable's length. When the snow ceased falling, we saw a large space of water before us, the land opposite being at least five miles distant. We sailed towards a strange looking islet in the middle of the gulf, very similar to the old mouldering figures of the fabled Sphinx,[8] but the snow becoming again almost incessant, only allowing us to see our way at intervals, while the wind was too strong for even a close reefed sail, we landed, and hauled the boat up on an island. I was in hopes of finding an opening which would lead me to the Skyring Water; and my boat's crew, being almost as eager as I was, cared little for the wind or snow. This night we made a larger tent than usual, with a top-gallant studding sail, and / the consequence was, we were extremely cold, as there was a sharp frost, and the snow was lying every where very deep. Next night we were wiser, and reduced our tent to the smallest dimensions.

'29th. Early in the morning we resumed our search. I had a chronometer with me, but as we never saw the sun, nor even a star, I should have been as well without it. We pulled and sailed towards the northernmost corner first, but found no opening, and went thence to the eastward, with a strong and favourable breeze. Passing Still-hope Point I felt sure of finding a passage, for before me were the tops of mountains seen from the Otway Water. I was, however, deceived, the gulf ended in two bights, or inlets, unconnected with other waters: so we returned to Still-hope Point and hauled up the boat. The night passed very well, in a snug place among trees, although the snow was falling thickly. Early next morning we left the shore, having employed a quarter of an hour in clearing the snow out of our boat. When we started, it snowed fast but without wind, and we steered by compass for the Sphinx. I sketched what I could see of the south side of this gulf, but did not consider it worth delaying longer, in such weather, for so unimportant a place, while anxious that the Beagle should reach Chilóe before her provisions were expended, and that I should fall in with the Adelaide before leaving the Strait. If ever a minute survey is made of this gulf, it should be after all others have been examined, as it is utterly useless. The temperature of the water within it we found to be 40° Fahr. We landed on St. Anne's Island, having run near thirty miles since the morning, and thence we sailed across the Strait, reached His Majesty's little vessel, and found that the Adelaide had not yet passed by. All looked cold and wintry, every thing being covered with snow; and our sails were hard frozen, for the first time.

'July 1st. After beating loose the sails, we stood out in the ship to meet the Adelaide, which was seen coming towards us. I went on board, and found every one well. They too, in attempting to anchor off Carlos Island, had, like ourselves, been driven / out: we compared chronometers, and supplied her with a few things not thought of before (keeping under all sail meanwhile to profit by an easterly wind); and the Beagle's officers lent the Adelaide their own stove.

'In the afternoon, we parted company; the Adelaide stood towards Upright Bay, and anchored at dusk, while we steered out of the Strait, with a freshening breeze from the east, which increased much as we made westing. At midnight, we were in the Pacific, and all our anxiety about weeks of beating to windward upon short allowance of provisions, vanished as quickly as the land astern. The glass falling, with the wind in the S.E. quarter, foretold unusually bad weather; we therefore shortened sail by degrees, making all secure.

'2d. At six o'clock in the morning, it was blowing a gale of wind, with so much sea, that it was necessary to steer right before it, – or heave-to, – which with a fair wind was not preferable; and we found the vessel scud extremely well, under close reefed fore and main topsails, and double reefed foresail. Our quarter boats caused anxiety, for the davits were low, and at every lurch the boats were risked. Frequently they dipped in the sea, and sometimes were half filled; but they hung fast till by a moment's neglect of the steerage, a sea broke over the whale-boat, and carried her away. The other, being much smaller and stronger, held on well, though frequently under water. Towards midnight the gale broke; by the next morning the weather was more moderate; and from that time it continued fine, until our arrival at Chilóe.

'On the 5th, at daylight, we saw land at a great distance, which afterwards proved to be the Island of Guafo, and in the afternoon the south end of Chilóe was seen.

'On the 8th, we were working towards the Port of San Carlos, being off Point Huapilacuy, and next day (9th) anchored in the port of San Carlos, which seemed to be well sheltered by a country, the appearance of which was very agreeable when contrasted with that of Tierra del Fuego.

'The town reminded me of a Cornish village. I thought, from their appearance and colour, that the houses were built / of stone, and roofed with slate; but afterwards found they were of wood, from their foundations, to the tops of their roofs. Except a few cleared spaces, the island is entirely covered with trees, even on the highest hills. The Captain of the Port (an Englishman) boarded us as we neared the anchorage, and was very obliging in his offers. From him I learnt that the Adventure had not yet arrived, nor even been heard of on the coast. We anchored under the lee of Barcacura Heights, in a good berth, and moored ship. I went on shore immediately, and paid my respects to the Governor, Don José Santiago Aldunate, a brigadier-general in the Chilian Service, whose kind manner, and friendly offers of every assistance he could render us, were very gratifying. From the master of a merchant ship, lately arrived, I was surprised and concerned to learn, that the Adventure had not reached Valparaiso before the time of his sailing thence (20th of June).*

'Refitting the Beagle, repairing and building boats, occupied most of the officers, and all the crew, while Mr. Stokes and I were engaged in the work of the survey, during our stay in the Port of San Carlos. Our ship required caulking, which, in so rainy a climate, was difficult to accomplish. So continually wet was the weather, that had we not dried our sails, and unbent them, during three fine days which we had (*r*) on our arrival, they would not have been dry during our stay.'

* The Adventure arrived on the 21st. – P.P.K.

(*r*) Por milagro (miraculously); as the inhabitants told me. – R.F. /

CHAPTER XV.

Extracts from the Journals of Lieutenants Skyring and Graves – Magdalen Channel – Keats Sound – Mount Sarmiento – Barrow Head – Cockburn Channel – Prevalence of south-west winds – Melville Sound – Ascent of Mount Skyring – Memorial – Cockburn and Barbara Channels – Mass of Islets and Rocks – Hewett Bay – Cypress trees useful – Adelaide rejoins Beagle in Port Gallant – Captain King's narrative resumed – Plan of future proceedings – Adelaide arrives at Chilóe – Abstract of Lieutenant Skyring's account of her proceedings – Smyth Channel – Mount Burney – 'Ancon sin Salida' – Natives – Kirke Narrows – Guia Narrows – Peculiar tides – Indians in plank canoes – Passage to Chilóe.

THE extracts from Captain Fitz Roy's first journal being ended, I shall now give some passages from the journals of Lieutenants Skyring and Graves, while employed in the Adelaide, exploring and surveying the Magdalen and Barbara Channels.

The reader will remember, that the Adelaide parted company with the Beagle, at the entrance of the Magdalen Channel, on the 19th of April; and steered to the southward under the direction of Lieutenant Skyring.

Lieutenant Graves says:–

'The east and west shores of the Magdalen Channel run nearly parallel to each other: but the east side is broken by a large opening, named Keats Sound, which runs into the land for eight miles, and appears very like a channel. *(s)*

'At the S.W. angle of the Magdalen Channel stands Mount Sarmiento: the most conspicuous, and the most splendid object in these regions. Rising abruptly from the sea, to a height of about 7,000 feet, it terminates in two sharp peaks, which seem absolutely in the sky: so lofty does the mountain appear, when you are close to its base.

'Two thirds of the height are covered with snow; and two enormous glaciers descend into the deep blue waters of the sea beneath. When the sun shines, it is a most brilliant and magnificent sight.

* *(s)* I do not think that there is any opening at the bottom of Keats Sound; which lies at the base of a chain of snow-covered mountains, whose southern side I have closely traced. – R.F. /

'Many days were almost lost to us, in consequence of heavy gales, accompanied by torrents of rain; but we profited by intervals of fine weather to move from cove to cove.

'On the 5th of May, while working out of Stormy Bay, we grounded, and remained fixed upon a rock several hours, but were lifted off again by the next tide, without having sustained material injury.

'To vessels navigating this channel, I should strongly recommend giving a preference to the south shore, where there are many openings, and I have no doubt good anchorages, which, as our time was limited, and the weather very tempestuous, we had not an opportunity of examining. If any such exist they would have a decided advantage over those on the north shore, from being generally to windward, and therefore easy to leave, as well as more secure. King and Fitz Roy Islands, lying in mid-channel, between Stormy and Park Bays, are of bold approach, as are also the Kirke Rocks, which lie further to the S.W.

'One morning, being anxious to obtain a more secure situation for the vessel, we started in search of a better berth, intending, if possible, to reach a bay on the other shore, near Barrow Head, apparently affording good anchorage; but after beating about, from nine until four o'clock, without being able to reach it, the breeze freshening, and sea increasing, we bore up, and again anchored under the lee of the same island. S.W. winds prevail in these parts throughout the year: in confirmation of which, besides the experience we ourselves have had, all the trees which stand exposed, are bent in an opposite direction; and on the S.W. side of all the land open to that point, not only does the vegetation commence much further from the water's edge, but it is scarcer, and more stunted. In sheltered places the trees grow to within a foot of high-water mark. / /

'May 11th. We remained at the above-mentioned anchorage; and while Lieutenant Skyring was examining a cluster of islands in the vicinity, I obtained observations for the latitude and longitude; and as it was the first fine day, indeed the only one since entering this channel in which we had a fair proportion of sunshine, it was taken advantage of to dry and air all our clothes and bedding, and clean out the vessel thoroughly.

'The next anchorage we took, was in a cove just large enough to hold the schooner, at the entrance of Dyneley Sound, on the north shore. In crossing over, we had a fine view of Mount Sarmiento; and looking to seaward, from the hill over this cove, the Tussac, and the Fury Rocks, at the entrance of Melville Sound, which are much resorted to by sealers, were clearly distinguishable.

'During our stay here, until May 15th, the neighbouring coast was examined, whenever the weather permitted. We also communicated with several canoes full of Indians, but gained no additional information respecting the habits of the natives.

'The next start carried us through the islands of Melville Sound, to an anchorage in a small cove, at the N.E. end of the largest of the Magill Islands,

C. Martens. T. Landseer.

MOUNT SARMIENTO.

upon which is Mount Skyring. Having resolved to ascend to the top, as it offered so commanding a view, and was so centrally situated, we remained for that purpose.' The weather, for several days, was very unfavourable, and it was not until the 21st, that there was any reasonable prospect of obtaining a view from the summit; when Lieutenant Skyring and Mr. Kirke had a most laborious excursion, and the latter was nearly frost-bitten in ascending the mountain; but they were fully recompensed for the trouble and difficulty they had experienced.

Lieutenant Skyring says:–

'We gained the summit after three hour's hard travelling. During the last five hundred feet of ascent, the mountain was almost precipitous, and we had the utmost difficulty in passing the instruments from hand to hand. Its formation is remarkable, although, I believe, the same structure exists throughout the hills around. The base is a coarse granite, but this solid / formation cannot be traced half the height; above is an immense heap of masses of rock, irregularly and wonderfully thrown together, many huge fragments overhanging, with apparently very little hold. This station was the most commanding we had chosen during the survey, and answered well for the object we desired; which being attained, we returned on board, and I rejoiced when all were safe, for it was neither an easy, nor a pleasant enterprise.'

A document, of which the following is a copy, was enclosed in a bottle and a strong outer case, and left at the summit of the mountain.

(Copy.)

This Memorial was left by the officers of H.M. Schooner Adelaide, while employed on a survey of the Magdalen, Cockburn, and Barbara Channels; and any person finding it is requested to leave the original document, and build the pile, under which it is placed, at least six feet higher.

Signed this 16th day of May 1829, by

W. G. Skyring, Lieut. and assist. surveyor of H.M.S. Beagle.

Thomas Graves, Lieut. of H.M. Schooner Adelaide.

James Kirke, Midshipman H.M.S. Beagle.

Alex. Millar, Master assist. H.M.S. Adelaide.

Benj. Bynoe, Assist. surgeon H.M.S. Beagle.

Jno. Park, Assist. surgeon H.M.S. Adventure.

God save the King.

'In the Cockburn Channel,* the flood-tide sets to seaward; / but it was not found to be of consequence to a vessel in working through. The rise and fall is not more than six, or at most, eight feet, at spring-tides.

* In the old Dutch charts, a passage was laid down near the place, and nearly in the direction of the Cockburn Channel, and named 'Jelouzelt:' but until some written authority can be

The first person known to have passed through the Cockburn Channel was the mate of the Prince of Saxe Cobourg, who went in a boat (see page 66). It was afterwards passed by Mr. William Low, master of the Mercury, and has since been used by several vessels.

'May 22d. We quitted this anchorage; and having worked to the westward, through the Adelaide Passage, took up a berth in a small bay, two miles and a half to the northward, where we remained during the night, and next morning; then, after examining the neighbouring coast sufficiently to carry on our triangulation, proceeded to an anchorage on the north side of Bynoe Island. From the summit of this place an extensive view was obtained of the island in Melville Sound, as well as of the entrance to the Cockburn and Barbara Channels. Such a complicated mass of islands and rocks, I never before saw; to lay them all down correctly would occupy a long time. Sufficient, however, has been done to take the navigator through this labyrinth; but I am well aware, that very much is still wanting to complete the survey.

'Fury and North Harbours, of which the former became more particularly known to us from the Prince of Saxe Cobourg having been wrecked there in December 1826, were laid down from an eye-sketch only; but the peaks of the island, and its extremes, were fixed by triangulation.*

'Melville Sound is formed by the islands which separate the Cockburn from the Barbara Channels. Generally speaking, they, as well as the coasts in the immediate neighbourhood which are exposed to seaward, present a most barren and desolate appearance.

'Until the 26th of May, we were much occupied among the surrounding islands; but time being short, we took advantage of a southerly wind to run up the Barbara Channel, and soon reached an anchorage in Hewett Bay. While securing the vessel, a canoe, containing only a man, woman, and child, and three dogs, was seen coming round the south point of the bay. As they seemed very unwilling to pay us a visit, remaining at a distance, and vociferating as usual, 'Ho-say,' 'Ho-say!' Mr. Bynoe and I communicated with them in the dinghy; but / finding they had not an article worth bartering for, we soon left them, and returned on board. It was suspected their companions were not far off, and indeed, the day after, Lieutenant Skyring saw several canoes; but the moment he was discovered, they were beached, and the men, taking to the woods, kept at a distance.

produced to prove that this passage was explored, or, at the least, discovered by the person who gave the name of 'Jelouzelt' to one of the almost innumerable openings in Tierra del Fuego, it does not appear that the inlet so called has any claim to our consideration, greater than that of the non-existing San Sebastian Channel, – or a number of other imaginary passages which must have been laid down, upon supposition only, in many old charts.

* Since surveyed by Capt. Fitz Roy in the Beagle, 1829-30.

'On the 29th, we left Hewett Bay, and, after threading the needle through a multitude of islands, islets, and small rocks, for more than three miles, reached an anchorage in a small cove, at the north entrance of Brown Bay, where we were detained, and confined to the vessel, by heavy gales, and stormy weather, until June 2d; when, having a fine day, we reached a spot (marked in the chart as North anchorage) sufficiently secure for a small vessel; but not to be recommended to any other.

'Between Hewett Bay, and the above anchorage, there are several rocks, among patches of kelp, which, as they only show themselves at half ebb, or near low water, render the navigation rather intricate. A good maxim in these channels is, 'Avoid kelp, and you avoid danger.' Forty-three days had passed since we left Port Famine; and in this interval, I find we had nine favourable days, twelve partially favourable, some hours of which we could employ in the work about which we were engaged, and the remaining twenty-three were days of rain and wind, far too unfavourable to serve our purpose in the least.

'June 4th. While turning to windward, we, for the first time, felt the influence of the tide, which, from the channel's narrowing, begins to be sensible: here it was sufficiently strong to prevent our gaining ground in beating to windward, although with a good working breeze; we therefore ran into a bay on the west side, and anchored. The country around had rather a pleasing appearance, the shores being partially covered with the evergreen, and deciduous-leaved beech, and a few stunted cypress-trees. These last are serviceable for boat-hook spars, or boats' masts; and, when seasoned, work up very smoothly, and wear well; the beech-trees do not equal those found further northward in the Strait, except here and there in sheltered corners. /

'With a leading wind, the next morning, we reached the south narrows of the Barbara Channel, through which we were carried by a strong tide, and anchored in Bedford Bay.

'Here, as well as throughout the Barbara channel, the flood tide sets to the southward. We obtained at this place angles which connected our triangulation with points fixed by Captain King during the previous year, and finished our examination of these channels within a very few days of the time allotted.

'On the 8th of June we attempted to pass through the Shag Narrows, but not saving the tide, were obliged to anchor for the night in Field Bay, which is small and much exposed to southerly winds; the bank also is very abrupt, and the water is deep close to the shore.

'On the 9th we succeeded in clearing the Narrows, and reached Port Gallant early in the afternoon, where we rejoined the Beagle.'

Having given these brief extracts from Journals kept on board the Beagle and Adelaide, during the time occupied by the Adventure about Cape Horn, or on her way to Chilóe, I will resume my own narrative.

As it was my intention to remain at this port** until the Beagle and Adelaide were equipped, the Adventure was made snug, and, by way of relaxation, such of the officers as could be spared from the duties of the ship, resided in turns at the town, where also the ship's company had frequently permission to amuse themselves.

The Hoxsley schooner arrived from Valparaiso[1] and brought me letters from the Admiralty, acquiescing in my request to return to England direct, instead of proceeding by way of New South Wales and the Cape of Good Hope, as was originally intended. I therefore determined to return to Valparaiso as soon as our consorts had taken their departure, proceed thence to Port Famine, where we were to be joined by the Adelaide, and afterwards repair to Rio de Janeiro to await the Beagle's arrival, when we should sail for England. /

On the 20th of September my anxiety for the Adelaide was relieved by her appearance, and by finding all on board her in good health. She had gone up the coast by the channels that communicate with the Strait of Magalhaens at Beaufort Bay, passing inside of Hanover Island and Madre de Dios; and Lieut. Skyring gave me a very interesting account of their discoveries, of which the following is an abstract.

It will be remembered that the Beagle left the Adelaide at anchor under Cape Upright. While there the wind freshened up from the eastward, and threw a swell into the bay, which rendered the anchorage very unsafe, as the schooner's stern was in the foam of the sea that broke on the rocky shore close to her. Much anxiety was felt for their safety, but the anchors held well. As soon as the weather permitted they sailed, entered Beaufort Bay, and steered towards a deep opening to the eastward of Cape Phillip, into which they ran with a steady S.E. wind, and found an anchorage on the west side in Deep Harbour.

On the 5th of July Lieut. Skyring and Mr. Kirke were absent in a whaleboat, exploring a deep opening eastward of Cape Tamar, which they found to terminate in two sounds, named by them Icy Sound and Glacier Bay; the first from its being covered with a sheet of ice, and the latter from its being full of large masses which had been detached from an extensive glacier occupying the bottom of the bay. The examination of this opening was made in search of a channel, through which, vessels had entered the Strait, and the schooner was to proceed to her rendezvous. The result proved that the Adelaide was already in the channel they were looking for, therefore they returned on board, and proceeded (7th) to the northward. In passing Mount Joy a strong tide was observed, the certain indication of a channel; for, as has been before remarked, within sounds the tide has no perceptible stream. To gain a better knowledge of their way they anchored early in Good's Bay; the course of the channel, from the intersection of points, and

* San Carlos, in Chilóe.

intervention of islands, being by no means distinct. Lieut. Graves made a plan of the bay, while / Lieut. Skyring, and his assistant,** completed the survey of the entrance to the passage, which was named Smyth Channel, as a compliment to Capt, W.H. Smyth, R.N., under whom, while surveying the Mediterranean, both Lieuts. Skyring and Graves had served.[2]

The best channel they found to the eastward of Renouard Island, and the Adelaide took that course, but stopped a night in a small cove on the eastern side of the island, and in passing Shoal Island next day struck on a rock; she was got off however without injury, and anchored afterwards, for a night, on the north side of the Island of the Narrows.

The two following days (10th and 11th) were spent in examining the coast, and exploring Clapperton Inlet, which had the appearance of being a channel. From the hills at the bottom Lieut. Skyring noticed a considerable tract of low land and open plain, extending to the northward. On the 12th, being Sunday, they remained quiet, and on the 13th the weather was so calm that they only reached Hose Harbour, on the east side; and the next day Oake Bay. Thence crossing the channel in a whaleboat they explored some distance along that shore; and on the 15th anchored in Otter Bay. This slow progress was unavoidable, owing to the calm state of the weather, and to the survey being principally, if not entirely, carried on in boats.

On the 16th the schooner was towed onwards, and passing over an extensive shoal flat of three fathoms, reached the Summer Islands, where she might have stopped, but, as the tide was still favourable, she proceeded to an anchorage under Long Island, the most northern in the Elson group.

The eastern shore of the channel was there very different in character from what they had so long been accustomed to, being nearly level; and, extending for some distance off every low point, there was shoal water.

For some days a lofty mountain, covered with snow, had been in sight; which, by angular measurement, proved to / be 5,800 feet in height. It was named Mount Burney, in compliment to the admiral.[3]

On the 17th the Adelaide reached Fortune Bay, situated at the east extreme of a headland, on each side of which is a channel, leading, apparently, towards Cape Isabel. The northern seemed to be the principal one, and therefore was followed next day (18th) as far as Welcome Bay.

Continuing the survey onwards they reached Victory Passage, which they entered, thinking they were in the mouth of the 'Ancon sin Salida,' as laid down from Sarmiento's journal by Admiral Burney.[4] The weather, however, became so bad, that they were obliged to take shelter in Island Bay, and the next day the wind setting in from the eastward, they gave up, for a time, their search

* Mr. Kirke.

for the 'Ancon sin salida,' and proceeded by Smyth Channel, as far as Hamper Bay, where they were again detained by bad weather. Here a few rock fish were caught, but at no other time during this cruise were the fishermen successful, although the channel was so filled by porpoises and seals, that it is probably well stocked with fish at the proper season: and there are many places where the seine might be shot. Proceeding slowly on the 25th, the Adelaide struck on a rock, and remained fast for a few hours, but as the tide rose she swung off without damage. Upon examining Rocky Bay they found it a complete bed of rocks; yet, bad as it was, the Adelaide was obliged to remain there five days, owing to the tempestuous state of the weather. On the 30th they reached the north end of Smyth Channel, and anchored in Narrow Creek.

On the 31st Lieut. Skyring went to a remarkable hill, which he called Mount Trafalgar, but thought it might have been the 'Monte Trigo'* of Sarmiento, so much did its appearance remind him of a corn stack. The day was most favourable: a round of angles, and an extensive view down Lord Nelson's Strait, were obtained from the summit. They remained on an island all night, sheltered by the boat, and next morning went to two points, called by Sarmiento 'Oueste,' and 'Mas / al Oueste,' (west and more west,) returning to the Adelaide in the evening.

The following morning was fine, and the Adelaide moved out of Smyth Channel, the survey of which was completed very satisfactorily, although their progress was slow, owing to constant northerly winds.

By towing the Adelaide during tedious calms, they reached Montague Bay in the evening, and next day anchored in Relief Harbour, on the S.W. side of Vancouver Island.

As it was evident that the 'Ancon sin salida' was within Piazzi and Ceres Islands, up the west coasts of which they had passed, Lieut. Skyring left the schooner moored in Relief Harbour, and proceeded, on the 4th August, to the southward, in a whale-boat with Mr. Kirke; but he took no more than a week's provisions, that time being all he could devote to this exploration.

The 4th, 5th, and 6th, Lieutenant Skyring employed in pulling or sailing to the southward and eastward, through winding and intricate passages; although strong winds and much heavy rain annoyed him, and impeded his progress.

On the 7th the weather was much more favourable than it had lately been. The boat pulled and sailed to the southward, and at noon Lieutenant Skyring ascended a height,† having on each side of it a deep opening, but he was disappointed in the view; and, after taking bearings, pulled round the adjacent bights, one of which was exactly opposite Artist Bay, in Smyth Channel, and so near

* A heap, or stack of corn.

† No doubt the Mount Oracion of Sarmiento, p. 144. – P.P.K.

it that the two waters were only separated by a few hundred yards;* the other,† eastward of the height, was large, and closed at the bottom by very low lands. It was directly supposed to be the 'Ancon sin Salida;'‡ but Sarmiento's description, and the chart compiled by Burney, / were insufficient to enable them to decide with any degree of certainty. After looking round this bay, they continued to the eastward, and passed a point beyond which there was apparently a wide channel; having run about six miles down it without discovering any termination, they hauled their boat up on the beach for the night.

On the 8th, two canoes were noticed on the west shore; but seeing strangers the natives, apparently much frightened, all landed, except an old man; and taking with them what they most valued, hid themselves among the brush-wood, leaving their canoes fastened to the sea-weed. By some Fuegian words of invitation, the men were, however, induced to approach and traffic, receiving for their otter skins whatever could be spared. In appearance and manner these Indians were exactly similar to the Fuegians; and by their canoes only, which were built of planks, could they be distinguished as belonging to another tribe.

After leaving the natives, the boat passed Cape Earnest, and Lieutenant Skyring observed a wide channel leading north and then N.N.W.;§* also, another opening to the eastward. The wind being easterly, he ran some distance to the northward, to gain more knowledge of the first inlet; and having gone ten or twelve miles from Cape Earnest, and observing the opening for eight miles beyond to be as wide as where they then were, he concluded it to be a channel, or else a deep sound terminated by low land, for there was evidently a division in / the mountains, such as to justify this belief. Returning, they entered the smaller opening to the eastward, and were almost assured of its being a channel; for when they were between the points, many porpoises and seals were observed, and a tide was found setting westward, at the rate of two knots. At dark, they

* This place is described in Sarmiento's journal, p. 144. – P.P.K.

† Ensenada de la Oracion of Sarmiento. – P.P.K.

‡ This bay is also described by Sarmiento as an 'Ancon sin salida,' p. 143; but it is evidently not the one that bears that name on the chart. – P.P.K.

§ Here is certainly the Ancon sin salida of Sarmiento, whose journal describes the inlet as terminating in a cove to the north, p. 142. The mountain of Año Nuevo cannot be mistaken; indeed the whole of the coast is so well described by the ancient mariner, that we have little difficulty in determining the greater number of places he visited. In all cases we have, of course, preserved his names. The chart compiled by Admiral Burney is a remarkable instance of the care which that author took in arranging it, and how ingeniously and correctly he has displayed his judgment; it is also a proof that our favourite old voyager, Sarmiento, was at least correct in his descriptions, although he appears to have been quite ignorant of the variation of the compass. – See Burney Coll. Voyages, p. 31; and Sarmiento, p. 162.

hauled their boat on the beach of an excellent bay, at the north side of the narrow reach, and secured her for the night.

On the 9th, shortly after daylight, they set out in a N.E. direction to ascertain the truth of their supposition; and before noon knew, beyond a doubt, that they were correct in their belief, being in the narrows of a channel before unknown, that had eluded Sarmiento's notice. These narrows, which Lieutenant Skyring felt assured would lead to a large opening, were upwards of three miles in length, and generally about one-third of a mile in breadth. A strong tide took the boat through; and at the N.E. extremity, where the narrows were reduced to four hundred yards in width, the water, although a neap-tide, rushed at the rate of four knots, forming whirling eddies, which were carefully avoided by Lieutenant Skyring. At spring-tide, the strength of these rapids would probably not be less than seven knots.

Having passed through them, a clear channel was seen, upwards of two miles wide, running to the N.b.E. for, at least, eight miles, and then turning directly eastward, between moderately high land. Another channel, nearly a mile and a half wide, trended to the S.E. for two or three miles, and then also turned to the eastward. Here they stopped. Lieutenant Skyring regretted extremely not being able to prosecute the discovery, and have one more view from the eastern point of the N.E. channel; but as only one day's provisions remained, it would have been imprudent to delay his return. It was evident, that they had passed through the range of the Cordilleras,* for to the eastward the country appeared totally different, the highest hill not being above seven hundred feet. The opening to the N.E. was thought to communicate with the waters' lately discovered by Captain Fitz Roy. The latitude / was obtained on Point Return; and in the afternoon, reluctantly but anxiously, they retraced their way, and passed that night at their former quarters, in Whale-boat Bay.

On the 10th, at daylight, they proceeded on their return. The wind was fair until they reached Cape Earnest, when it drew right against them; and they had the unpleasant prospect of a tedious pull to the schooner, with very little provision.

The 11th was a thoroughly wet day, and the wind was so strong from the northward, with a very heavy sea running, that it was impossible to proceed.

On the 12th, they left the bay soon after daylight, and having pulled along shore a few miles, crossed Union Sound, and gained the Narrows of San Benito, the wind being still fresh from the northward; thence they continued pulling until they hauled up, after dark, in a bay, opposite Point Benito, and waited till the morning of the 13th, when with a fresh S.W. wind they made good progress, which was of the more consequence, as their provisions were expended, although they had eked them out with corvorants and muscles. At last, the sight

* Cordillera Nevada' of Sarmiento.

of the Adelaide rejoiced them, and they soon afterwards reached her. Their appearance was a relief to all who were on board, as they were becoming very anxious, and Lieut. Graves was preparing to send the other whale-boat in search of them. During their absence he had made the necessary astronomical observations, and finished the examination of those shores adjacent to the harbours.

From the 13th to the 17th, the schooner was detained by bad weather, and the following day only succeeded in reaching Escape Bay, in San Estevan Channel, which was found to be a good and well-sheltered anchorage, although small.

On the 19th, after angles had been taken on each side of the Channel, the Adelaide got under weigh, and steered up the Channel. At noon she passed the mount which they supposed to be Sarmiento's Monte Trigo, and soon after, nearing Esperanza Island, they sought for some mark by which to recognise the Mountain of the Fox ('Monte de la Zorra'). In the white part of a cliff, they fancied some resemblance to an / animal, and noticed a harbour opposite, in which they anchored. They had such trouble in getting to the northward, that this day's run, though only eighteen miles, was a cause of much satisfaction.

On the 20th, at daylight, the boats were employed around the anchorage, and at nine o'clock the vessel was underweigh, and working to the northward, although it rained hard then, as well as throughout the whole day, after beating until the evening, she anchored on the west shore.

Constant rain fell through the whole night, and during the 21st; it was therefore impossible to make any progress to the northward.

On the 22d the Adelaide weighed, and the weather being calm, was towed during the whole forenoon. At noon a southerly wind sprung up, and by the evening she was in the Guia Narrows (of Sarmiento). They tried for anchorage in Unfit Bay, conceiving it to be Sarmiento's Port Ochavario; but none being found, the vessel was towed into a cove, and securely moored.

Next day the boats surveyed the Guia Narrows. Although long, they did not appear hazardous to pass, for the tides are not very rapid. The ebb tide runs to the northward, but at the south entrance of the San Estevan Channel, the ebb sets to the southward; which difference in direction, within so short a distance, is extraordinary, and difficult to account for without knowing more of the coast. Certainly there is a meeting of tides between the two entrances; probably, all the land westward of San Estevan is a collection of large islands, and water flows into this channel, from the Pacific, through many openings, which may be the cause of this peculiarity.

24th. With light breezes from the eastward, the schooner weighed and stood through the Narrows; passed Point San Juan, and continued along the eastern shore of Concepcion Strait to Guard Bay, where she was moored.

25th. Rainy weather until near noon, when the boats were employed.

On the 26th the schooner was towed out, and, as it was calm, / kept a boat a-head the whole day. She anchored in a small bight, formed by Chance Islands, about seven miles from Guard Bay.

The 27th was rainy, but the boats went to different points, and angles were taken before the schooner weighed and worked northward. At noon she came to an anchor in a small bay, northward of the Hocico de Caiman. Constant rain during the remainder of the day.

On the 28th it rained too incessantly the whole morning, to allow the party to work, even in boats; and the day was passed in laying down former observations.

29th. After angles had been taken near the anchorage, the schooner was moved, and worked along the coast. A strong wind from the N.W., with a heavy sea, brought the vessel under close-reefed sails, and obliged her to anchor in Walker Bay.

On the 30th, the Adelaide anchored in Molyneux Sound. To give a clearer idea of the delays experienced in making progress to the northward through these intricate channels, I shall now extract part of Lieutenant Skyring's Journal, in his own words:

'31st. Wind N.N.W. with a heavy swell in the Strait; the boats at daylight went north and south of the anchorage, and angles were obtained. At nine, ready for starting; but the weather was too unfavourable, and continued so until the 4th of September, when, at seven o'clock in the morning, we weighed. At nine, squally – obliged to double-reef; but the tide serving, we gained a few miles to wind ward, and at one P.M., stood among a mass of islands on the west side, and moored in Tom's Bay, steadying the vessel with the stream anchor. In the afternoon the survey was continued, and from the heights a view was obtained of the Gulf of Trinidad, and of several points observed last year. Another detention of two days, owing to bad weather.

'7th. Cloudy; weighed at daylight, and stood for the narrows. At eight, squally, with thick snowy weather; but, being once under weigh, we refrained from returning, until compelled. It certainly was not a favourable day for working / through; but the wind moderated, and our attempt succeeded. No anchorage being found by the boats on the north side of the narrows, we made for the weather-shore of the gulf, and anchored early in Windward Bay. In the afternoon, angles were taken on Middle Island, and east and west of the anchorage. The time of our departure drawing near, it became doubly necessary to work constantly, that we might join this survey with that of last year, in the Beagle.

'8th. Weighed at daylight; wind light from N.W.; but, falling calm, boats were detached for continuing the angles, and the latitude was observed on Red Beak Rocks. At five o'clock, we gained an anchorage, close to the eastward of the Ancon del Morro, on the S.E. side of Division Isle, in a bay which answered our purpose, although it was rather a confined place. Some angles were taken on Point Candelaria, preparatory to continuing our course next morning.

9th. At daylight weighed and stood over to the northern shore, and at eleven, anchored in Neesham Bay, in eleven fathoms. Boats employed in the afternoon, on the survey. While at anchor, two canoes, containing together thirty-two Indians, came alongside; they were chiefly men, a finer race of people, better formed, and better featured than the Fuegians, and much less noisy. Their canoes were made of planks, the longest upwards of twenty-three feet in length: they appeared exceedingly buoyant, and pulled quickly.

'10th. At daylight, we sailed out of the bay, with a light breeze from the eastward; at seven, the wind increased, and a heavy sea rose in the gulf. It was my intention to get an anchorage under Mount Corso; but, as that was now a leeward coast, with a heavy sea setting upon the shore, it would have been improper to attempt seeking for one. If it had answered our purpose, we might have gone to Port Henry, and, indeed, this was the only safe course we could have pursued, if our object had been to remain in the gulf; but no time was left to wait for favourable weather; therefore I chose in preference to leave the gulf, and take advantage of the fair wind to gain an offing, the time of our return being so near. /

'We left the gulf two days before I had expected to have done so; but we all rejoiced at our departure. No crew could have performed their duty more willingly than the Adelaide's; but such lengthened fatigue as they had undergone, was sufficient to make any men feel happy at the prospect of a respite.

'It was a pleasing reflection to Lieutenant Graves and myself, that the orders had been fully executed; that the coast we had passed was throughout well connected; and that this service was concluded without any illness or accident among the crew, without any damage to the vessel, without any loss of boats, or even the slightest misfortune.'

During the Adelaide's passage to Chilóe, Lieutenant Skyring and his companion were assiduously employed in transferring their observations to paper, notwithstanding the violent motion of their little vessel, during ten days of rough weather. /

CHAPTER XVI.

Chilóe – Its probable importance – Valdivia founds seven cities; afterwards destroyed by the Indians – Migration of Spanish settlers – Province and Islands of Chilóe – Districts and population – Government – Defence – Winds – Town – Durability of wooden buildings – Cultivation – Want of industry – Improvement – Dress – Habits of lower classes – Morality – Schools – Language – Produce – Manufactures – Exports and imports – Varieties of wood – Alerse – Roads – Piraguas – Ploughs – Corn – Potatoes – Contributions – Birds – Shell-fish – Medical practitioners – Remedies – Climate.

As the Island of Chilóe was formerly shrouded from notice, by the policy of its master, the King of Spain, and therefore little known to the world;[1] I have considered it not irrelevant to the narration of the voyage, to introduce a short account of its present state, particularly as since the trade of the whole coast has been opened, a new era has dawned upon this interesting island; and although it has been, as yet, the least frequented of the South American States, I think the time is not far distant, when it will become an important part of the Chilian territory.

After the foundation of the city of Penco, or Concepcion, by Don Pedro de Valdivia,[2] in the year 1550, he passed on towards the south in search of convenient situations for other cities; and crossing the river Bio Bio, which separates Concepcion from the territory of the Araucanian Indians, successively founded Imperial, Valdivia, Villa Rica, Angol, Cañete, and Osorno; the last being effected in the year 1558. The necessary distribution of the Spanish forces, to protect so many points, made them comparatively defenceless, in a country inhabited by a large population of Indians, who contemplated the hostile occupation of their native land, by the invading army, with a deep dissatisfaction. They had for some time endured, with sullen patience, the yoke of the Spaniards; but at last, incensed by the servility and bondage to which they were reduced, and, probably, by no small portion of ill-treatment; the whole population rose simultaneously, and waged a most destructive and / harassing war against the Spaniards, in which the above-mentioned cities were all destroyed, and the greater number of their inhabitants put to death.

The destruction of the city of Osorno caused the province of Chilóe, or, at least, the adjacent districts of Calbuco and Carelmapu, to be occupied. This town, being more distant from the seat of war, where the main body of the Indian army was actively employed, was enabled to hold out for some time; but, at last, cut off from assistance, prevented from communicating with friends, and utterly destitute of supplies, the inhabitants retired to the fort, or citadel; which they maintained, until compelled, by absolute want of provisions, to abandon their position, and proceed to the south, with a view of establishing themselves in Carelmapu and Calbuco; where they hoped to be safe from attack.

Their retreat was attended by much suffering; many died from fatigue, and many were cut off by the Indians, who hovered about them and murdered all who fell into their hands.* At last they reached their destination, and established themselves first at Carelmapu, which is on the main-land, on the north side of the Boca de Chilóe, opposite to San Carlos; and afterwards at Calbuco, on an island at the entrance to the Gulf of Reloncavi. The latter position by its insularity, was effectually protected against any attack from Indian tribes, who, for many years, continually harassed the inhabitants of Carelmapu.

At what date this journey was made does not appear; nor is it certain that these places were occupied before the foundation of the city of Castro, in 1566, by the Licentiate Lope Garcia de Castro, in pursuance of an order from the Viceroy of Peru, Marshal Don Martin Ruiz de Gamboa.†

The island of Chilóe, from its situation, is a place of considerable importance, and may be termed the key of the Pacific, / It is the northernmost of that vast archipelago, which borders the coast from latitude 42° south to Cape Horn.

The province of Chilóe, one of the eight divisions of the Chilian Republic, includes several islands, and extends on the main-land, as far as the south bank of the River Maullin;‡* which takes in the districts of Carelmapu and Calbuco. Its southern extent is not defined; but as the existence of Chilian authority is not known, to the southward of the Chonos Archipelago, certainly not farther south than the land of Tres Montes, the parallel of 47° may be considered its southern limit. The country thence, to the Strait of Magalhaens, is known by the appellation of Western Patagonia.

* A very full and detailed account of this journey is given by Agüeros, in his 'History of the Province of Chilé,' pp. 50 to 56, as well as in the 'Chronicles of the Province of Lima, by Padre Fr. Diego de Cordova,' Salinas, chap. xvii. p. 485.[3]

† Agüeros, l.c.p.57.

‡ Agüeros describes its boundary thus:– It is situated between the latitudes 41° 30' and 44°; from Point Capitanes to Quilan. On the north it is bounded by the territories of the Indian tribes Juncos and Rancos, which extend to Valdivia; on the N.E. by those of the ancient but destroyed city Osorno; on the south by the archipelago of Guaitecas and Guaianeco, and others which extend to the Strait of Magalhaens; on the east by the Cordillera; and on the west by the sea. (Agüeros, p. 61.)

Besides the Isla Grande, as Chilóe is called, the following islands are inhabited:– Achao, or Quinchao, Lemuy, Quehuy, Chelin, Linlin, Llignua, Quenac, Meulin, Caguach or Cahuache, Alao, Apiao, Chaulinec, all in front of Castro; the Chaugues Islands, opposite to Tenoun; Calbuco, Llaichua, Quenu, Tabor, Abtao, Chiduapi (on which is the fort); Huar in the neighbourhood, and district of Calbuco; and, to the South, Tanqui, to which may be added Caylin, which is also called El fin de la Cristiandad.*

Of the above, next to the Isla Grande, the principal are Quinchao and Lemuy, both of which are very populous, and / almost entirely cultivated. The other islands are small, and very close to each other; but separated by navigable channels, which offer many dangers to the frail vessels in which the islanders move about.

The province is divided into ten districts, or Partidos, as follows:–

1. San Carlos, containing the northern coast of the island, as far as Chacao.
2. Chacao. The N.E. part of the island.
3. Carelmapu and Maullin.
4. Calbuco.
5. Dalcahue, extending from Chacao to Tenoun.
6. Quenac.
7. Quinchao.
8. Castro.
9. Lemuy.
10. Chonchi, which extends from Castro to the south extremity of the island.

By the census of 1828, the population of the large island, and those in its neighbourhood would appear to be, comparatively, very considerable; the number of souls being 43,131:† particularly as the greater portion of the interior, and much of the sea-coast, are quite uninhabited. The population of the district of San Carlos is confined principally to the town; for between it and Chacao, there are very few inhabitants. At Chacao there are only about two hundred houses, and Dalcahue is but thinly occupied: but Castro, Quinchao, and Lemuy, are very populous. These three districts are the most fertile and productive part of the island, particularly for seven or eight miles round Castro. The peninsula

* When the Yntendente, or governor of the province, visited Castro for the purpose of taking a census of the population, a family of Indians waited upon him to render an account of their property; who, upon being asked whence they came, replied, 'Del fin de la Cristiandad.' The name being new to the Yntendente, it was explained to him that they belonged to Caylin, which was more generally known by the above name, because there existed no Christian population beyond, or to the southward of, that island.

† In the year 1783 there were 23,447 (Agüeros): and in 1832, 43,830.

opposite to that town, which is entirely cleared, would abundantly repay its cultivators, were industry more common among them.

Chilóe is governed by an 'Yntendente,' or civil governor, who exacts obedience to the constitutional laws, as well as to the orders of the executive powers, and the resolutions of the provincial assembly, which is composed of members, elected / by the people, at the rate of one deputy for 7,500 souls; but whatever the number may be, short of 90,000, twelve deputies are to be elected. The duration of the assembly is biennial, and its business is to superintend the civil regulations of the province.

Under the Yntendente each province has a local governor, whose principal duties are to maintain order, preside in the municipal meetings, see their regulations carried into execution, and obey the orders of the Yntendente of the province. Whilst we were at Chilóe, the duties of Yntendente, and military commandant, were at performed by one person, Brigadier-general Don José Santiago Aldunate; but, upon his resignation,[4] the offices were separated: the military commandant retaining the charge of the treasury. The duties of the military chief, are to dispose of the troops under his command, as he sees occasion, so as to ensure the quietness, and subordination of the province, for which he is responsible; and to render the Yntendente such assistance as he may require; but, for all ordinary purposes, the Militia, who are under the immediate control of the Yntendente, are employed. For the administration of the law there is a Judge (Juez de letras), who tries all civil as well criminal actions. The province sends two deputies to the Chilian congress, one from San Carlos, and the other from Castro. At the beginning of the year 1829, the Militia amounted to more than seven thousand men, and the regular troops to three hundred and thirty, which was quite sufficient for the province.

The port of San Carlos is capable of being well defended, and, during the time of the Spaniards, was in a good state of defence. The entrance was protected by a battery on the highland of the Corona, and by the castle of Aguy, which effectually commands it. Farther in, on the same side of the port, was the small, but well-placed, two-gun battery of Barcacura; close under which is the anchorage. On the town side there are several batteries; but, towards the Pudeto it is weak, although capable of being made very strong. Fort San Carlos, which, for some years past, has been used as a cemetery, was well-selected as to position, and constructed in a manner very / creditable to the engineer. It was surrounded by a deep and wide ditch; and under it lay two small batteries: one, San Antonio, commanding the passage between the small island of Cochinos, and the Main; and the other flanking the anchorage off the town. At the Mole were two guns, and opposite to it, under the governor's house, was the battery, Del Carmen, mounting twelve or fourteen guns. In the town, in a convenient situation, there were excellent barracks, capable of containing more than one thousand men.

The original establishment was at the Sandy Point, on the western side of the port, where the situation is better sheltered, and, perhaps, equally capable of being well defended. It is, also, on the windward side of the harbour, and close to the safest anchorage which the port affords; but the inconvenience of water-carriage was found to be so great, that the establishment was removed to its present site. A still better situation might have been selected opposite to Sandy Point, at Leche Agua; where the anchorage is perfectly safe, and the communication with Castro could be more advantageously made.

Northerly and westerly winds prevail, and the town is exposed to all their fury, which, at times, is extreme. The anchorage nearest to it, for the sake of convenience, and expedition in loading and unloading cargoes, is often taken up, but is very unsafe, many vessels having been lost there, from the bottom being shoal, and rocky; and the swell, during a northerly gale, is so short and deep, that anchors will not hold.

The town is built on two rising grounds, and in the valley that separates them; through which a rivulet runs into the bay, at a mole which affords sufficient protection to the boats and piraguas frequenting the port. The houses, which are all of wood, are generally small, and have but little comfort. The plaza, or square, without which no town in Chile of the least importance is to be found, is situated on a flat piece of ground at the summit of the southern hill, and commands an extensive view. It is about one hundred and eighty yards square, with a flagstaff in the centre.

On the north side there is a strong, well-built stone store-house, / / and opposite to it is the church, also built of stone. On the side next the sea is the Yntendente's residence, a low range of wooden buildings, erected without regard to taste, convenience, or comfort; and opposite to this are two or three dwellings, very little superior to common huts, or ranchos.

Within the last few years, however, some substantial buildings have been erected by the more wealthy people in the town, an example which is likely to be followed. During our visit, several were built equally creditable for strength and convenience; and not a little remarkable for the rapidity, with which they were completed.

Wood, being abundant, and cheap, as well as easily worked, is the only material used in the construction of houses, which, with the exception of the provision-store, and the church, are all built of it; and notwithstanding the perishable nature of the material, which is not protected by paint, or any external coating, from the humidity of the climate, they are of extraordinary durability. The treasury, one of the oldest houses in the place, has been built upwards of seventy years; and is even now tight, and dry, and by no means unserviceable: but its removal has been ordered, and, probably ere this, it has been replaced by another. In Chacao, where, in former days, the Yntendente resided, the greater number of

SAN CARLOS DE CHILÓE.

C. Martens. S. Bull.

SAN CARLOS DE CHILÓE.

the government-buildings, not less than sixty or seventy years old, are still standing. This durability can only be accounted for by the nature of the wood, and the practice of charring the ends of the timbers before they are inserted in the ground. The lower frame is of 'Roble;' (*t*) the beams are of laurel, and the floors and partitions, as well as the weather-boarding and shingles, of 'Alerse:' the latter forms an excellent substitute for tiles, or slate, being much lighter, and almost as durable. Some of the houses are thatched with reeds; but this shift is only used by those who cannot afford the expense of shingling.

The inclosures, round the houses, are fenced with stakes of / Luma, three or four yards in length, fastened above and below to cross-rails, by ligatures of creeping plants, of which there is an abundance in the woods close to the town: the general name for them is Buque.

The land in the vicinity of San Carlos, which is a peninsula, is cleared of timber, and partially cultivated. In the valley, through which the rivulet runs into the sea near the mole, there are a few attempts at gardens; but the extent to which the inhabitants cultivate, seems to be confined to a rood of potatoes and wheat, which, with a litter of pigs, and an inexhaustible store of shell-fish on the coast, are the principal support of their families. It is not surprising, when so little personal trouble is necessary to provide subsistence, that the Chilotes (*u*) should not be an industrious race. Byron, in his narrative of the loss of the Wager, has given a most excellent and correct account of the inhabitants of this island; which, excepting for those about San Carlos and Castro, may well serve at the present time.[5] In the town, trade, a free communication with other parts of South America, and the residence of several Europeans, have introduced approaches towards refinement; and besides the articles of luxury that occasionally make their appearance, such as chairs and tables, crockery-ware, and similar domestic comforts; shoes and stockings are now, on feast days, in common use among the females; although in many instances one can easily observe, that the wearer is actuated by vanity, rather than by any comfort or pleasure she derives, from a confinement to which her feet have not been accustomed.** This is one of the steps towards civilization, which the Chilote peasantry are making, and among the higher classes 'el ultimo modo' (the latest fashion), is not less the theme of conversation than it is in other parts of the Republic.

In style of dress, among the upper ranks, the men are more advanced than the women, many having been in other countries. / They have given up the use of

(*t*) A kind of beech, found every where on these shores. The literal meaning of Roble, is oak. – R. F.

(*u*) Native of Chilóe. – R.F.

* Agueros says, 'both men and women go generally with the foot and leg uncovered; with the exception of the principal families; but even those do not all wear shoes.' – (Agüeros, p. 108.)

the poncho, and in this particular, they say they are before the gentry at Concepcion, who wear it on all occasions: and probably are quite right, for, with respect to comfort, there is much to admire in the poncho, as, of all cloaks, it is the most generally convenient, and the best adapted for protecting the person, especially on horseback, where it is indispensable: its use, however, offers the wearer such an opportunity to neglect the other part of his dress, which it effectually conceals, that sometimes, beneath the poncho, the body is very ill-clothed.

The dress of men in the lower orders, consists of a pair of trowsers, and a shirt, over which is thrown the all-concealing poncho. The women are as slightly clad; but instead of a poncho, they wear a rebozo, or shawl, which, however, is very often dispensed with, and their persons are left too much exposed.

These lower classes, or Indians, as they, with much reason, are termed, are scarcely superior to the uncivilized savages of the southern coasts; and live principally upon shell-fish, with what little they are enabled to procure besides by the sale of a few pigs, or poultry, which they rear on the scanty store of potatoes and-wheat, that remains after their new crop comes to maturity. One roof shelters a whole family. Father and mother, sons and daughters, dogs and pigs, all live and sleep in their only room, in the middle of which, a fire is made; whence the smoke escapes by numerous apertures in the roof and sides of the dwelling.

As to their morals, within the precincts of their habitations, I have reason to believe they have not much to boast of, although they are described, by Agueros and other writers, as most innocent, and well-conducted. Agüeros speaks highly of their character; and cites Padre Ovalle, who, writing upon Chiloe, between the years 1629 and 1636, says: 'The natives of these islands are the most docile and noble (dociles y nobles) of all Chile, and are the least given to drunkenness, and other vices; therefore they are best disposed to be edified by the light of the Gospel.'

Since the province became subject to the Chilian Republic, / the government has made several attempts to improve the condition of the inhabitants; among which, the instruction of public schools, was not the least important. From an official report[6] there appear to be ninety schools, in which 3,840 children receive an education, according to the abilities of the masters, who are employed; but these, from the small salary attached to the situation, cannot be expected to be superior.

The language in common use, is Spanish; the original Indian tongue being almost forgotten: but it is supposed to be the same as that spoken by the Indians of Madre de Dios; for, on a late occasion, a whaler which had been upon the coast of those islands, and had taken on board an Indian, as a pilot, called at Castro; and during her visit, the Indian communicated with those who understood the language of the Chonos, and by them was tolerably well understood. This Indian has been frequently embarked on board American or English sealers, which frequent those coasts, to serve as a pilot to the seal-rookeries.** He is known by the name of Dan.

The products of the island, for the year 1828, according to the census, and returns, officially made, were –

Wheat	64,935 fanegas (175 lbs. in a fanega) about 200,000 bushels.
Barley	21,645.
Potatoes	194,805.

and the muster of stock, and apple-trees, as follows:–

Horned cattle	5,411 head.
Sheep	86,580
Swine	21,645
Apple trees	75,754

The manufactures of the province are Carro, a coarse woollen cloth, two and a half, or three yards long, and three quarters of a yard wide, used for men's garments, and of very durable quality.

Ponchos – both these and the carro are manufactured by women, in a rude sort of loom, of wool dyed of various colours from plants that are found in the island, or imported for the / purpose. Of the latter indigo is much used, and it is the general colour for the ground-work of the ponchos.

Frezadas, bordillas, sabanillas, mantillas de lana, blankets or rather counterpanes of different textures, are also among the manufactures: none of the above are exported, being made merely for their own use.

Cables, hawsers, and rope, they make of a plant, called Quilineja, which is supposed to be the root of a species of *Callixene*.

No wine or spirit is made in the province, but Chicha (a very good cyder) is manufactured from apples. The only other fruit produced is the 'Frutilla,' a kind of strawberry.

The exports must very nearly amount to the value of foreign imports, which consist principally of sugar, wine, brandy, salt, wearing apparel, and household furniture. The import duty on European and North American produce is twenty-seven per cent.; from which, however, some articles, such as arms and munitions of war, instruments of music, and other things of less importance, are exempt. Spirits of all kinds, foreign wines, tobacco, tea, and cards, are monopolized by the government, and sold at an immense profit. The unauthorized sale of these goods is declared illegal, and is punishable by a heavy fine, and sequestration of goods.

* Places where seal congregate – so called always by the sealers.

The exports, during the year 1828, consisted of wood in beams, planks, and boards; hams, wheat, a small quantity of dried fish, fire-wood, and brooms,** to the amount of 52,320 dollars, of which 35,683 dollars were for wood, and 10,887 for wheat. These articles were exported in sixteen vessels under national, and eight under foreign flags. The exports are said to be increasing very much. In the year 1791, Agüeros describes the exports of alerse planks (tablones) to Lima, to be between fifty and sixty thousand in number; and some years previous to have been in a much greater quantity. The number of alerse boards exported, during the last year, was 328,928, but of planks only 2,623.

The island, and neighbouring part of the main land, produce / a great abundance, as well as variety, of wood fit for exportation, as well as home consumption. The following is a list of the principal trees, with their qualities, and the use to which they are most adapted.

Avellana *(Quadra heterophylla)*, a handsome tree, in appearance like the ash of Europe, of a light wood, which shrinks very much when dry, and may be used with advantage for oars, being light, strong, and springy, as well as for planking small vessels below the water, and for the ceiling within; it is bad for firewood, being too light. The seed is a nut, about the size of a cherry, the kernel of which is roasted and eaten. The tree abounds at Concepcion, and in the country to the south, and grows on the Peninsula of Lacuy.

Roble *(Fagus obliqua*, Mirb.), a large tree; and, from the durable quality of its timber, considered the best in the island, for ground-frames of houses, planks for vessels, and beams. The piraguas are built chiefly of this wood. There are two sorts, one an evergreen, and the other a deciduous-leaved tree. It is evidently a beech, and the same that grows in all parts of the Strait of Magalhaens; the smooth-leafed sort is *F. obliqua* of Mirb. – see Bertero, in Mercurio Chileno, No. 14, p. 640.[7]

Tiqui, heavy wood; but esteemed strong and durable. Piraguas are sometimes built of it.

Laurel, used for house building in-doors, for beams and rafters, and posts; durable when not exposed to damp, in which it soon perishes.

Mañu, a tree of great dimensions, tall and straight, the leaf is like that of a yew; it is a very useful wood in ship-building, for planks, and, next to alerse, is the best for spars which the island produces; but the large trees have a great tendency to become rotten at the heart, owing possibly to the humidity of the climate, and to the very wet soil.

As the Adelaide wanted a mast, I sent her round to Castro for a mañu spar, for which I agreed to pay eighty dollars; but of twenty trees that were cut down, not one was sound at the heart. The wood is heavy, with large knots, which pen-

* Potatoes are not mentioned in the report, yet they must have been exported in considerable quantities.

etrate into the trunk to a great depth. A great deal of this timber grows in the Gulf of Peñas. /

Muermo. There is no wood produced on the island more useful than the muermo. It is used for timbers, and knees, and all other purposes of ship-building: and is excellent for the planks of boats, as it bears wet and dry without suffering from either. It is abundant, and much used as firewood, for which it is well suited.

Luma *(Myrtus Luma)*, a very tough and useful wood, used for tree-nails, for stakes in fencing, for rafters in the roofs of houses; and is exported in large quantities to Lima, for shafts and poles of carriages. The fruit is sweet, and might yield a strong spirit; it is called cauchao.

Ciruelillo, a small tree, used only for washing-bowls and boxes; it is of little value.

Quiaka. Of no value.

Tapu, a very crooked tree, growing along the ground in swampy places. It might serve for floors, and timbers for small vessels; but it is not used, from its being so very hard.

Tenu, something like muermo, and considered a good wood.

Peta, a species of *Myrtus*, of which hoops for barrels are made.

Ralral, considered to be like the wood of the walnut-tree, and of general use, on account of its toughness and durability; it is made into blocks for ships.

Meli, more tough than luma: of this the country people make pick-axes, for cultivating the ground (Agüeros, p. 127).

Pelu, also tough; useful for axle-trees and gun-carriages (Agüeros, p. 127).

Mayten, useful for turning; and lasts long under water.

The above mentioned are produced on the island; but the two following, alerse and cypress, are from the main-land, in the neighbourhood of the Cordilleras. They are not only in general use in Chilóe, but are exported in large quantities to all the ports to the northward. The alerse, near Chilóe, is of better quality than that which comes from Concepcion.

The Cypress is brought to the island in 'tablones' (or planks), seven or eight feet long, two inches thick, and nine or ten inches wide, as is also the alerse; but the latter, from the facility with / which it splits, is brought in boards also, four feet long, half an inch thick, and six inches broad, which, as I have before remarked, are the principal articles of barter.

The Alerse is found in great quantities near Calbuco; but at so great a distance from the beach that it cannot easily be conveyed thither for embarkation, except in the above form. The tree is cut down and squared, then hewn by the axe into as many logs of seven or eight feet long as it will afford; and these, with the assistance of iron wedges, are split into planks and boards, in which state, without being further trimmed, they are tied together in bundles, and carried on men's backs, or dragged over the ground to the beach.

The extraordinary straightness of the grain of this tree enables the natives to split it, so as to make it appear as if it had been dressed with an adze, or even with a plane; but, as I have said, the axe is the only instrument used. So great is the difficulty of obtaining a spar of this wood, that when I wished to procure a new mast for the Adelaide, I offered four times the value of an alerse spar to the natives, besides the assistance of twenty men, and tackles, &c. to assist in conveying it to the beach. The temptation was almost too great to be withstood; but the man to whom I applied, who had before been employed to get masts for a schooner in the Chilian service, and a flag-staff for the town, said that it would take his own party two months to bring one to the beach: with the assistance of our people, however, it might be done in a month. The trees were distant, and there were two or three ridges of heights to cross, that would cause much delay. The facility with which these people usually handle timber was a sufficient proof to me that such a task, if refused by them, must be very difficult indeed, and I gave it up, as the Yntendente was so obliging as to give me the flag-staff, which had taken the same party two months to procure.

The Hoxsley, a national schooner, built at Chilóe, for the government, was masted with alerse spars, which proved to be very strong.

Alerse is used principally for the floors, partitions, and / weather-boards of houses, also for shingling the roof; for which purpose it is very superior and durable: after exposure to the weather it turns blue, and has the appearance of slate. It does not shrink or warp; and though brittle, is of a very close grain, and well adapted for furniture. Of this wood the country people make staves for casks; and the bark of the tree is used for caulking the seams of vessels, for which it answers remarkably well, being extremely durable when constantly wet, though it soon decays when exposed to the sun and air.

Spars of alerse, eighty or ninety feet in length, may be procured; and from eight hundred to a thousand boards are frequently obtained from a single tree. I was told that as many as one thousand five hundred have sometimes been cut out of one trunk. Alerse is found on the island, but not of any size. It is also common in the Strait of Magalhaens, in all those parts west of Cape Froward; but there, from the poverty of the soil, it is of very stunted growth.

The cypress is thought to be a different tree, but I rather imagine it to be only a variety; the wood being white, whilst that of the alerse is of a deep-red colour. As the trade of the island is principally carried on by water, roads are seldom used for that purpose, for which, indeed, the few that exist are far from being convenient. Between San Carlos and Castro there is a road cut through the forest, forty or fifty feet in width, in the middle of which is a causeway, four or five feet wide, formed of logs of wood, laid transversely. This is the only way of communication, unless, which rarely occurs, the weather has been dry during some days; for, off the causeway, there is a mere bog, in which a horse frequently sinks up to the girths in mud. In many parts of the causeway, indeed, where the logs have decayed, and have not been repaired, the passage is equally bad, so that in wet weather, only persons without a load are able to pass. For the greater part of the way, the trees on each side prevent an extensive view; but on approaching within five or six miles of Castro, the country becomes more open, having been cleared by cultivation, and there, of course, the road improves. /

There is a track branching off from the main road to the district of Dalcahue; but on it, I believe, there is no causeway.

As the only mode of supplying the town of San Carlos with provisions is by water-carriage, it is frequently ill supplied during winter, when N.W. winds prevent the arrival of the piraguas. A southerly wind for two days, at that season, brings from fifty to a hundred piraguas from Dalcahue and Castro, laden with hams, potatoes, pigs, grain, fowls, calves, dried fish, and charcoal, which are sold at a cheap rate, paying one-tenth to the government.

The arrival of so many piraguas at San Carlos creates no slight bustle in the neighbourhood of the mole; and a stranger happening to arrive at the time would think it a place of considerable trade; the return, however, of the N.W. wind, with all its attendant 'vapours, clouds, and storms,'[8] very soon dispels the illusion: the piraguas depart, one after another, and in two days all is dull and monotonous.

These piraguas, the boats used by the natives of the archipelago of Chilóe, are all similar in form and material; but vary much in size, according to the voyage they have to perform. The largest are from thirty-five to forty feet long. The head and stern are alike, and resemble those of a whale-boat, being sharp at both ends. The transverse section is that of a thick wedge, so that they have no bearings, and must be extremely unsafe, *(v)* particularly with so lofty a sail as they hoist; and yet these vessels have made long, and even dangerous passages, as is fully attested in Agüeros's account of the missionaries' visit to the archipelago southward of Tres Montes. These boats are literally sewn together, there is not a nail used in their construction; every portion of the hull is of a vegetable nature. The lower, or garboard strake, is sewn to the keel by strips of the stem of a creeping plant, called Pepoi,* and the seam is caulked with bark of the alerse, which, while under / water, is admirably adapted for the purpose. The upper planking consists of three or four broad boards on each side, sewn together, and their seams caulked. The wood of which they are made is the roble, or sometimes tiqui.

Agüeros's description of the construction of a piragua cannot be improved. 'They are constructed of five or seven planks, each of which is from two to four

(v) When moderately laden they are stiff under sail; and are not such very bad sea-boats, if properly managed. – R.F.

* Molina, i. 167. A species of 'Dolichos.'

fathoms long, half or three-fourths of a yard wide, and two or three inches thick. These are fashioned, or worked, narrow at each end, so as to form the bow and stern, and afterwards are exposed to the fire, in order to burn the outer surface on both sides. To unite these planks, they bore or burn holes, two inches from each other, along the edges of the planks, through which they sew them together with a rope of solid reeds (soquillas), or twisted cane (coligues), forming a junction as close as a seam of cloth. To prevent water from passing through the seams, they apply along the plank, within and without, pounded leaves of trees, over which they pass the stitches, and with the same preparation of leaves the holes are filled up. Thus constructed, it is in appearance a perfect boat, or vessel, but without keel or deck. That they may resist the pressure of the water, and retain their shape, curved pieces (curbas) of wood, called 'barrotes' are fitted inside, and fastened by wedges of wood, instead of nails. For all this, they are dangerous; and, since their sails, oars, and other furniture are very inferior to what boats require, they are much exposed to be easily sunk, and the risk is greatly increased by want of care and management in those who navigate them.'

In the above description Agüeros has given a very good account of the rude manner in which they are built, and has not in the least magnified the danger attendant on their use. It is, indeed, a miserable and unsafe vessel; and for the rudeness of its construction, and the poverty of its equipment, is a perfect prototype of the crew which it conveys.

The largest have from eight to ten people, each of whom furnishes one poncho, and the 'patron,' who steers, and directs / the course and all their movements, provides two ponchos, all which are sewn together to form their sail, which is hoisted by 'lazos,' or thongs of bullock's hide.

These sails are generally in a wretched state, the name Santisima is applied to them all by the crews, with the hope of securing the protection of their patron saint. The anchor is of wood, formed of four crooked pieces, in the shape of a grapnel with four flukes, at the bottom, or crown of which a large stone is fastened, to increase its weight. The crews are exceedingly timid, and instead of making exertions to extricate their vessel from any impending danger, they throw themselves on their knees, beating their breasts and calling loudly upon their saint, for 'misericordia.'

I was given to understand that very few of them can swim, which seems extraordinary, since they are born and bred in the immediate vicinity of the sea, and depend chiefly upon its productions for subsistence. The fact speaks strongly for the indolence of their character, even although the rigour of the climate forms a bar to bathing as a mere amusement. Several piraguas were lost while we were at Chilóe, and, as may be inferred, their crews were all drowned.

With regard to the cultivation of land, they are very far behind, and, comparing the present state with the description of Byron (1740), and of Agüeros

P. P. King. T. Landseer.

BREAST PLOUGHING AT CHILOE.

(1791), very little improvement seems to have been made. The ground is prepared by make-shift ploughs, of a very rude construction. Two poles of hard wood (luma), about three yards long and proportionally large, trimmed to a sharp point at one end and rounded at the other, are held by the middle, one in each hand, and pointed very obliquely into the ground; in this direction they are forced forward, by pressing against the blunt end with the abdomen, which is defended by a sheepskin, suspended in the form of an apron. After these have penetrated twelve or fourteen inches into the soil, a second person, generally a woman or a boy, places a stout stick under the poles, or 'lumas,' as they are called, close to the earth, to form a solid support for them. / / The large ends are then forced down, the ground turned up, and the lumas pushed forward again, which the woman uses her stick to turn the clods over, to the right and left, alternately. These clods are afterwards broken up by a wooden tool, in the shape of a pick-axe, called 'hualate,' made of the wood named meli. Rude as this process is, the operation is rapidly performed, and I have seen a field, ploughed in this way, that would not do much discredit to an expert ploughman with a European plough.

The soil is a rich, sandy loam, of a dark red colour; and although rarely, if ever manured, produces fair average crops. According to the usual allowance of 175lbs for a fanega of wheat,* the weight of a bushel would not be more than 51 1/2 lbs., which shews that the grain is but poor. Wheat is sown in the month of April, and cut in the same month of the following year; but from the humidity of the climate, and constant rain, particularly at that season (the commencement of winter), it is frequently reaped before it is ripe, and almost always gathered in wet. Every subsequent sunny day is taken advantage of, to dry the grain, but a part must be spoiled by mildew. The evaporation, however, is so great, that merely moving it about, and keeping it thinly strewed in granaries, will effect much. It is trodden out by oxen, and to clean it, the grain is thrown up in the wind by means of broad wooden shovels, and effectually separated from the chaff. This rude winnowing takes place frequently in the principal streets of San Carlos, and even at the mole, where one would suppose that a great deal must be lost; but from the adroitness of the operation, it is not only well cleaned, but suffers no diminution.

Potatoes are planted in September, October, and November, and are fit to dig up in May.

Of the proceeds of harvest, one-tenth is paid as a tribute, or / tax, to the government; but forced contributions may be required, when the necessities of the state demand them. These contributions are sometimes unfairly levied in Chile;

* The fanega weighs 175lbs. and contains twelve almudes, which being cubic measures of eight inches and a half, contain each 614.125 cubic inches; therefore a fanega contains 7369.5 cubic inches, and as an English bushel contains 2150.4 cubic inches 2150.4 x 175/7369.5 = 51 6/10 lbs. the weight of a bushel.

for the subsidy is only taken from those who possess grain, or some equally tangible article which can easily be turned into money; so that persons who are rich enough to live without cultivating land, or trading for their support, contribute nothing towards the emergency of the State. How does this accord with republican principles? or how can a republican government, so conducted, expect to become respectable among nations?

I am not aware that such contributions have yet been levied in Chilóe. From the character of General Aldunate, I do not for a moment think he would commit such an act of injustice; but it is in the power of any Yntendente to call for them, and I afterwards witnessed an example of this, during my visit to Conception. A considerable quantity of wheat, purchased by a Russian vessel, for the use of their settlements on the coast of California,[9] was brought down to the port, at a time when the government was much in want of money, and knew no just way of obtaining it. They therefore very unceremoniously seized the wheat, and applied its value in dollars to their own use, giving only an uncertain, almost a nominal security to the owner for the recovery of his money. The only way of accounting for such an arbitrary proceeding is, that the country was distracted by civil war, and that the person who owned the property was opposed to that party, which at the time happened to have the upper hand, and which held, by main strength alone, the reins of government.

Among the birds of Chilóe, the most remarkable are the 'Cagge,' the 'Cancania,' or 'Canqueña,' and the 'Barkingbird.'*

These birds are very common in the Straits of Magalhaens, and every where on the west coast between the Strait and Chilóe; also at the Falkland Islands.

The *Cancania* (or *Canqueña*) is the *Anas Magellanica, Anser Magellanicus* (Ency. Méth. p. 117). From Buffon's description, and a well-drawn but badly-coloured figure, in the Planches Enluminées, No. 1006, I have no hesitation in assigning it to that kind. The colour of the head, however, instead of being

* Molina notices the 'Cagge,' or 'Chilóe duck,' (*Anas antarctica*) vol. i. p. 268, and calls it *Anas hybrida*. M. Lesson, in his 'Manuel d'Ornithologie,' ii. 409, has taken great pains to describe it, and remarks, with reason, that much obscurity exists in the specific descriptions of the goose kind in the Malouine (Falkland) Islands, and the extreme southern land of America. The male, Lesson says, is white, the feet and beak of a bright yellow colour. All the specimens that we saw, and numbers were killed by us, had a black beak with a red cere – otherwise M. Lesson's description is correct. In many specimens, however, we found the tip of the primary wing feathers black, which is not to be wondered at when the colour of the female is considered, but which it is not an easy task to describe. M. Lesson, I think, has done it justice in a note to his vol. ii. p. 409:– 'Anas antarctica. A capite griseo, genis gulo colloque albo et nigro acuti-striatis; oculorum circuitu nudo: pectore abdomineque omninò atris, atque vittis niveis notatis: tectricibus alarum nigris; dorso uropygio caudâ et ano albis; alis niveis cum speculo lato virescente, brunneo marginato; pennis longis aterrimis; rostro et pedibus, aurantiacis.'

'reddish purple,' is cinereous with a reddish hue; the feathers of the sides and thigh covers are white, with five black bars, the extremity being white; the central portion of the abdomen is white; the speculum of a splendid shining green. This bird is common to the Strait as well as to Chilóe, and is probably Byron's 'Painted Duck,'[10] and the *Anser pictus* of the Ency. Méth., p. 117. M. Lesson considers *Anas leucoptera*, Gmel. as the male of *Anas Magellanica*, which may be doubted. The 'Barking Bird,' as our sailors called it, was first brought to me by Capt. Stokes, having been shot during the Beagle's visit to Port Otway, in the Gulf of Peñas. It was an imperfect specimen; but Mr. Tarn afterwards obtained for me several others. It seems to have a great affinity to the genus *Megapodius*; but no specimens of that genus being in England when I was last there, and the Barking Bird differing in essential points from M. Tenminck's description of the genus, and from the figured specimen of *Megapodius Freycinettii*; – particularly in the length and form of its wings, which are rounded, and so short as not to reach beyond the base of the tail; – also in the emargination of the upper mandible; – I have been induced, by Mr. Vigors' advice, to form it, provisionally, into a new genus, termed *Hylactes*.[11] (See Proc. Zool. Soc., vol. i. p. 15.) There is another specimen in our collection (now in the Zoological Society's Museum), which will probably be placed in this genus, but there existed some uncertainty in essential points, which prevented my describing it before I left England. /

The shell-fish,* for which this island is justly famed, are principally brought from Calbuco, and consist of the finest / muscles, of which there are two sorts: the Choro (*Mytilus Choras*, Molina), and Cholgua (*Mytilus Magellanicus*, Lamarck), Picos (*Balanus psittacus nob. Lepas psittacus* Molina, 1, p. 223), a large barnacle,† and the oyster (*O. Edulis*), which is exceedingly well-flavoured. Besides which there are several kinds of shell-fish of less value, but equally abundant, such as Navajuelas (*Solen sp.*); Caracoles *(Turbo*); Cornes (*Pholas Chiloensis*, Molina); Campaña (*Calyptræa*); Lapas (*Crepidula*); Tacas (*Chama Thaca*, Molina); Locos (*Concholepas Peruviana, Murew Loco* of Molina); Quilmagues; Piures (*Pyura sp.* Molina); and others.

The apparently inexhaustible abundance of shell-fish with which nature has provided the inhabitants of these islands, the facility with which they are

* Among the numerous testaceous productions is a small shell, which constitutes a new genus. *Marinula*, nob. in Zool. Journal, vol. v. p. 343. It was found on the wooden piles which support the mole in the bay of San Carlos, below the wash of the high water. The mole stands out into the sea, and there is no fresh water near it, save a very little rill, which discharges its tiny stream more than fifty yards off. This shell was named *Marinula Pepita*, Zool. Journal, l.c. No. 43. The following is its generic character:– 'Testa ovato-producta, sub-solida; apertura ovata, integra; columella bidentata et basin versus uniplicata; dentibus magnis subremotis conniventibus, superiori maximo; operculum nullum.'

† Zool. Journal, vol. v. p. 333.

obtained, and their consequent cheapness, is the principal cause of that want of industry which is so remarkable in the Chilotes.

Of the above-mentioned shell-fish, those deserving more particular notice are the large muscle, the oyster, and the pico.

Molina has described the choro of Concepcion, which is not at all different from that of Chilóe. It is often found seven or eight inches long. The fish is as large as a goose's egg, and of a very rich flavour: there are two kinds, one of a dark brown; and the other of a yellow colour; but the last is most esteemed. There is also another sort, much larger than the choro, yet equally delicate and good, the fish of which is as large as a swan's egg: it is called cholgua; but as the shells seem to be of the same species, I think the distinction can only be owing to size. In Febres's Dictionary[12] of the Chileno language, the word / cholchua is rendered into Spanish by 'cascara de choros blancos,' or shell of the white muscle. Cholhua, or cholgua (the letters g and h are indiscriminately used), must be a corruption; for it is now used in Chiloóe to distinguish the large from the small choros.

The manner in which the natives of these islands, both Indians and descendants of foreigners, cook shell-fish, is very similar to that used for baking in the South Sea Islands, and on some parts of the coast of New Holland. A hole is dug in the ground, in which large smooth stones are laid, and upon them a fire is kindled. When they are sufficiently heated, the ashes are cleared away, and shell-fish are heaped upon the stones, and covered, first with leaves or straw, and then with earth. The fish, thus baked, are exceedingly tender and good; and this mode of cooking them is very superior to any other, as they retain, within the shell, all their own juiciness.

The oyster, which is a true *Ostrea edulis*, is found in beds, at low water, or taken with the dredge. It is about the size of the native oyster of England, and not at all inferior to it in flavour. In Agüeros's account of Chilóe, he notices this excellent shell-fish; but remarks, that the islanders are ignorant of the value at which it is appreciated. It is rather curious, that, excepting in the neighbourhood of Chilóe, the oyster is very rarely to be met with on the South American coast, while there it is in the greatest abundance. We have never observed any shells of this fish anywhere between the river Plata and Chilóe; nor is it known elsewhere upon the western coast, I believe, to the southward of Guayaquil, which is very near the equinoctial line.* The oyster-shells at Port San Julian are fossilsOf the Linnæan genus, Ostrea, there are many sorts, on all parts of the coast, both east and west, but they are what we call the pecten or scollop. All Coquimbo, a species of scollop is much used as an article of food, and called oyster; but it has no further right to the name than because Linnæus classed them all as *Ostrea*, and Molina describes this to be *Ostrea edulis*.

* Some have since been found on the north-east side of the Guaytecas Islands. /

The pico, which is a barnacle, grows to a very large size; at Concepcion, however, it is still larger, being six or seven inches in length. It has, when properly cooked, very much the flavour of a crab, and by the inhabitants of this Archipelago is considered preferable to any other shell-fish.

Before concluding this imperfect description of the shell-fish of Chilóe, the piure claims some consideration, if it be only for its peculiar and disagreeable appearance. It was considered by Molina as a genus allied to *Ascidia* (Mol. i. 214), none of the varieties of which are inviting in their look, as an edible substance, but the piure is still less so. It is thus described by Molina: 'The piure, scarcely deserving the name of a living animal, is as remarkable for its figure, as for the manner in which it is lodged. The body is about the size and shape of a small pear, an inch in diameter; or it may be described as a small, conical, fleshy bag, of a red colour, filled with saline liquor, and provided with two trunks or processes in the upper part, one of which is the mouth, similar to that of the Tetias; and between these processes are seen two small, black, and shining points, which are supposed to be the eyes. I could distinguish no other organs, nor any viscera in the fleshy substance of which it is composed, which is smooth without and spongy within They are extremely sensitive, and when touched, spout water out of both apertures. These small animals are shut up in a firm, but glutinous case, of various shapes; one case often contains eight or ten distinct bodies, separated from each other by cells, formed of a strong membraneous substance. They are attached to rocks or stones, under water, excepting when left uncovered by a low tide. The natives cat them boiled, or roasted in their shells. They also dry them for exportation to the province of Cusco, where their flavour is much esteemed, and considered equal to that of the lobster.'

At Chilóe, the piure is said to be a remedy for barrenness; and to such an extent has this idea prevailed, that a Chilote woman, eating this fish, literally says, if asked what she is doing, that 'she is making children.' One would not, however, suppose, from the number of children which are seen / crowding round the doors, that the Chilotes had any necessity for such food.

If one may judge from the few applications made to our medical men for advice, the climate is either very healthy, or the natives prefer their own mode of cure. They have very few medical advisers, and those few are not held in much estimation, being people of little or no education. A prejudice against medical men has been, even in late years, extended to foreign practitioners, and carried to great lengths. This illiberal feeling is, however, fast wearing away; but, among the lower orders, the application of herbs and other simples is yet wholly resorted to for the removal of their complaints. One day, when I was employed in making some astronomical observations, at Sandy Point, a woman passed me, and forcing her way through a thicket of thorny plants, began to gather branches of a species of arbutus (*A. rigida.*), a small shrubby plant, which is every where abundant, especially to the south, and in the Strait of Magalhaens. My curiosity prompted me to inquire her reason for collecting it with such apparent anxiety. She replied, with a desponding air, 'It is chaura* for a poor, sick child. These branches,' she said, 'are to be put into the fire, and, being green, will produce a thick smoke, and yield a very strong aromatic smell. The child, who is only five months old, is to be held over it, which, as they say, is a good remedy; but,' she added, with an air of doubt, 'I know not (dicen que es bueno, pero yo no sé).' 'Who says so?' I asked. 'Los que saben (those who know),' replied the half-credulous mother, with a deep sigh, partly doubting the efficacy of the remedy, but unwilling to lose the advantages of whatever virtue it might posses, for the benefit of her sick infant.

The climate of Chilóe is considered, by those who live in other parts of Chile, to be 'rigorous, cold, and damp.' Certainly there is much reason for such an opinion, particularly in the winter months, when it almost always rains, and the wind, with little cessation, blows hard, from N. to N.W., and, / by the W. to S.W.; but notwithstanding the great quantity of rain that falls, the evaporation is great, and it cannot therefore be called unhealthy; indeed, from experience, it is considered quite otherwise. Agüeros, to whose excellent account of Chilóe I have so often referred, dilates much upon this subject, and from having resided there a considerable time, may be taken as the best authority. Those who now reside upon the island speak very much against it, and all whom I met, previous to my visit, condemned it, as being 'the worst in the world.' Perhaps we, who had lately been experiencing a much more disagreeable climate, went to Chilóe with the expectation of finding it exceed in severity that to which we had been accustomed in the Strait of Magalhaens, but we found ourselves agreeably mistaken. Our visit certainly was in the better season, and we had, perhaps, no right to form a decided opinion upon the other part of the year. I shall, therefore, first quote Agüeros, and then describe what we found the weather from September to December; yet as these months were considered by the inhabitants to be finer than is usual at that season, we can only form a vague idea of the spring and summer. For the autumn and winter I must depend upon the accounts of others.

After explaining the contra-position of the seasons, to what is experienced north of the equator, with regard to the months of the year; Agüeros says, 'Chilóe has also its four seasons, but does not enjoy the benefit of those changes, as do other parts of Chile; for there is neither that abundance of fruit, nor are its fields adorned with so many and such beautiful flowers, and useful medicinal plants. The summer is the best time; for in the month of January, from ten o'clock in the morning till three in the afternoon, the heat is excessive. Between these hours, however, a sea-breeze, which is called 'Vira-zon,' refreshes the air. In

* Chaura. Una murta que no se come.[13] Febres, Dict. of the Chileno language. It is, however, edible, and has rather a pleasant flavour.

the winter the temperature is very cold; but the frosts are by no means so severe as in Europe. I have never seen ice, even in the small streams, nor does snow lie any length of time on the ground.

'In the winter months, as well as in other parts of the year, there are falls of rain, and heavy gales from N.N.W., and west, / which last frequently for the whole moon, with scarcely a cessation, and the wind, at times, is so furious, that the houses are not secure, and the largest trees are torn up by the roots. The weather, when it is fine, cannot be depended upon for any length of time; not even in summer; for in the month of January I have frequently experienced gales, and rain, as severe and copious as in the winter. During the summer months southerly winds are more prevalent, and, while they last, the weather is fine, and clear, and the air particularly dry.

'Although the winter months, and a considerable part of the other seasons, are very disagreeable, owing to the severity of the winds, and exceeding quantity of rain, it cannot be denied that the climate is healthy. In Chilóe no epidemic diseases are experienced. The small-pox and measles are not known;** nor have tertian fevers,[14] so common in the north, ever been experienced on the island. Spotted fever (tabardillo),[15] and acute pains in the stomach, are the only disorders to which the inhabitants of this archipelago are subject. Thunder and lightning are rarely experienced; but earthquakes have occurred at intervals. In the year 1633 the church and houses were destroyed, and in the year 1737 much damage to the village of Isla grande was caused by earthquakes.'

So far Agüeros. On the whole, the climate is not so unfavourable as we had been led to expect from all that we had heard.

Captain Fitz Roy arrived there in July, during the latter part of which, and the month of August, the weather was very wet, with some heavy gales from the N.W.; but in his Meteorological Journal for those months there is no record of the thermometer falling below 38°, and it is recorded to have fallen to that degree only on one occasion, the general height being from 45° to 50°. The first part and the middle of September were boisterous and wet; but towards the end of the month the wind was chiefly from the southward, and the weather dry and / extremely fine. In October it was rather changeable; but for the last ten days, with the exception of one, on which there was a fresh gale with a heavy fall of rain, it was fine and dry, and the winds were moderate.

The month of November was generally fine, but the first half of December continued tempestuous and wet. The mean temperature of the months, and other meteorological remarks, are as follows:

* The small-pox was introduced into the island, in the year 1776, by a ship from Lima; but it was confined to San Carlos, and was soon eradicated. The measles also were introduced by similar means, in the year 1769; but did not re-appear after once ceasing.

1829.	Temperature.					Hygrometer (Daniells').					No. of Days.		Rain.		
Months.	Mean at 9 A.M.		Ex. of Temp.		Pressure reduced to 32°.	Dew Point.	Dew Point less than Air.	Expansion.	Dryness by Thermo. Scale	Weight of a cubic foot of air.	Fine.	Rain.	Quantity fallen.	Quantity evaporated.	Remaining in the gage at end of month.
	Air.	3 P.M. Water at anch.	Max.	Min.											
July 22 days	46.9	47.9			29.927	–	–	–	–	–	–	–	–	–	–
Aug. 31	–	–	–	–	–	–	–	–	–	–	–	–	–	–	–
Sept. 30	47		64.5	35.2	30.061	40.9	6.18	296.9	806.2	3.3854	7	7	1.68	0.48	1.2
Oct. 31	50.9		73	37	29.979	45.8	5.14	349.8	845.0	3.9575	21	10	4.22	2.25	1.97
Nov. 30	53.5		68.5	42	29.898	48.4	4.79	416.2	844.5	4.3361	14	16	4.89	2.28	2.61 /

This table partly shows the state of the weather during three spring months. The greatest quantity of rain in the gage at the end of the month of November did not exceed 2.6 inches. At St. Martin's Cove, near Cape Horn, after thirty days' observation, the rain-gage contained eight inches; so that although Chilóe bears the character of being a very wet place, it is not one-third so bad as Cape Horn. The time of our visit to San Carlos was certainly the finest part of the year; and I believe that the weather we experienced was unusually dry even for the season; therefore, the above table does not present a fair criterion of the climate: I do not, however, think it is by any means so bad as has been represented. /

CHAPTER XVII.

Chilóe the last Spanish possession in South America – Freyre's Expedition – Failure – Second Expedition under Freyre and Blanco – Quintanilla's capitulation – Chilóe taken – Aldunate placed in command – Chilóe a dependency of Chile – Beagle sails to sea coast of Tierra del Fuego – A delaide repaired – Adelaide sails – Adventure goes to Valparaiso – Juan Fernandez – Fishery – Goats – Dogs – Geology – Botany – Shells – Spanish accounts – Anson's voyage – Talcahuano – Concepcion – Pinoleo – Araucanian Indians – Re-enter the Strait of Maghalhaens – Fuegians.

THE island of Chilóe was the last place the King of Spain possessed in South America; and even to this day he is not without friends there, who would gladly restore his absolute monarchy, notwithstanding the advantages that are acknowledged to have been derived from the change of masters, and the consequent opening of trade, which has added very much to the comfort, as well as civilization of the inhabitants.

During the struggle for independence, this island was too distant from the seat of war to render it important; but when all other parts of Chile were freed from the king's troops, the new government dispatched an expedition, consisting of between three and four thousand men, commanded by the Director-General Freyre, to attack it. Upon the appearance of this expedition off the harbour of San Carlos, the Spanish governor, Quintanilla, was inclined to capitulate; but, instead of anchoring in the roads, the squadron proceeded to Chacao, landed troops there, and dispatched some of their forces to Castro, where they were repulsed by the Spanish and native troops, and obliged to re-embark. In this interval, one of the ships left the squadron, and returned to Valparaiso, whence she was immediately ordered back; but meanwhile the Director had embarked his troops, and returned to Conception. Not long afterwards, in January 1826, a second expedition, under the same general, sailed from Valdivia, convoyed by a strong squadron, under the command of Admiral Blanco. /

'Upon this occasion the troops landed, on the 8th, at the little inlet of the Bay of Huechucucuy; and Fort Corona was immediately taken. On the 10th, the disembarkation of the troops was completed. A battalion was left to mask Fort Aguy, while a force, under Colonel Aldunate, passed on, and took the battery of Barcacura. On the 10th, Admiral Blanco shifted his flag; and, leaving the O'Higgins outside, stood into the bay with the rest of the squadron, which anchored off Barcacura.

The governor, Quintanilla, with upwards of three thousand Royalists, took up a strong position on a hill, at the S.E. side of the bay, flanked on the left by an impenetrable wood, on the right by the shore, and supported by three gun-boats in shallow water. These were taken by the boats of the squadron, under Captain Bell, and turned against the Royalists. Their position was thus enfiladed, and they retired. Freyre then advanced: some skirmishing took place: Quintanilla capitulated; and the territory of Chile was no longer sullied by the Spanish flag.

Colonel Aldunate, Majors Maruri, Asagra, and Tupper (a native of Jersey); and Captain Bell, of the navy, greatly distinguished themselves.' – Miller's Memoirs.[1]

Colonel Aldunate was afterwards invested with the government of the island; but, owing to the disaffection of the troops, who were urged on by the King of Spain's agents, a revolution took place, Aldunate was imprisoned, and afterwards sent to Valparaiso, and the Spanish flag once more waved in Chilóe. It was, however, for a short time only; Aldunate was despatched once more, and with a small force of three hundred veteran troops, headed by Colonel Tupper, and accompanied by the Aquiles, brig of war, again obtained possession of the island, which he has since kept, though not quietly, for the Royalists were constantly on the alert, and made several futile attempts to recover the place for their king. Time has now reconciled the greater number to the change; and, I believe, Chilóe may be considered a contented dependency of the republic of Chile. /

The Beagle being ready to resume her voyage, sailed on the 19th of November to survey the southern coasts of Tierra del Fuego; after which, she was to rejoin the Adventure at Rio de Janeiro.*

As the Adelaide had received some damage in getting aground, it was requisite to lay her on the beach for examination and repair. Her mainmast, also, was found to be sprung so badly, as to render a new one necessary; which we should have found much difficulty in obtaining, but for the kindness of General Aldunate, who, finding that we were at a loss, proposed to give us the flag-staff of the town, a beautiful spar of alerse, that was in every way suitable. Previously, however, to accepting his offer, being aware that such an act might expose him to much reproach from the people of the town, who were all very proud of it, I caused inquiry to be made whether a spar of the necessary dimensions could be brought from Calbuco; and in the meantime we proceeded with the repairs.

A creek behind Sandy Point offering every convenience for heaving her down, the Adelaide was moved into it, and laid on the beach. On stripping her copper off, the injury proved to be considerable; but not beyond our means

* See orders to Captain Fitz Roy, in the Appendix.[2]

OLD CHURCH AT CASTRO.

NEAR P^T ARENA.

P^T ARENA – SAN CARLOS CHILOE.

to repair. Upon examination, the foremast was found to be in a bad state, but could be rendered effective by fishing it with the sound portion of the other mast, therefore our only real difficulty was to get a mainmast. From the account I received from Calbuco, I found that, without a great delay, not less than two months, and sending a portion of our people with ropes and tackles, there was no chance of procuring a spar: it could only be obtained at a considerable distance from the shore, and when felled must be dragged over several high ranges of hills, which might be called mountains, before it could be got to the water-side. General Aldunate, through whom this inquiry was made, then renewed his offer of the flag-staff, which I accepted most thankfully; and by his order it was taken down, and conveyed to the ship, soon after which it was converted into an excellent mainmast for the schooner. Before it was moved, a new, but / / / shorter staff, with a topmast, was fitted for the flag; notwithstanding which, many unpleasant observations were made, and absurd reports circulated, which spread to Chile, and even to Peru, that the English were about to take possession of Chilóe, and had already removed the flag-staff of San Carlos.

By Lieutenant Mitchell's activity in superintending the Adelaide's repairs, she was got ready for sea at the beginning of December, and sailed on the 8th, under the command of Lieutenant Skyring, with orders* to survey those parts of the Gulf of Peñas which had not been examined by the Beagle; particularly the River San Tadeo, in San Quintin's Sound; the openings behind Xavier Island; the Channel's Mouths; and the Guaianeco Islands, where the Wager was wrecked: and then to proceed down the Mesier Channel, behind the Island Campana, which was supposed to communicate with Concepcion Strait, by the Brazo Ancho (or Wide Channel) of Sarmiento. He was then to go to the Ancon sin Salida, examining all the openings into the main land, on his way, and search for a communication with the large waters, discovered by Captain Fitz Roy, through which he was to try to enter the Strait, and join the Adventure, at Port Famine, during the month of April.

Lieutenants Skyring and Graves again took with them, by Captain Fitz Roy's permission, Mr. Kirke and Mr. Bynoe, of the Beagle; Mr. Alexander Millar and Mr. Parke also accompanied them.

Having thus dispatched our companions, we prepared, on board the Adventure, to return to Valparaiso; intending to proceed to Rio de Janeiro; by way of Concepcion, Port Famine, and Monte Video; for the sake of adding some links to our chronometric chain: with a view to which, I had taken the opportunity of having the chronometers cleaned at Valparaiso by Mr. Roskell, agent for Messrs. Roskell chronometer-makers at Liverpool.[3] General Aldunate being on the point of returning to Valparaiso, I had an opportunity of obliging him, and showing

* See orders to Lieut. Skyring, in the Appendix.

my sense of the assistance, and essential kindness we had / received, by offering him and all his family a passage in the Adventure, which he accepted; and on the 17th we left Chilóe. In our way we touched at Concepcion, and anchored at Valparaiso on the 2d of January.

We remained there until the 11th of February, and then sailed on our return to Rio de Janeiro,[4] with the intention of passing though the Strait of Magalhaens, and taking that opportunity of completing some few parts, which our former surveys had left unfinished. As the breeze, which, on this coast, blows with the constancy of a trade wind, would carry us close to the island of Juan Fernandez, I determined upon visiting it, for a few days; and then proceeding again to Concepcion.

We reached Cumberland Bay, on the north side of Juan Fernandez, on the 16th, and anchored, within two cables lengths of the beach, in ten fathoms.

I have seldom seen a more remarkable and picturesque view, than is presented by the approach to Juan Fernandez. When seen from a distance, the mountain of the 'Yungue' (Anvil), so called from its resemblance to a blacksmith's anvil, appears conspicuously placed in the midst of a range of precipitous mountains, and is alone an object of interest. It rises three thousand feet above a shore, which is formed by an abrupt wall of dark-coloured bare rock, eight or nine hundred feet in height, through whose wild ravines, broken by the mountain torrents, views are caught of verdant glades, surrounded by luxuriant woodland.

The higher parts of the island are in general thickly-wooded; but in some places there are grassy plains of considerable extent, whose lively colour contrasts agreeably with the dark foliage of myrtle-trees, which abound on the island.

The Yungue is wooded, nearly from the summit to its base; whence an extensive and fertile valley extends to the shore, and is watered by two streams, which take their rise in the heights, and fall into the sea.

This valley appears to have been formerly cleared and cultivated by the Spaniards, who had a colony here; for the stone / walls, which served to divide their enclosures, still remain. From Walter's account of Anson's voyage,[5] and the view given with it of the commodore's tent, there is no difficulty in determining this valley to be the spot on which his encampment was placed.

The island is now (1830) occupied, or rather rented from the governor of Chile for a term of years, by Don Joachim Larrain.[6] The establishment consists of a superintendent (mayordomo), there called, 'the governor;' and forty persons, who are employed in the seal and cod fishery, and in drying fish for the Chilian market. Their dwellings are erected on the flat land, at the north side of the bay, where the soil is richer than in other parts; and where it is more sheltered from the squalls, which, during strong southerly gales, rush down the valley of the Yungue, the situation of the former establishment, with great violence.

The remains of a fort, called San Juan Baptiste, are yet in a tolerable state; and from an inscription on the wall, it appears to have been repaired, or completed, in the year 1809. It is situated on a rising ground, about one hundred and thirty feet above the sea, at the S.W. part of the bay, and overlooks the village; there are now no guns mounted, but, with a few, it might be made very effective in a short time; and, from its situation, would command the bay.

In the middle of the beach are some ruins of a four-gun battery, and there are also traces of a fort at the N.W. end of the bay.

At present, except wild-goats, wild peaches, figs, abundance of fish, and excellent fresh water, no refreshments can be procured. An establishment of forty persons, with very little to do, might naturally be expected to cultivate the land, raise vegetables and fruit, and rear poultry and pigs, to supply the vessels, which frequently touch here for wood and water; but it is not the character of the Chileno to take any trouble, unless obliged, although his own comfort and advantage may be materially concerned.

The mayor-domo, however, told me that their attempts to / cultivate the soil, and raise potatoes, had been defeated by the destructive ravages of a worm.

By sending a boat to the east point of the bay, to fish in forty fathoms water, a most delicious kind of cod-fish may be taken, in such numbers, that two men, in half an hour, could fill the boat. Craw-fish, of large size, are almost equally abundant; they are taken with a hooked stick: one of our boats caught forty-five in a very short time. The inhabitants catch them, and cure their tails, by exposure to the sun, for exportation to Chile, where they are much esteemed, and fetch a high price.

Wild-goats are very numerous among the inaccessible parts of the island, but are not easily obtained; they are sometimes shot, or taken with a lazo. These animals, according to Woodes Rogers,[7] and other writers, were originally left on the island by Juan Fernandez, who, for a short time, lived there. According to the 'Noticias Secretas,'[8] p. 50 to 56, they are supposed to have been landed by the Buccaneers, who frequented this island. Certain it is, that, without such refreshments, the Buccaneers would not have been able to carry on their harassing war of plunder against the Spanish possessions on the American coast to such an extent; nor should we, perhaps, have heard anything more about Commodore Anson, and the crews of the Centurion and Gloucester, who were, on their arrival at this island, in the last stage of scurvy.

To prevent Juan Fernandez from being so tempting a resort to Buccaneers, the Viceroy of Peru caused a great many dogs to be landed, which hunted down and destroyed the goats in great numbers: this in some measure has prevented their subsequent increase. The dogs however drove the goats to places where they could not follow them, and were then obliged to destroy seals for food. Large troops of these dogs still range about the lower grounds; but the heights are in the undisturbed possession of wild-goats; which may be seen in numbers browsing on elevated and almost inaccessible places, where they live in safety.

The geological character of this island, according to Mr. Caldeleugh, who accompanied me in this trip, is of basaltic / green-stone, and trap, which appears, at first sight, to be volcanic; but, on a more particular examination, the lava-like appearance of the rock does not seem to arise from an igneous origin.

The green-stone is full of crystals of olivine, which, as they decompose, leave hollows, resembling those of scoriæ. Mr. Caldcleugh communicated an account of the structure to the Geological Society.* In Captain Hall's interesting journal,[9] there is a list of Geological and Mineralogical specimens, of which one from Mas-a-fuéra† is named 'Vesicular Lava.' May it not be this same rock in a decomposed state?

The late Signor Bertero, whose botanical collections from Chile have enriched many of the principal herbaria in Europe, accompanied me to make a collection of the Flora of the island;[10] and he considered that the character of the vegetation was very little allied to the Chilian, but partook more of that of California. The sandal-wood, which has been described as indigenous to this island, was not found by us, growing, but a large quantity was collected about the hills and vallies, in a dry state, and apparently very old. It is of the red kind, and still preserves a strong scent. The mayor-domo told me there were no sandal-wood trees in the island; but we had reason to think his information was incorrect, for one of the inhabitants would have taken us to a place where he said they were growing in large quantities, had not our arrangements for sailing interfered.

The island produces several kinds of grass; but the most abundant herbaccous plant is a species of oat, which grows very luxuriantly, and towards the westward covers the ground for many miles. The neighbourhood of Cumberland Bay is over-run with strawberry plants, wild radishes, mint, and balm, besides peach, apple, cherry, and fig trees, which are / found wild every where, and remind one of Lord Anson's visit.‡

Not only in its botanical productions does this island differ from the Chilian coast, but also in its shells: the shell fish being extremely scarce, and dissimilar in character. On the rocks we found a patella and a small chama, but we saw no mytilus. From the deep water I fished up some coral, and attached to one fragment was a new species of arca.§ The fishing-lines brought up, from the depth of eighty fathoms, a branch of coralline, to which an infinite number of a species of caryophyllia were attached. The existence of coral is mentioned in Mr. Barry's

* Phil. Journal, and Annals of Philosophy, for March 1831 (new series x.), 220.

† Juan Fernandez is called 'de Tierra,' because it is nearer the mainland than another adjacent island, which is called 'Mas-a-fuéra' (farther off, or more in the distance).

‡ Anson's Voyage, p. 118.

§ Arca angulata. See Zool. Journal, vol. v. p. 336.

translation of the 'Noticias Secretas de America; por Don J. Juan, y Don A. de Ulloa,' a work which contains a long and, generally speaking, good account of the island; but their description of the anchorage does not agree with ours. They say, 'The distance between the two points, which form the bay, is two miles, and its depth about half a league; and, although the depth is nearly the same in all parts, the best berth to moor ships is in the front of the 'Playa del Este;' but it is necessary to be close to the stones of the beach, for at one or two cables' length there are fifty fathoms water, and the outer anchor is in the depth of seventy or eighty fathoms; but if the vessel is three or four cables off, it will be necessary to drop the outer anchor in one hundred fathoms, which, even with two cables an end, will scarcely secure the ship.' Now, at three cables' length from the beach, we had only ten fathoms, our outer anchor was dropped in seventeen fathoms, and in a line between the two points of the bay there is not more than fifty fathoms.

If the accounts of those Spanish officers were correct, the earthquakes, which certainly affect these islands, must have caused a considerable uprising of the base of the island; but, on referring to the plan in Anson's voyage, the soundings in 1741 do not appear to have been different from ours. The innermost ship, whose berth we occupied, is, in that plan, at anchor in / nineteen fathoms, and the depth between the points of the bay is shown to be about fifty fathoms.

There are few persons who have not read, with much interest, Mr. Walter's account of the Centurion's voyage, and who are not well acquainted with his description of this island, which we found exceedingly correct.[11] The views of the land, although old-fashioned in execution, are most correctly delineated, and the plan of the bay is quite sufficient for every common purpose of navigation; but as we had an opportunity of fixing its latitude and longitude more correctly, it became desirable to make a more detailed plan than Commodore Anson's.

The seals and sea-lions, which were so abundant formerly, are now reduced to such a small number, as to make the sealfishery scarcely worth notice. They have been destroyed by taking them indiscriminately, without regard to age or sex, leaving none to propagate the race but those who by chance escaped. At present the island is let to a tenant, who is not permitted to kill them until the young have taken to the water, by which means an opportunity is given for them to increase.

I am not aware that there are any indigenous animals. Dogs, goats, and rats, have been imported. Land birds are not numerous; some pigeons, said to have been imported, and a few hawks, are occasionally seen, besides three species of humming-birds, two of which are new to science.* Of sea-birds we saw very few; but were informed that the 'Goat Islands,' at the south-west end of Juan Fernandez, are completely covered by them at the breeding season.

* Trochilus Fernandensis, nob. *Troch: ferugineo-rufus; capitis vertice splendento-coccineo; remigibus fuscis.* Long. 5 uncias.

Trochilus Stokesii, nob. *Troch: corpore supra viridi-splendente, subtus albo, viridi-guttato; capite supra, guttisque confertis gulæ lazulino-splendentibus: remigibus fusco-alris; romigum omnium, mediis exceptis, pogoniis internis albis.* Long. 4 1/2 uncias. Proceed. Zool. Society, vol. i.; also Phil. Magazine, for March 1831, p. 227.

During our stay, several excursions were made, in various directions, from the village, and much facilitated by beaten paths, one of which leads up a valley, westward of that of the / Yungue, and thence to a pass over the principal range, communicating with the other side of the island. This pass, called the Puertozuela, is 1,800 feet high, and was visited several times by the officers. On one occasion, they went to the western part of the island, to hunt wild goats. The party set out in boats with the mayor-domo, or governor, as their guide; but before they reached the proper landing-place, became so impatient that they landed, intending to walk back. The governor, however, persevered, and returned, in the evening, with five fine she-goats, which he had taken with 'lazos.' Our pedestrians found their return by no means so easy as they had contemplated, being obliged to pass the night in a cave, which they fortunately found at sunset, and they did not reach the ship until the following afternoon, fatigued, but much pleased by their ramble.

The thermometer on board ranged, during the day, between 63° and 82°, and the barometer between 29.98, and 30.16. On shore the thermometer stood higher, in fine, unclouded weather, and lower when the summits of the hills were covered with clouds.

We put to sea on the 22d, anchored at Talcahuano on the 3d of March, and sailed again on the 17th, to proceed through the Strait of Magalhaens.

While at Concepcion I had an opportunity of seeing Pinoleo,* the Indian chief, from whom Captain Basil Hall endeavoured to obtain the release of a captured Araucanian female, whose husband had been murdered in cold blood before her eyes.[12]

Mr. Rouse, our consul,[13] procured for me the necessary introduction, and, with one of the governor's aides-de-camp, accompanied us to the Indian quarters, situated on the out-skirts / of the town, towards the river Bio-Bio. We found the chief's residence (little better than a rancho, or hut of the country), surrounded by Indians, some of whom were armed; and at the door were his two daughters, young, and rather good-looking, whose persons and dress we had leisure to examine, whilst waiting the chief's pleasure to receive us. They were clothed with a mantle, or wrapper, of green baize, enveloping the body from the

* Pinoleo (from 'Pino,' pisando; and 'leo,' rio; or, pisando sobre el rio, living close to the banks of a river), is the Chief of a small tribe, whose territory is near the River Imperial; but he generally lives in the confines of Concepcion. He has four wives in the interior (la tierra) and three in the town.

neck to the feet, and fastened at the breast by a toup, or tupu* (a silver pin, or skewer, headed with a round silver plate, three inches in diameter), over which hung a string of beads. Their hair, which was remarkably fine and clean, as well as neatly dressed, was divided into two plaited tails ('trensas'), and their foreheads were ornamented with a broad fillet, worked over with beads.† They also wore necklaces, bracelets, ear-rings, and anklets of similar manufacture.

Our names having been announced to Pinoleo, he came to the door to receive us, and invited us to enter. Some of our party he recognized, and seemed pleased at their visiting him. We were early, and found him sober; but from his bloated and haggard appearance, it seemed that he had not been long so. On entering the hut, we observed a number of Indians, scarcely sober, seated round, near the walls. Some turbid wine was presented to us, in a silver cup, which we sipped as it passed round; but the last of our party knowing that to return the cup without emptying it, would be an offence, was obliged to drink the contents, and a bitter potion they were. Pinoleo was then stout and rather corpulent, five feet ten inches in height, of a fairer complexion than the generality of his countrymen, and had lost much of his hair. He had laid aside the Indian / dress, and wore the deshabille of a Spaniard, a shirt and pair of trowsers, in a very slovenly manner. He spoke Spanish with great facility, and appeared to be quite at his ease in conversation. He has the rank of lieutenant-colonel in the Chilian army, and receives pay, as a retainer for his friendship.

A very short visit was sufficient to satisfy us, and we took the first opportunity of retiring, for fear of a second cup of wine. While leaving the hut, we were beset by some of his followers, asking for money. The Indian quarter is a scene of drunkenness the whole day; the women, however, are prevented from thus injuring themselves; they are industrious and cleanly, and are principally occupied in the manufacture of ponchos. These Indians are frequently at war with other tribes, who live on the south side of the Bio-Bio river, and who have never yet been conquered by white men, of which they are not a little proud. (*w*)

These Araucanians are by no means to be despised. The Cacique Mariloan,‡ who resides near San Carlos, on the Bio-Bio, has three hundred fighting men

* In Febres 'Arte de la lengua de Chile' they are thus described 'Ahujas grandes con una plancha redonda de plata como una hostia, ò mayor, con que prenden las mujeres sus mantas – Certain large bodkins, with a round silver plate, as large as, or larger than, an oyster, with which the women fasten their mantles.'

† The ornament on the forehead, which is worn only by unmarried women, is called Trarelonco, from the old Chilian words trarin, to fasten, and lonco, the head. The bracelet is called Anello cure; the anklets, Anellco.

(*w*) Not since the first Spanish conquest, perhaps. – R. F.

‡ From 'Mari,' diez, and 'loan,' huapo: whence Mariloan means 'huapo como diez,' or, 'equal to ten men.'

under his own command; and from the influence he holds over neighbouring Caciques, could bring upwards of one thousand men into the field. Upon the occasion of a late revolution in Chile, a deputation of chiefs was sent by the Araucanian Caciques to inquire into the cause of those disturbances, of which they had received intelligence. They first asked for an interpreter, whom they cautioned to give a true and literal translation of their speech; and then they made a long harangue, in which they explained the cause of their visit, and declared their willingness to assist their friends, if their aid should be required, to expel a foreign foe; but if the troubles were caused only by the quarrels and dissensions of parties, they would not take an active part. They were then given to understand that an attempt had been made by one party to put down another, upon which they declined assisting either. The conference being ended, some horses were / slaughtered and skinned. Large holes were dug, and the skins put into them, to form substitutes for vessels, into which barrels of wine were poured, and the Indians commenced their feast of horses' flesh and turbid wine, which threw them rapidly into a state of excitement and intoxication, that lasted some hours after the wine was all drunk.

In this neighbourhood, the Araucanian pine (*Araucaria imbricata*) is found, but very few of the trees grow near the sea. One beautiful specimen which I saw in a garden was, at least, forty feet in height, with branches sweeping the ground. The cones of these trees, called piñones, are brought to the town from the mountains where they grow, and are roasted, to be sold in the streets.

On the 31st of March, the land about Cape Lucia was seen, and at noon it bore E. b. N., distant twelve miles, when the wind ceased, and a heavy swell setting us towards the land, made our situation an anxious one. A breeze, however, sprung up, and by carrying a press of sail, we succeeded in gaining an offing before dark. The night was very equally, but next morning (1st April) the weather was better, so we stood in, and made the Evangelists, which were seen from the masthead, at a distance of twenty-two miles. Between these islands and Cape Pillar we found a most turbulent sea; yet no sooner had we entered the Strait, than the water became perfectly smooth. I intended anchoring in the Harbour of Mercy; but the night proved fine, and the wind was so favourable, that we proceeded by the chart, using a patent log,[14] and passing within two miles of the headlands. Sail was reduced as much as possible, to give us space sufficient to run on during the night, steering E. 3/4 S. by compass. Towards midnight the weather became cloudy, and occasionally the land was concealed from our view.

Abreast of Cape Tamar, and as far as Cape Providence, some sharp squalls raised a sea, rather heavy, considering we were in the Strait; but afterwards the water became smooth again. Off the latter cape, the patent log indicated a distance run equal to that shown by the chart, which proved that we had /

experienced no current. At daylight we were in the entrance of the 'Long Reach,' abreast of Cape Monday.

While passing the opening opposite to Playa Parda, a schooner was observed at anchor, and a boat was seen coming out to us. It contained the mate of the schooner Industry, of New Bedford,[15] who informed us that she had been lying there, weather-bound, for nearly a month. He came to make inquiries about good anchorages to the westward (having already lost two anchors), and to learn in what part of the Strait he was; his own idea being, that the vessel was under Cape Monday. Having given him the required information, we proceeded; but the wind fell light, and we were glad to anchor in the cove of Playa Parda. With our chains we found it safe; but the bottom, being rocky, would probably do much injury to hempen cables.

The opening opposite to us, where the schooner was lying, was evidently Sarmiento's 'Abra.' It appeared to us to be a mile and a half wide, with an island in the entrance. Within, it seemed to take a south, then a south-west direction, and afterwards to trend round a low hummocky point of the eastern shore, under a high, precipitous ridge, on the opposite or western shore, towards the S.E.; beyond this its course could not be observed. When passing through this part of the Strait, Captain Stokes found the weather so bad, that although the distance across was only two or three miles, the shores were often concealed by clouds and rain, so as to render it impossible for him to make any survey of them.

We were detained the two following days by bad weather. On the 5th we proceeded, but before we got abreast of Snowy Sound, heavy rain set in, which lasted all day.

As we passed Borja Bay, a schooner was observed at anchor in it, so like the Adelaide, that we altered our course to communicate with her. From a boat which came off to us, we learned that it was a sealing-vessel, called the Hope, of New York, going through the Strait, from Staten Land. She had seen nothing of the Adelaide.

When abreast of Bachelor River, a canoe, containing two / men and two women, came out to us; but we did not delay long, and at five the anchor was dropped in Fortescue Bay.

As it did not appear that the Adelaide had preceded us, I determined upon remaining, to make a chronometric measurement from Port Gallant to Port Famine; and the next morning Lieutenant Graves landed, and obtained a set of sights for time.

In the early part of the day, two canoes, containing eight or ten Fuegians, entered the bay. They came from the westward; but we did not recognize among them any of those who visited the ship as we passed Bachelor's River. Several had red baize shirts, and some had 'Union caps,' such as are supplied to our men-of-war; which they must have procured from the Beagle or Adelaide, or from the

Chanticleer, at Cape Horn. (x) After hanging about us all day, they landed at sunset, and took up their quarters in some old wigwams in the inner harbour.

The canoes of these natives were very different in their construction from any we had seen to the eastward. Instead of being paddled, they were pulled with oars; one of which was an ash oar, probably obtained from some sealing-vessel. The canoes were large; at the bottom was a plank, twenty inches wide, to which were sewn the sides, in the manner of the piraguas, and they were caulked with bark, in a similar way.

We did not remark any thing peculiar among these people which we had not perceived in other natives of Tierra del Fuego, except that they frequently used the word 'pecheray,' a word particularly noticed by Bougainville, who thought that it meant the name of the tribe; and, in consequence, the Fuegians have been often called Pecherays.

On one of the officers cutting a lock of hair from a woman's head, the men became angry, and one of them taking it away, threw half of it into the fire, and, rolling up the other portion between the palms of his hands, swallowed it. Immediately / afterwards, placing his hands to the fire, as if to warm them, and looking upwards, he uttered a few words, apparently of invocation: then, looking at us, pointed upwards, and exclaimed, with a tone and gesture of explanation, 'Pecheray, Pecheray.' After which, they cut off some hair from several of the officers who were present, and repeated a similar ceremony.

From this fact, one might suppose the word to be connected with their ideas of divine worship; but we had heard it used for so many opposite things, that I could not consider it of so much importance as some of the officers were inclined to think it.

The next day a party ascended the Mountain de la Cruz, to deposit a pewter plate, on which were cut the names of the ship and officers. At the summit they found the pile of stones made by Captain Fitz Roy, which they left undisturbed; but made another, in which a bottle was placed, containing the little Spanish coin, and copies, on vellum, of the memorials we had formerly taken from it, also several English coins, and some medals. The bottle was corked, covered with resin, and enveloped in sheet lead. Our party returned in the evening, having been seven hours in going up and descending.

The next day I obtained an angular measurement of the Mountain de la Cruz, with a thedolite, having measured a base of 2,608 feet, which gave for its elevation 2,364 feet, 74 feet more than Captain Fitz Roy's barometrical determination.

During the day several Fuegian families had arrived, and, by the evening, ten canoes, containing altogether about sixty natives, were collected. I landed

(x) I believe that the natives who have canoes of the kind described, above, do not go near the Hermite Islands, on which Cape Horn is situated. – R. F.

to visit them, for I had never before seen so many assembled. We entered all the wigwams but one, which was said to be occupied by a woman in labour. In the opening stood her husband, painted all over with a red ochrous earth, and his head and breast ornamented with the white down of birds. The other Fuegians called him 'Pecheray;' and appeared to consider him, while in the character he had assumed, as a being superior to themselves. /

Hence, there evidently is something of a superstitious nature connected with the word; but our frequent attempts to find out its precise meaning, were unsuccessful. On repeating this expression to a group of natives, one of them immediately coughed up a piece of blubber, which he had been eating, and gave it to another, who swallowed it with much ceremony, and with a peculiar guttural noise; then, looking up, and pointing with his finger to the skies, solemnly pronounced the talismanic 'Pecheray.' This word is also used in pointing to the sun.

On the 10th April, I went to Charles Islands, and surveyed them. There is very good anchorage for a small vessel, in eighteen fathoms, at the north end of the passage which separates them; and at the bottom, or elbow, under the eastern island, in thirteen or fourteen fathoms. The next day, a fresh arrival in two canoes increased the number of Indians to eighty; rather a formidable body for a small vessel to encounter. They conducted themselves, on the whole, very peaceably, but seemed determined that our curiosity should not be gratified by finding out the contents of the 'tabooed' wigwam. It was always guarded by the 'Pecheray,' who seemed ready and determined to dispute all access to it, by means of a heavy club. One of the midshipmen, however, with a little coaxing, persuaded the man to let him put his head in; but those who were inside, having received their lesson, threw ashes in his face, and nearly blinded him. After this, seeing they were determined on the point, I desired that no further attempt should be made to ascertain what was really going on inside the wigwam.

We sailed the next day (11th), not without some apprehension that the Adelaide might meet this large concourse of Indians before they separated; as Port Gallant was a place rarely passed by vessels without stopping, and the natives being all housed behind a point of land, could not be seen until too late.

We were abreast of Cape Froward at noon; in the evening we anchored in French Bay, and next day (13th) reached Port Famine. As I purposed remaining until the Adelaide should arrive, the tents were set up, the boats landed for repair, and the transit instrument was set up, in the hope that a comet / might be visible, which we had seen in our passage from Concepcion to the Strait; but the weather was at first too cloudy, and afterwards the comet itself was too faint to be discerned.*[14]

* The same comet was seen at the Mauritius; and its orbit calculated. See Ast. Soc. Proceedings, and Phil. Journal.

On the 21st, nine canoes arrived in the bay, containing a large party of Fuegians, principally those who frequent the Magdalen Channel, and probably the sea-coast. They had generally shown themselves disposed to be mischievous, and I determined upon preventing their encamping near us; for their presence would greatly impede our watering and wooding parties, by distracting the attention of the people. I, therefore, went to meet them at the watering-place, under Point St. Anna, where they had landed, near one of our boats which was on the beach. Among them we only recognised three who had visited us before, and those three were brought to our remembrance by their former misconduct. I had always made it a rule to treat them kindly, with the view of obtaining their good-will; but I found it was the wrong way to gain their respect, for it only made them expect more from me, the consequence of which was, that when we separated, neither party was pleased with the other. I used on this occasion a more dictatorial tone than I had hitherto done; for, seeing several with slings in their hands, and a collection of large, round pebbles wrapped up in the corner of their mantles, I desired them to throw the stones away, which they did not hesitate to do. The Indians were now all landed, and evidently presuming upon their numerical strength, upwards of eighty being assembled, began to make themselves very familiar.

I though it best to check their advances, by desiring them not to visit the side of the bay where our tents stood, but to go round Point St. Anna, to an adjoining cove. They seemed to understand me perfectly, and soon afterwards embarked, while I returned on board. The natives, however, landed again, in the middle of the bay, at the north side, and there encamped.

Next morning, the men of the tribe visited our tents, but found them surrounded by a rope I had caused to be fixed, / and which they were not permitted to pass. At noon, after observing the sun's transit, I went to the barrier, and while the people were at dinner, endeavoured to amuse our visitors, who were from fifteen to twenty in number, by showing them several trifles; among the rest, a pocket set of coloured glasses, belonging to the transit. They looked through them at the sun, but handled them rather roughly, and broke the frame; upon which I expressed my anger, and turned them away. Soon afterwards, however, I walked towards them, and selecting the Indian who had offended me, gave him a bunch of beads, and thus restored peace; but desired them, at the same time, to go to their wigwams, which they did. In their way, they mischievously broke down a part of my meridian mark; seeing which, I sent a carpenter, attended by a marine, to repair it, and went myself to inspect its being again set up. The natives were collected round it, evidently in expectation of my being angry, and awaited my approach. Upon my coming near, I showed them that I was much displeased, and ordered them into their canoes; when one of the party, muttering a few words, picked up a stone from the ground, and was fixing it in his sling, when I took the marine's musket, and presented it at him, upon which the whole took to their heels; the principal offender and another ran along the beach, and the rest to their canoes. I could not resist the opportunity of letting them know we were prepared for them, by firing over the heads of the two who were running near the water.

The report of the musket attracted the attention of Lieutenant Mitchell, who was on board on the look-out, expecting some fracas would, sooner or later, take place; and seeing four or five canoes paddling across, and the two Indians running along the beach, he manned a boat, and pulled towards the canoes, which tried to evade him, and stones were thrown at him as he approached. A musket fired over their heads, soon quieted them, when he pulled round their canoes, to show them they were in his power, but did not molest them, and then allowed the party to proceed.

This affair alarmed the women at the wigwams, and hastily / gathering up their effects, they hurried into their canoes, and joined the others, who all paddled round Point St. Anna. The men, however, landed there, and remained on shore, armed with slings, spears, and bows, ready to defend themselves, and, by their gestures, defying us to land. No attention was paid to them, and, after a short time, they went over the hills to the coves on the north side of the point. As we had now openly quarrelled, I thought it better that they should keep at a distance; and therefore, taking two boats, pulled round the point, to tell them to go five miles farther, to Rocky Bay; but the canoes were already beached, and the women had taken up their quarters. As we approached, the hills echoed with the screams of the women and the shouts of the men; all of whom, stark naked, armed, and daubed with white paint, their heads being stuck full of white feathers, hastened down to the point of the bay. The place, from its nature, offered a good defence, as the beach was lined by large rocks, behind which they could conceal themselves from our view, and yet assail us with stones. When within a few yards of the beach, we held a parley – the object of which was, that they should go farther to the northward; to this they vociferously replied, by desiring us to leave them. Seeing there was no chance of enforcing our demand, without shedding blood, I ordered the boats away; and on getting about a musket-shot from the beach, one of the Fuegians threw a stone, which fell close to us. In an instant, every one of them was concealed behind the rocks; but we returned their fire, and another large stone fell within two feet of the boat. A second musket was fired, and another stone was returned, with equal precision. After the interchange of a few more stones for bullets, they ceased throwing them, and we returned on board, It was very unlikely that any of our shot took effect; for we were at a long distance, and could only see their heads above the rocks. Fortunately, none of the stones struck us, for they were large enough to have caused a severe bruise. It is astonishing how very correctly they throw them, and to what a distance. When the first stone fell close to us, we all thought ourselves out of musket-shot. /

The next morning, five or six natives were seen crouching down among high grass, on the hill over our watering-well, waiting for the people to go for water; probably with the intention of assailing them, for it appeared afterwards that their slings and bows were in readiness. To show them they were not out of our reach, I caused a six-pound shot to be fired over their heads, which, as it went high above them, made no impression. The gun was then pointed lower, and another ineffectual shot fired. A third, however, fell close to them, when they jumped up, shook their mantles in the air, with the most violent gestures, and, apparently in a furious rage, scampered off; but the last man, before he disappeared, threw an immense stone, which did not reach one quarter of the distance.

We saw nothing more of the natives until the evening, when Lieutenant Mitchell, who went to look for them, found they had moved away to Rocky Bay, where they had encamped on the open beach. The next day, I sent him to endeavour to make peace, which he very easily effected, by the interchange of a few trifles.

After this we had much bad weather, during which most of the Indians kept close to their wigwams; but a few occasionally communicated with our watering party, quite peaceably, as if nothing had happened. A day or two after, the weather improved, and the Fuegians dispersed, probably for want of food, some going to the northward, but the greater part along shore to the southward. These people pointed upwards to the sky, when they were going away, repeating the word 'Pecheray.'

This was our last interview with the wretched Fuegians. Naturally petulant and quarrelsome, they are also ever intent upon mischief; the fear of punishment alone restraining them. Weakly-manned vessels passing through this Strait should always avoid them, if they are numerous; for unless they are given what they want, they try to steal it, and any consequent punishment probably brings on a quarrel. Their conduct, and servile bearing, at our first seeing them, gave them an appearance of being timid and inactive; while, in reality, they / are the very reverse. Had we attempted to land on the last occasion, I do not think we should have effected our object, without receiving some severe contusions from their stones, which they sling with such extraordinary precision and force: so much so, that I consider the sling, in their dexterous hands, to be equal to a musket in ours. Indeed, with many of us, a native would have had the advantage. It has been too much the practice, when obliged to fire upon them, to fire over their heads; by which proceeding the savages are led to consider our weapons as so uncertain in their effect, that they become much depreciated in their estimation. It would be almost preferable to inflict a slight wound, in order to show the nature of our arms, and as a warning against further hostilities.

When the Uxbridge, sealer, was at anchor in a harbour in the Magdalen Channel, some Indians, who were on board, angry at being ordered out of the vessel at sunset, threw stones at the person who was walking the deck, as they returned to the shore. Several muskets were fired over their heads, at which they expressed neither fear nor concern; but paddled leisurely away, and the next morning came off again to the vessel, as if nothing had happened. At Port Famine, Duclos Guyot had a skirmish with natives, the particulars of which are described in Dom Pernetty's History (ii.653).[15] Three of the Indians were killed, and three of the French were severely wounded. It may be here remarked, that the chief's name, according to M. Duclos Guyot, was 'Pach-a-chui,' which is not unlike 'Pecheray;' the women were called 'Cap, cap,' probably a mistake for 'Cab, cab;' which evidently means 'no, no!' for it was an expression we frequently used, and was never misunderstood. Their cunning is sufficiently proved by the theft of the Adelaide's boat, in St. Simon's Sound (page 142).

The absence of the Fuegians permitted us to move about a little; and among other places, we visited their late encampment at Rocky Bay, our approach to which was offensively indicated by a most sickening smell. On our way, I found two fossils; one was very interesting, bearing the appearance of a / large orthoceratite:* the other was a Venus. From Rocky Point we descried a strange sail, which, by her movements, we thought must be the Beagle: I returned, therefore, and sent Lieutenant Mitchell out to her. She arrived in the evening, but proved to be a ship belonging to the Hudson Bay Company, called the Dryad,[16] bound to the Columbia River, and last from the Falkland Islands. She came to wait for Mr. Low, of the Adeona, who had promised to pilot her through the Magdalen Channel. The Adeona arrived on the 3d of May; and the following day, to our great joy, the Adelaide hove in sight: and being becalmed, was towed to an anchorage.

The result of her cruise proved to be very interesting, although no communication had been discovered between the 'Ancon sin Salida,' and the Skyring Water. The only loss they had sustained was, however, a severe one; Mr. Alexander Millar having died of inflammation in the bowels. The death of this promising young man threw a damp over the happiness we felt at meeting again, after having so nearly completed this long and tedious voyage.

We had, for some days, been getting ready for sea, and now hastened to complete our preparations. The Dryad, after receiving some assistance from us, sailed in company with the Adeona, and passed out to the Pacific, by going through the Magdalen Channel. The day afterwards we took our final departure – crossed the shoal that extends off Magdalena Island, in five fathoms, sailed on rapidly, and passed Gregory Bay at noon. Seeing us approach, a large party of Patagonians, at least a hundred in number, assembled at the usual place of communication; but as both wind and tide were in our favour, and we could derive no novel information from them, we continued on our course. The Indians were probably much

* They are deposited in the Museum of the Geological Society.

mortified and disappointed; but all on board were delighted by avoiding the anticipated delay. We showed our colours to them, but I dare say our friend, Maria, was not very well pleased with my want of courtesy, in passing by so old an acquaintance / without a salutation; or, what she coveted much more, such presents as she had always received when we anchored.

Just before entering the First Narrow, we passed through a furious 'tide-race,' which broke over the Adelaide, and not a little impeded her progress. No accident, however, was the consequence; and a rapid tide, running at the least nine knots an hour, swept us through the Narrow, and round the reef off Cape Orange: after which we proceeded rapidly, and rounded Cape Virgins at ten P.M., not a little elated by leaving behind us, with no expectation of ever seeing it again, the famous Strait of Magalhaens.

Our voyage to Monte Video was rather long; but we delayed there only to water the ship, in the usual place, off Cape Jesu Maria, and then proceeded to Rio de Janeiro, where we awaited the arrival of the Beagle. Our anxiety for her safety, during so hazardous a survey as that of the sea-coasts of Tierra del Fuego, was soon removed, by hearing that she had touched at Monte Video; and, on the 2d of August, our consort was seen entering the harbour; when we were delighted by finding all well on board, and the little vessel quite ready for sea, having refitted on her passage. /

CHAPTER XVIII.

Adelaide's last cruise – Port Otway – San Quintin – Marine Islands – Unknown river or passage – San Tadeo – Isthmus of Ofqui – San Rafael – Sufferings and route of the Wager's party – Channel's Mouth – Byron – Cheap – Elliott – Hamilton – Campbell – Indian Cacique – Passage of the Desecho – Osorio – Xavier Island – Jesuit Sound – Kirke's report – Night tides – Guaianeco Islands – Site of the Wager's wreck – Bulkeley and Cummings – Speedwell Bay – Indigenous wild potato – Mesier Channel – Fatal Bay – Death of Mr. Millar – Fallos Channel – Lieutenant Skyring's illness – English Narrow – Fish – Wigwams – Indians – Level Bay – Brazo Ancho – Eyre Sound – Seal – Icebergs – Walker Bay – Nature of the country – Habits of the natives – Scarcity of population.

I WILL now relate the principal incidents of the Adelaide's last cruise. The following pages contain extracts from Lieutenant Skyring's journal, and also notices obtained from other sources.

The Adelaide sailed from Chilóe on the 8th of December 1829, made Cape Tres Montes on the 14th, and anchored in Port Otway the same evening. Of this place Lieutenant Skyring writes: 'Good anchorage, wood, water, and shell-fish (such as muscles and clams), Port Otway affords: but no more. Excepting in one or two sandy bights, a landing is hardly to be effected; walking along shore is impossible, and it is scarcely practicable to enter the country, the land being so thickly wooded, from the summits of the hills down to the water-side. No soil is to be discovered; the shrubs, and even the trees, which are of large growth, rise out of moss, or decomposed vegetable substances. The climate is very wet; none but amphibious animals were seen, among which hair-seals were numerous. There were very few birds, excepting turkey buzzards; and not a trace of human beings; indeed, I do not believe Indians ever go there – (*y*) they rarely leave the direct channels; as a proof / of which, some articles left by the Beagle, in a conspicuous place, were found by us untouched.' During the Adelaide's stay at Port Otway, the openings on the east side of Hoppner Sound were explored, yet they proved to be only small inlets. Mr. Kirke examined some, which appeared to communicate

(*y*) For evidence that Indians have been thereabouts, see Byron's account of the cave entered by the surgeon of the Wager.[1] I believe that curious place was either in, or close to, Port Otway. – R.F.

with San Quintin Sound; but found them to be merely channels dividing the group of the Marine Islands,* excepting the most southern, which is the entrance of Newman Inlet, a deep bight, without anchorage, but abounding with hair-seal.

From Byron's Narrative it would appear, that there is a channel somewhere hereabouts communicating with the Gulf of San Rafael, to the east of the Peninsula of Tres Montes; for the Indian guide wanted to conduct the Wager's barge through it, but was prevented by the strength of the current.

The Adelaide sailed from Port Otway on the 18th, and the same evening reached San Quintin Sound, anchoring opposite an opening northward of Dead-tree Island, that proved to be the mouth of the River San Tadeo, by which Byron and his unfortunate companions effected their escape to Chilóe.

The sufferings of this party, which are so affectingly described in Byron's narrative of the loss of the Wager, made so deep an impression on our minds, that I thought it not irrelevant to the object of this voyage to endeavour to trace their steps. Among the numerous incidents that occurred to them, the passage of the 'Desecho,' or carrying-place over the Isthmus of Ofqui, is, from all the circumstances connected with it, one of the most interesting. It may be remembered, that, upon the departure of Captain Cheap, and his shipwrecked crew, from the place of the wreck (Byron's Narrative, p. 69), they proceeded round the shores of the Gulf of Penas, with an intention of tracing the Coast of Chilóe.[3] They first attempted to steer for Cape Tres Montes, which headland they had seen, in one of the intervals of fair weather, from the summit of Mount Misery, and which appeared to be twenty or thirty leagues distant. The wind, / however, freshened to a gale, and they were obliged to run before it, and throw all their provisions overboard to lighten the boat.

At night they took refuge in a small opening, which led to a secure harbour, and next day advanced a little farther, till they reached some small islands, where they were detained three or four days by bad weather.

After leaving that place, they found an opening, into which they rowed, flattering themselves it would prove to be a passage; but, being disappointed, they were obliged to return. This was probably the inlet, called 'Channel's Mouth.' Xavier Island was the next place they went to, named by them Montrose Island. Byron describes this island so exactly, that there cannot be the least doubt of its identity. 'The next morning,' he says, 'being calm, we rowed out; but as soon as clear of the island, we found a great swell from the westward: we rowed to the bottom of a very large bay, which was to northward of us, the land very low, and we were in hopes of finding some inlet through, but did not; so kept along shore to the westward. This part, which I take to be fifty leagues from Wager Island, is

* The Marine Islands were so called, in remembrance of the four marines who were put on shore from the Wager's boats, and left behind.[2] See Byron's Nar., p.85.

the very bottom of the large bay it lies in. Here was the only passage to be found, which (if we could by any means have got information of it) would have saved us much fruitless labour. Of this passage I shall have occasion to say more hereafter.' – Byron's Nar. p.74. This is evidently San Quintin Sound. They proceeded to the westward and northward, entered a larger bay (Holloway Sound), and discovered another headland at a great distance to the westward (Cape Tres Montes), which they reached with much difficulty; but being unable to get round it, and losing the boat that accompanied them besides being obliged to leave four of the marines behind, they became quite disheartened, and returned to Wager Island, to linger out their miserable lives, without the least prospect of again seeing home. This expedition occupied two months, during which they lived principally upon sea-weed, called 'tangle;' but sometimes passed whole days without eating anything at all. While they / were absent, some Indians had visited the wreck; and, about a fortnight after their return, they arrived a second time, in two canoes. Among them was an Indian Cacique of the Chonos tribe, who live in the neighbourhood of Chilóe. It was supposed that a report of the wreck had reached that place; and that this Cacique, and another Indian, had come to derive some advantage from it. As the Cacique spoke Spanish, the surgeon, Mr. Elliot, made himself so far understood, as to let him know that they wished to reach some of the Spanish settlements; and eventually bargained to give him the barge, and every thing in it, if he would conduct them to Chilóe. The party consisted of Captain Cheap; Mr. Elliot, the surgeon; Mr. Campbell, Mr. Hamilton, and Mr. Byron, midshipmen; and eight men, besides the two Indians; in all fifteen. The first night they slept on an island, and the next laid upon their oars, to the westward of Montrose Island, not being able to land.

They then pulled, 'to the bottom of a great bay, where the Indian guide had left his family, a wife and two children.' There they staid two or three days; after which, taking on board the family, they proceeded to a river, 'the stream of which,' Byron says, 'was so rapid, that after our utmost efforts, from morning to evening, we gained little upon the current; and, at last, were obliged to desist from our attempts, and return.'

This was probably a river, or channel, to the westward of San Quintin Sound, which eluded our search; and, if so, it must communicate with channels north-eastward of the Peninsula of Tres Montes. The Indians, anxious to get the barge to the Chonos, had no other way to effect their purpose; for the usual route was over the 'Desecho;' to pass which, it was necessary to take a boat or canoe to pieces, and carry her, piecemeal, over a high mountain.

After losing the barge, they crossed the Peninsula of Forelius, by hauling canoes over a narrow neck of land, and reached the water of San Quintin Sound; where they met another native family, with whom they proceeded to the River San Tadeo, 'up which they rowed four or five leagues; and then / took to a branch of it that ran first to the eastward, and then to the northward.' There they landed, took the canoes to pieces, and carried them over the isthmus; then putting them together again, re-embarked, and proceeded through the Chonos Archipelago to Chilóe.

When at Chilóe, I saw an old man, Pedro Osorio, who had been in two of the last missionary voyages (in 1769 and 1778), to the Guaineco Islands; where the Wager was wrecked. He related to me the particulars of these voyages, and gave me an account of the 'Desecho,' over which the missionaries transported their piraguas. He also remembered Byron and his companions; and described them by the following names:– Don David (Captain David Cheap); Don Juan (John Byron); Hamerton (Hamilton); and Plastà. The name Plasta is not once mentioned in Byron's Narrative; but on referring to Bulkeley's and Cumming's account, one Plastow is described as the captain's servant; and perhaps he was one of the number who remained with Captain Cheap. (*z*) Pedro Osorio must have been upwards of ninety years of age, in 1829.(*a*) A detailed account of these voyages is given in Agüeros's Historical Description of the province of Chilóe, p. 205.

Captain Stokes's 'Dead-tree Island,' in the entrance of San Estevan Gulf, is near the 'Cirujano Island' (Surgeon Island) of those voyages. Pedro Osorio told me that it was so called, because the surgeon of the Wager died there. From Byron's Narrative it would appear, that the surgeon died, and was buried, just before they embarked to cross the sound. – See Byron, p. 147.

As the examination of the River San Tadeo, and the discovery of the 'Desecho,' formed a part of Lieutenant Skyring's instructions, he proceeded up it, in a whale-boat, accompanied by Mr. Kirke. The entrance of the river is blocked up by a bar of sand and stones, which, at low spring-tide, must be nearly dry; and a heavy swell breaks upon its whole length, joining the surf of the beach, on each side; so that there is / no deep channel; and, except in very fine weather, an attempt to cross is hazardous.

At its mouth, the breadth is about a quarter of a mile, but within the entrance it increases for a short distance: at three miles up, it is three hundred yards, and thence gradually diminishes. The shores are a mixture of clay and sand; and the country, on both sides, is low and marshy, abounding with brant-geese, ducks, teal, and snipe.

The land, near the mouth of the river, is studded with dead trees (a species of pine, about twenty feet high), which appear to have been killed by the sea overflowing the banks;(*b*) as it does at high-water for several miles.

Three miles from the entrance this river divides into two branches, one leading N.W., and the other eastward. Considering the latter, from Byron's description, to be the proper course, Lieutenant Skyring followed it. At nine miles from the

(*z*) Could 'Plasta' refer to Alexander Campbell? – R.F.

(*a*) Pedro Osorio died at San Carlos in 1832. – R.F.

(*b*) Or by an earthquake wave. – R.F.

mouth, a stream was found falling into the river from the north, in every respect differing from the principal stream; the water being fresh, dark, and clear, and the current constantly running down, uninfluenced by the tide; while the water of the river was brackish and turbid, and affected by the ebbing and flowing of the tide, although, at that distance, its effect was much diminished.

The shores of the Black River, as this new stream was called, are thickly wooded, which is not the case with the principal stream. They had entered it about a hundred yards before they discovered that they had left the main river; but being desirous of proceeding, they followed its windings, the next day, for three leagues; during the greater part of which distance, they found a strong current against them, and were also much impeded by fallen trees lying in the bed of the river. In many parts they dragged their boat along by the help of overhanging branches, or projecting roots; and the width, generally, was not more than fifty yards. As no piragua could pass there, Lieutenant Skyring felt assured that he was not in the right stream; therefore, returning to the main river, he proceeded / up it during the next two days. At two miles above the junction, the tide ceased to be felt; and a rapid current met them, which increased in strength until they were unable to stem it; and as they were prevented from tracking the boats, by trees growing on the banks, they could ascend no farther.

This place was not more than eleven miles from the sea; although, from the tortuous course of the stream, they had gone double that distance, and were about two miles from the foot of a mountain, whence the river descends. The mountain was very high, and the vallies, or ravines, were filled with glaciers. From Byron's description, it seems probable that Lieutenant Skyring was near the carrying place; but as further delay could answer no good end, he very prudently returned, looking carefully about, as he proceeded, for some signs of a landing-place, but without success. He re-crossed the bar, reached the Adelaide without accident, and the next day went on in her to Xavier Island. On the way they passed Dead-tree Island; where, observing seal on the rocks, a boat was sent ashore, and her crew succeeded in killing a few sea-elephants, twenty feet long.

Favoured with fine weather, they were enabled to land on the north side of Xavier Island, to improve the former survey; and in the evening anchored in Xavier Bay, where they remained four days; during which, Jesuit Sound was explored, and found to terminate in two narrow inlets. Being a leewardly opening, it is unfit for any vessel to enter.

The name Jesuit Sound, and those of the two inlets at the bottom, Benito and Julian, are memorials of the missionaries, who, in the expedition of 1778, entered and explored it.* (Agüeros, p. 232.)

* Mr. Kirke, who examined them, says, 'There are two openings opposite Xavier Island, on the main land: the northernmost runs through high land, and is terminated by a low sandy beach, with a river in the middle, running from a large glacier; the southern inlet is ended by high mountainous land.'

The Adelaide anchored the next night in Ygnacio Bay, at the south end of Xavier Island, which Lieutenant Skyring / recommends for small vessels; the depth of water being six or eight fathoms, and the anchorage well sheltered from the wind.

On the 31st they anchored under the Hazard Islands, in the Channel's Mouth: 'preparatory,' writes Lieutenant Skyring, 'to commencing new work with the new year; for since entering the gulf, except while examining the San Tadeo, we had followed the Beagle's track, and only completed what she left unfinished; but from this place all would be new. This was the last wild anchorage she had taken; and although now fixed in the best situation, and in the height of summer, we found our position almost as dangerous as hers.

'Early on the 1st of January 1830, Mr. Kirke went in a whale-boat to examine the openings, at the mouth of which we had anchored: he returned on the 9th, having traced to the end, all which had the least appearance of being channels. The two largest, the south and the east, penetrated into the Cordillera for thirty miles. All these inlets are narrow but deep arms of the sea, running between ranges of very steep hills; their sides affording not the least shelter, even for a boat, and apparently deserted; for neither seal, nor birds of any kind were seen, nor were there even muscles on the rocks.'

Mr. Kirke, in his report, says: 'The three northernmost of the inlets of the Channel's Mouth end with high land on each side, and low sandy beaches at the head, beyond which there rises a ridge of high mountains, about two miles from the beach. The S.E. inlets end in rivers rushing down from the mountains, and a rocky shore: not the smallest shelter could I find, even for the boat. Two days and nights I was forced to keep her hauled up on a rock, just above high-water mark, in a strong gale, while the williwaws were so violent, that we were all obliged to add our weight to that of the boat, to prevent her from being blown off: and twice we were washed out of our resting-places, on the beach, by the night tide rising about fifteen or sixteen inches above that of the day.'

This opening in the coast is noticed by the pilot Machado (Agüeros, p. 210); but by whom the name of Channel's Mouth was given, does not appear. It is by no means descriptive of / what it has been proved to be; but as Lieutenant Skyring thought that a change in the name would not answer any good purpose, he very properly left it unaltered.

The day after Mr. Kirke returned, very bad weather set in, and detained the Adelaide nine days, during which nothing could be done, out of the vessel.

'January 19th,' Lieutenant Skyring writes, 'with moderate weather, and an easterly wind, we left the Channel's Mouth, and, standing for the Guaianeco Islands, passed those of Ayautau (between which and the mainland are several rocky reefs, though the passage seems to be sufficiently clear for any vessel); and skirting Tarn Bay, we distinguished the Mesier Channel, and could see many leagues down it. The entrance of the Mesier Channel is very remarkable, from having two high and singular peaks on the islands at its mouth: the northernmost very much resembling (although higher than) Nelson's monument, near the Strait; and the other, more to the southward, and much higher, resembling a church with a cupola, instead of a spire. Both are easily made out from the westward, at a distance of twenty or thirty miles.

'We reached the Guaianeco Islands in the afternoon. The two largest are divided by a narrow passage, on the west side of which we anchored, in ten fathoms, in a spacious and secure haven, which proved to be Speedwell Bay of Bulkeley and Cummings; the boats were employed next day, and, while the examination of the coast was pursued, I sought to ascertain the exact spot of the wreck of the Wager, but never could discover it: not a fragment of that ill-fated vessel was seen in any of our excursions. A few pieces of the boat lost by the Beagle last year were picked up; but nothing more that could tend to denote the misfortunes which have occurred near these islands.

'From the description of the Wager's wreck, in Bulkeley and Cummings, there seems to be little doubt of the place being at the N.W. end of the eastern Guaianeco Island, near my Rundle's Passage, which is the place so often mentioned in their account as the 'Lagoon.' /

'Being well supplied with powder and small shot, the people provided themselves plentifully, during our stay at Speedwell Bay, with a variety of wild-fowl, namely, geese, ducks, redbeaks, shags, and the ibis; curlew, snipe, plover, and moorhens, were also met with, and fish were observed in shoals near the vessel, but, as we had no seine, they escaped. With hooks and lines our fishermen had no luck; the baits were no sooner at the bottom, than they were taken away, and for a day or two the cause of their loss was unknown; but being accidentally ascertained, small trap-nets were made, and great numbers of crabs were taken, about a pound each in weight.

'In almost every bay we noticed the potato, growing among wild celery, close above high-water mark: but in so unfavourable a situation, choked by other vegetables, its produce was very small.

'The trees are not of large growth in these islands, neither is the land thickly wooded; but above the beach, and almost round the coast, there is a breast-work of jungle and underwood, from fifty to one hundred yards broad, and nearly impenetrable; beyond which is a great extent of clear, but low and swampy ground.

'On the 25th, we left this port, and ran to the S.E., through what I have named Rundle's Passage. This small channel, where the islands approach each other, is about a quarter of a mile wide, perfectly clear in the whole extent, and also at its southern entrance; but at the northern there are many detached rocks, which are obstacles to entering Speedwell Bay, except in daylight. Rounding the islets, at the S.E. extreme of Byron Islands, we anchored in Muscle Bay, which lies on the northern side: by no means a secure place, – but the only one that could be found, by the boats, after many hours' search. I selected this situation in order that the entrance to the Fallos Channel, and the whole outline of these islands, might be laid down, and properly connected with the land of Port Barbara; which was thoroughly executed by Mr. Kirke and Mr. Millar, although delayed in the completion of their / work until the 1st of February. (*c*) On that day we sailed, and entered the Mesier Channel, anchoring in a small open bay, the only stopping-place we could perceive; which, from the loss we sustained shortly after our arrival, was called Fatal Bay. It is insecure, and the anchorage ground confined: the only convenience was, that wood and fresh-water were near. During our stay we had much rain, which retarded us. Mr. Kirke went away in a boat, whenever the weather permitted, and, on the 8th, we sailed for an anchorage, about ten miles to the southward, where he had previously been; but a sad event happened before our departure.

'On the afternoon of the 3d, we had the misfortune to lose Mr. Alexander Millar, who died in consequence of a severe attack of inflammation of the bowels, which carried him off, after an illness of only three days.

'On Thursday afternoon he was buried, close to the shore, near the anchorage, and just within the edge of the wood.

'That our progress had been so slow during the last month, was a great disappointment; but we had had many causes of detention. All the early part of January the weather was stormy: eighteen days we were anchored within the Channel's Mouth; yet during two only could our boats leave the vessel.

'Among the Guaianeco islands we had moderate weather, but also much wet: still the chief cause of our delay, I fear, was my own illness. From the beginning of January, I had been confined to my bed, with a tedious and obstinate disease; and from that time most of the angles were taken, and all the observations were made, by Mr. Kirke, who was ever exceedingly willing and indefatigable. After the loss of Mr. Millar, not only almost the whole duty of surveying fell upon him, but much of the duty of the vessel.

'At noon this day (8th), we moored in Island Harbour, a small but excellent landlocked anchorage, with good holding / ground, and abundance of wood and water. The two following days, Mr. Kirke was away examining the coast; the third we were confined by bad weather; and, indeed, during our whole continuance at this place, we had very much rain.

'We sailed early on the 12th from Island Harbour, and by night reached Waterfall Bay, an anchorage about fifteen miles to the southward: the wind all

(*c*) During much of this cruise, Lieutenant Skyring was so ill that he was unable to leave the Adelaide; and for a month he was confined to bed. His illness was caused by fatigue, and by sitting too long while constructing charts. – R. F.

day was light, and the tide, the greater part of the time, against us; so that, with every exertion, we scarcely gained anchoring ground before it was quite dark: the strength of the tide was upwards of a mile an hour, at neap-tides: the ebb and flood were of equal duration, the former running to the S.b.E., the latter N.b.W. Thirty miles within the Mesier Channel it is as wide as at the entrance, and for several miles to the southward appears clear: so that no one is liable thus far to mistake its course.

'The land on the west side appears to be a number of large islands, with here and there wide passages leading to the S.W., rendering it probable that there are many (although not direct) communications between the Mesier and the Fallos Channels. Our anchorages were chiefly on the eastern shore, that the openings on that side might be more readily examined; but all which appeared to run far inland were found to be merely narrow inlets, or sounds ending abruptly. On each side the land is hilly, but not high; and this distinguishes the Mesier Channel from many others, whose shores for miles are formed by ranges of steep-sided mountains. Here, in many places, there is much low land, which is generally thickly wooded, yet with no greater variety of trees than is to met with in the Strait of Magalhaens. The beech, birch, pine, or cypress, Winter's-bark, and a kind of red-wood, form the forests; but none were observed that could be at all serviceable for the larger spars of a vessel.

'(16th). Left Waterfall Bay, and with a N.W. breeze passed Middle Island, entered Lion Bay, and moored in White Kelp Cove. The coast survey was soon finished, but we were confined at our anchors here four days; not by bad, but by extraordinarily fine weather. During such intervals, so very rare / in these regions, the wind, if there is any, is almost always southerly, and light.

'At every anchorage we had found Indian wigwams, but as yet had not met with any natives. Here we took a great number of fish; and, among them, one like the ling, found on the east coast of Patagonia, off Cape Fairweather, but of smaller size, for the largest did not weigh more than two pounds. Very few water-fowl were seen; steamers and shags were the only ones shot; but in the woods we noticed king-fishers, wood-peckers, barking-birds, parroquets, and humming-birds.

'(21st.) With a light northerly wind we left this cove, and about ten miles to the southward the appearance of the channel changed greatly. Instead of sailing through unconnected land, of moderate height, we were confined between two mountainous ridges.* At noon we were obliged to anchor in Halt Bay, no opening appearing to the right or left, and being apparently embayed. On the west side,

* On the west shore Mr. Kirke noticed what appeared to be a channel, about twelve miles N.W. of Halt Bay, in the mouth of which was a considerable tide-ripple; an almost certain indication of such an opening. 'I thought the inlet about twelve miles north-west of Halt Bay much like a channel. I also noticed a distinct tide ripple, which I did not remark

the high land was skirted by several low islands, among which our only way of proceeding seemed to lie. This day and the next Mr. Kirke was away, seeking a passage; and having found one, and noticed the tides, we sailed through on the 23d, and gave it the name of the English Narrow. It is long and intricate, chiefly formed by islands; and in three places, where the shores approach each other, the distance across is less than four hundred yards, yet with a fair wind and slack tide, there is no hazard in passing. In the afternoon, we moored in ten or twelve fathoms in Level Bay, a spacious anchorage near the southern entrance of the Narrow; the bottom mud and sand, and the depth of water equal throughout. Mr. Kirke, who was among the islands opposite this bay, saw numerous shoals / of fish in many of the bights; with a seine, therefore, an abundant supply might be obtained.

'The woodland eastward of our anchorage had very recently been on fire, and the conflagration must have been extensive, and very destructive; for throughout a space of ten or twelve miles along shore, all the trees had been consumed, the dead trunks of the larger ones alone remaining. We left Level Bay on the morning of the 25th, and passed a canoe full of Indians; but they pulled to the shore, and ran into the woods; therefore, since they avoided us, and we had a fair wind, I did not seek their acquaintance. We had noticed traces of them in the neighbourhood of the Narrow, on each side of which many wigwams, that had been recently occupied, were seen.

'For the next ten or twelve miles we went through a fine reach, whose shores were low, and whose channel was interspersed with several islands, affording probably excellent anchorages; but to the southward the hills became more steep, and, except in the ravines, were destitute of vegetation. At four or five leagues to the E.S.E., beyond the English Narrow, an opening, apparently a channel, presented itself, and the reach in which we were sailing seemed to end. Doubtful which course to follow, we anchored the vessel in Rocky Bight, and despatched the boats to examine both passages. That to the E.S.E. was found to run direct nearly ten miles, and to communicate with a fine clear channel, trending to the S.S.W., which proved afterwards to be the Wide Channel (Brazo Ancho) of Sarmiento. At the junction, a considerable arm extended to the N.N.E., apparently a continuation of the Wide Channel.

'On Mr. Kirke's return from examining the passage in which we were sailing, I learnt that the same width continued about five miles southward of our present anchorage, and that there the shores approached closely, forming the intricate passage called Rowlett Narrow; which, after a S.E. course of many miles, also joins Wide Channel. The island formed by the two channels was named Saumarez Island, in honour of the gallant admiral.[4]

near any other opening. To me this appeared the southernmost inlet, of any depth; or at all likely to be a channel.' – Kirke MS.

'It rained hard and blew strongly the whole day, which / prevented our moving; but on the 27th we shifted our anchorage to Fury Cove, in Wide Channel.

'Mr. Kirke, on the 28th, examined an opening to the northward, called Sir George Eyre Sound, which terminates in a wide fresh-water river, running through low land from a large glacier. The low grounds extend two or three miles from it, and then the land becomes high. Behind the glacier there is a ridge of high mountains, covered with snow, which we had seen twice before; first, from near White Kelp Cove, and again from Halt Bay. In the sound, we saw three whales, and being the first we had observed, since leaving the Gulf of Peñas, they inclined us to think we were near the Gulf of Trinidad. A great number of fur seal, besides two of their rookeries, or breeding-places, were also seen. Several icebergs were floating out of the sound, some of which were dark-coloured; and upon one I found a quantity of rock that had come down with it from the mountains. There was serpentine and granite, specimens of which were collected, and given to Captain King. One of the bergs, which was large, was aground. It was nearly seven fathoms above the water, and bottom could not be found by sounding round it with twenty-one fathoms of line.

'Fury Cove is diminutive; there is not more than sufficient space for two small vessels; but the ground is good, and in every other respect it is a secure haven. We sailed on the 3d of March with the expectation of soon recognizing some known points in the Gulf of Trinidad; but as the wind failed, we were obliged to anchor for the night in Sandy Bay, in eight fathoms.

'As we proceeded to the southward, the appearance of the country gradually changed: the mountains seemed more barren, the trees and shrubs more stunted, the land rose more suddenly, and the shores of the channel became bolder, and presented an uniform rocky line of coast.

'(4th.) We again steered southward, and at noon an opening appearing on the east side, which ran several miles inland, I sought an adjacent anchorage, in order that it might be explored. Our boats were examining the shore all day, and / sounding in the coves, but no fit spot was found; therefore we were forced to stop in an ill-sheltered nook, termed Small Craft Bight, which just served us (having fair weather) as a resting-place until morning (5th), when we set out again to find a better anchorage; for I still desired to ascertain whether the opening to the eastward was a sound or a channel. In our course to the southward we traced both shores in search of a stopping-place; but there was neither bight nor cove where it was possible to anchor, until we arrived at Open Bay, which lies near the entrance of Wide Channel. Even this was such a very insecure place, that although I remained the next day, to examine the neighbouring coast, it was far too exposed an anchorage for the vessel to continue in while the boats were away at a distance.

'Disappointed by not finding a place for the schooner near the opening I wished to explore, I was yet averse to leaving it unexamined, having traced every inlet to its extremity for upwards of two hundred miles along the continent. I wished to continue so sure a mode of proceeding; and although I felt certain that this opening terminated like the rest, and Mr. Kirke held the same opinion, I would gladly have prevented any doubt by following its course in the boats, could we have gained a safe anchorage for the vessel. The nearest harbour that could be found was thirty miles from the opening, and it would have detained us too long to send the boats such a distance; so considering that we had yet a great extent of coast to examine; that my state of health did not permit me to undertake any very exposed or arduous service; and that Mr. Kirke was the only person to whom such duty could be entrusted, I was induced to relinquish our former practice of exploring every opening to its end.

'We left Open Bay on the 7th, and soon entered Conception Strait, keeping along the east shore, and sending a boat, at every opening, to seek a situation for the vessel. In the afternoon, a tolerably sheltered bay was found, at the south end of the North Canning Island, open only from S.E. to S.W.; but those winds being frequent and violent, and the / bay exposed to a long reach of sea from that quarter, it cannot be accounted a safe harbour; yet it was very far preferable to many places in which we had been obliged to anchor.

'This bay (Portland Bay) is on the north side of an opening called by Sarmiento 'Canal de Tres Cerros,' and from the broken state of the interior high land, one is led to imagine a channel might be found there. His conclusion, I have no doubt, was drawn from this appearance, since the view down the opening is very limited, and, at the distance of three or four miles within the entrance, is interrupted by several small islands. Mr. Kirke passed between those islets, and followed an opening to the S.E., for upwards of eight leagues. On his return, he reported that he had found a fine channel, of which the principal entrance was the opening of Sarmiento's 'Canal San Andres.'

'On the 12th, in full anticipation of making some interesting discovery, we sailed into the 'Canal San Andres,' anchoring in the afternoon in Expectation Bay, where we remained until the 15th. During that time, Mr. Kirke was employed examining the different openings, and tracing this supposed channel farther. At his return, he said that he had found a termination to every opening, even to that in which we then were, which he had previously thought to be a channel. Like the rest, it extended only to the base of the snowy Cordillera, and then was suddenly closed by immense glaciers.

'This information caused great disappointment, as all hope of passing through the Cordillera, thus far northward, was now given up; and I was fearful we should be delayed many more days before we could extricate ourselves from this (as we then supposed) false channel. We were many miles within the entrance; in that distance there were no anchorages, and the wind being generally from the westward, I anticipated much labour before we could effect our return; but the very next day we were so fortunate as to have a slant of fair wind, by which we cleared this opening, and a second time entered Conception Strait. Knowing, by our former survey, that there was no anchorage along the coast to the southward of Cape San Andres before reaching Guard Bay, I ran over to Madre / de Dios, and brought up in Walker Bay. Fortunate we were, too; for before midnight the weather became so stormy as to oblige us to strike the topmasts and yard, let go a second anchor, and veer a long scope of cable. At few places in these channels where we had anchored, could we have veered even half a cable. We remained the following day, and on the 21st, the weather being moderate, ran for the Guia Narrow, and having a favourable tide, passed through easily.

'It was my wish to have anchored among the islands to the southward of Cape Charles, since that would have been the most convenient place for the Adelaide, while examining the opening beyond Cape San Antonio; but hauling round the headland into a bay formed by those islands, no soundings could be gained; and not perceiving any bight at all likely to afford shelter, I continued my course for Puerto Bueno, where Sarmiento thought there was good anchorage. In the evening, with the assistance of the boats, we moored in Schooner Cove, Puerto Bueno, and the next day, Mr. Kirke went to examine the opening north of San Antonio.

'While we remained, a plan was made of this port, which lies five miles S.E. from Cape Charles and three and a-half from Bonduca Island. The shore is steep, and without any indenture. To the southward is Lear Bay, a mile in extent, affording anchorage, but not to be chosen when such an excellent haven as Puerto Bueno is near. The south extreme of this bay forms the north point of Puerto Bueno, and a few hundred yards south of that point is Rosamond Island, which is low and pointed; four hundred yards S.S.E. of this, is a small round islet, bold to on every side; and between this islet and a low point, a quarter of a mile to the S.E., is the widest channel to the anchorage. Sarmiento, indeed, most appropriately named it Puerto Bueno. It has both an inner and an outer port, the depth of water throughout is from nine to six fathoms, and any position in either I consider safe; but excepting that it affords better shelter, it differs in no respect from other anchorages in these regions. Wood and water are generally found in abundance near them all; fish may be caught; geese, ducks, shags, and / steamers may be shot; and shell-fish gathered. The country, also, has the same appearance, and is of a similar nature; for if you force a passage through the woods, it is over fallen trees and moss; if you walk over clear flat ground, the place is found to be a swamp; and if you ascend the hills, it is by climbing over rocks, partially covered with spongy moss.

'Mr. Kirke returned on the 24th, having found that the opening beyond San Antonio led to the N.E., and at ten miles from the cape communicated with that called the Canal San Andres.

'At daylight we left Schooner Cove, and in passing down Sarmiento Channel I tried, though unsuccessfully, to reconcile some of his remarks with our own observations. South of San Marco and San Lucas there are two extensive bays, which we afterwards found communicated with an opening between San Mateo and San Vicente, separating the greater part of the eastern shore of this channel from the main land.

'I wished to anchor near Cape San Lucas, but around that opening no place could be distinguished likely to afford shelter, the shore in every part being bold, steep, and rocky. A like uniformity of coast presented itself as far as Cape San Mateo; but on the west side, along both Esperanza and Vancouver Island, lie many bays that are well adapted for vessels. Sailing, however, under Cape San Lucas, we stood for San Mateo, and succeeded in anchoring in a small port, formed by Weasel Island, scarcely large enough, but perfectly safe, when once we were secured. From this place the boats were despatched. An opening east of our present station was to be traced, and this part of Sarmiento Channel, with the entrance between San Mateo and San Vicente, was to be laid down. These operations, which in moderately fair weather would not have occupied three days, were not completed before the 31st, from our being delayed by violent winds, and almost continual rain. We had also had exceedingly bad weather during our stay in Puerto Bueno, and those employed in the boats had undergone very severe fatigue, and had suffered much from wet and cold. A short distance within the entrance of the / opening, between Cape San Mateo and San Vincent, it turns suddenly to the south and S.b.E., continues in that direction for nearly thirty miles, washing the base of the Cordillera which rises from it precipitously, and is closed by a low isthmus, two miles across, dividing this inlet from Stewart Bay, and over which Mr. Kirke passed to take the bearings of several points that he recognised in Collingwood Strait.

'In the prosecution of the survey northward of our anchorage, those passages were discovered which separate so much of the east coast of Sarmiento Channel from the main land; and the islands thus made known I named after Commodore Sir Edward Owen,*[5] the channel of separation being called Blanche Passage.

'One of the boats met with a canoe containing eight Indians; this was only the second that had yet been seen during our cruise.

'An interview, which two of the schooner's men had with these people, is so characteristic of the habits of the natives who wander in canoes, that I add the account, as given by one of those men: 'When we arrived at the wigwam, there were two women and five children inside, and a dozen dogs near it. At our entrance, the children crept close to one side of the wigwam, behind their mothers, who made signs for us to sit down on the opposite side, which we did. The women, seeing that we were wet, and meant to do them no harm, sent the two eldest children out to gather sticks, and made up a large fire; so we cut some pieces of bread from a loaf which we had, and distributed them. They all appeared

* At the request of Lieutenant Mitchell, of the Adventure.

to like the bread, particularly the youngest, which was sucking at the breast; for it eat its own slice, besides one we gave its mother. After we had been there about half an hour, and had given them some beads and buttons, a man came in from behind the wigwam, where he had concealed himself when we entered, and sat down beside us. By signs, he asked where our boat was, and how many men there were with us. We told him the men and boat were a little way off, and made signs that we wanted to / stay all night with him. We then gave him some bread, which he smelt, and afterwards eat. He offered us some sea-elephant blubber, about two inches and a-half thick; we took it, and making signs it was not good, flung it on the fire. As soon as it began to melt, he took it from the fire, put one part in his mouth, and holding the other drew it back again, squeezing out the oil with his teeth, which were nearly shut. He put the same piece on the fire again, and, after an addition to it, too offensive to mention, again sucked it. Several more pieces were served the same way, and the women and children partook of them. They drank large droughts of water as soon as they had done eating. As it grew dark at about eight o'clock, the man began to talk to the women about our 'sherroo' or boat, and our men, who he thought were near. They seemed to be alarmed, for the women shortly after left the wigwam, and did not return. They were quite naked. The man took the youngest child in his arms, squatted down with the rest, and making signs that he was going to sleep, stretched himself by the fire, the children lying between him and the side of the wigwam. Soon afterwards another man came in, who seemed to be about twenty-two years of age, younger by ten years than the first we saw. He had a piece of platted grass round his head, in the form of a band. After talking some time with his companion, he talked and laughed with us, ate some bread, and would have eaten all we had, if we had not kept it from him. He ate about two pounds of blubber, broiling and squeezing it, as the other had done, and drank three or four pints of water. We had only one case knife, which he was very fond of borrowing now and then, to cut the blubber, pretending that the muscle shells, which he broke for the purpose, were not sharp enough. He examined all our clothes, felt our limbs and breasts, and would have taken our clothes off, if we had let him. He wanted a knife, and was continually feeling about us for one, as we did not let him know that we had only one. He opened a rush basket, and took out several trifles, such as fire-stone,** feathers, spear-heads, a sailor's old mitten, part of / a Guernsey-frock, and other things, some of which he offered for the knife.

' 'About midnight it rained very hard, and the inside of the wigwam became soaked with wet; so they all roused up, and made a large fire; then ate some blubber, and drank some more water. They always carried a firebrand with them when they went out in the dark to get water, or for any thing else they might want. When they had well warmed themselves they lay down again. The young

* Iron pyrites.

man lay close to us, and, when he supposed we were asleep, began to search the man who had the knife, but we kept watch and he could not get it. About two hours afterwards he made up the fire, and went out, as we thought, for firewood: but for no other purpose than to take away bushes from the side of the wigwam, that he might have a clear passage for what he intended to do. Returning, he took up a piece of blubber, and asked for the knife to cut it. As soon as he had cut a slice, and put it on the fire, he darted through the part of the wigwam, which he had weakened, like an arrow. The other man seemed to be very much vexed, and thinking, perhaps, that we should do some mischief in consequence of the loss of the knife, watched an opportunity, when he thought we were asleep, to take out all the children, and leave us quite by ourselves. About two hours after, he returned, and pulling down dry branches, from the inside of the wigwam, made up a large fire. We had no doubt that the younger man was at hand watching us, and just at daybreak, as we were preparing to start, he jumped into the wigwam with his face streaked almost all over with black, and pretended to be quite a stranger. When we asked for the knife, he would not know what we meant, but took up one of our shoes that lay on the ground, and gave it to us. The band of grass was taken off his head, and his hair was quite loose. There were neither skins, spears, nor arrows in the wigwam, but no doubt they were in the bushes; for when we threatened to take the canoe he jumped into the wood, resting on one knee, with his right hand on the ground, and eyed us sharply till we were out of sight.' /

'The other family seen in the Mesier Channel we did not communicate with, and it may be remarked that in this passage, although between four and five hundred miles in extent, we did not meet twenty human beings; a strong evidence that these regions are very thinly inhabited, particularly when it is considered that we made no rapid progress, and that our boats traversed, through different channels, at least twice the distance run by the vessel.' /

EDITORIAL NOTES

Preface and Introduction

1. *successors in office*: Fitzroy, possibly reflecting his own firmly Tory views, names only Robert Dundas, Viscount Melville (1771–1851). The First Lord of the Admiralty was a Cabinet position, and with the change to a Whig government in 1830 it went to Sir James Graham (1792–1861). Graham held the post until 1834; he was succeeded by Whig aristocrats George Eden, Earl of Auckland (1784–1849) and Gilbert Elliot, Earl of Minto (1782–1859), to whom the *Narrative* was dedicated.
2. *do it justice*: King had returned to his estates in New South Wales in 1832; on publication delays, see General Introduction.
3. *has been disappointed*: apparently a reference to the botanist Robert Brown (1773–1858), a protégé of Sir Joseph Banks. Brown was a leading figure of the Linnean Society and a curator at the British Museum; he was notoriously cautious and slow to publish.
4. *On board ...were*: see Table of Ship's Companies.
5. *unfortunately drowned*: see pp. 61–2.
6. *very much regretted*: see p. 274.
7. *nine Government Chronometers*: instruments supplied officially and verified (that is, compared to standard instruments based at Greenwich Observatory).
8. *Commander-in-chief, on that Station*: then Rear-Admiral Robert Otway (1770–1846), who was commander-in-chief of the South American station from 1826 to 1829.
9. *belong to the Public*: Collecting curiosities of natural history from distant locations had become increasingly lucrative during the eighteenth century. By the time of the first of these voyages, Admiralty policy was clear: ship's officers contributed to natural history collections on behalf of the expedition and the nation, rather than acting as private individuals. In contrast, Fitzroy told Darwin in 1837 that the second voyage was the first in which the ship's officers were *not* under Admiralty instructions to collect on behalf of the nation, and therefore were at liberty to act in their own interests and profit – a point he raised while reprimanding Darwin for insufficient acknowledgement of the assistance the young gentleman naturalist had received from naval officers. See *Darwin Correspondence Project* (Robert Fitzroy to Charles Darwin, 17 November 1837).
10. *G. Cockburn*: Sir George Cockburn (1772–1853); the highest-ranking naval officer at the Admiralty, a position also known as the First Sea Lord.
11. *J. W. Croker*: John Wilson Croker (1780–1857), a respected secretary to the Admiralty for over two decades beginning in 1809.

Chapter I

1. *quitted the river Plata*: the narrative starts when the work of surveying began, but the ships had left Plymouth on 22 May 1826. Their voyage across the Atlantic is briefly described in the preface.
2. *for our purpose*: Alessandro Malaspina (1754–1810), a Spanish naval officer and hydrographer of Italian origins, made two scientific expeditions on behalf of Spain, in 1786–8 and 1789–94. He was in this region in the *Astrea* in late 1786 and early 1787, and again in late 1789 in the *Descubierta*. The latter ship was named after the *Discovery* in tribute to the Pacific explorations of James Cook, and the reference is probably to Malaspina's second and more scientifically ambitious voyage. It was surveying practice to compare data with a prior and trustworthy point of observation, especially on this difficult coast.
3. *Decatur of New York*: The American whaling industry boomed during the first decades of the nineteenth century, growing from its established base in Nantucket, Massachusetts to include several other ports along the north-east coast, including New York. By the 1820s and 30s, the focus was shifting to the Pacific, and the American 'golden age' of whale fishery exploited waters west of Peru as well as up to the north Pacific and Bering Sea. Accordingly, both King and Fitzroy mention many encounters with American whalers throughout their journeys, but the *Decatur* is not recorded in the index of vessels in the standard reference, Alexander Starbuck's *History of the American Whale Fishery* (Washington, 1878).
4. *herds of guanacoes*: this animal, like a small llama, was the most common supply of meat in the southern plains for the ships' crew.
5. *when the world was formed:* Sir John Narborough (*c.* 1640–88) visited the region in 1670–1; his description of the expedition was published in *An Account of Several Late Voyages and Discoveries* (London, 1694). The remark regarding oyster shells comes from this work, p. 46, and was widely repeated in subsequent English accounts of the region.
6. *as well as other works*: referring to two standard French works of reference, Bory de Saint-Vincent and Jean Baptiste Genevieve Marcellin (ed.), *Dictionnaire classique d'histoire naturelle*, 17 vols (Paris, 1822–31); and *Encyclopédie méthodique*, 206 vols (Paris, 1782–1832). The former was an item in the *Beagle*'s library during her voyage of 1831–6 and was likely in King's collection as well.
7. *Byron, and Wood*: John Byron (1723–86) was shipwrecked in the *Wager* in the region in 1741; King and Fitzroy frequently cite his work, one of the most famous and dramatic accounts of the region. John Wood accompanied Sir John Narborough in his 1670 passage through the Strait.
8. *ascertaining the fact*: King pulled out all the authorities here, perhaps an example of the kind of commentary that Charles Darwin so disliked. (See General Introduction.) Anselm-Gäetan Desmarest (1784–1838) was a French zoologist and author of many articles in reference books identified above, n. 6. Felix D'Azara or Azara (1746–1811) was a Spanish surveyor and naturalist who spent many years in South America; he was considered the best zoological authority for the region by English writers. Thomas Pennant (1726–98) was a prominent eighteenth-century British naturalist, author of the widely read *History of Quadrupeds* (London, 1781) and, incidentally, the recipient of the letters from Rev. Gilbert White that evolved into one of the most famous works of natural history, *Natural History of Selbourne* (London, 1789). George Shaw (1751–1813) was a botanist and zoologist at the British Museum, and author of a standard reference, *General Zoology, or Systematic Natural History*, 16 vols (London, 1809–26).

9. *Falkner, the Jesuit missionary*: Thomas Falkner (1707–84) was an English Catholic and Jesuit, a missionary in Patagonia from his conversion in 1732 to 1768. His work was the chief ethnographic source for the region; see later discussions by Fitzroy of the South American indigenous tribes, Volume 3, pp. 109–33.
10. *they can collect*: See the discussion of burial customs on pp. 112–13 and by Fitzroy, Volume 3, pp. 132–6.
11. *in Anson's voyage*: the fleet of George Anson (1697–1762), an English naval officer, circumnavigated the globe in 1740–44; the voyage had a shockingly high mortality rate from disease, and Anson's account of difficult conditions in the southernmost reaches of the continent was published in 1748.
12. *Viage de los Nodales, p. 53*: Garcia Jofre de Loyosa (d. 1526), a Spanish explorer, sailed for South America in 1525, shortly after Magellan's return, in a disastrous expedition which saw his fleet face storms, shipwreck and military action due to the outbreak of war between Spain and France. Only two of six ships crossed the Pacific, and de Loyasa himself died after passing the Straits. The careful remarks here show the importance of visual recognition of headlands in navigation. Naval officers were trained to sketch accurately and rapidly, and charts often included inset 'views' or sketches to help a navigator identify his location.
13. *a question of some interest*: chalk – a major European geological feature – held a critical place in contemporary debates about how the strata of earth's crust were formed, and whether changes occurred unidirectionally or cyclically, gradually or suddenly. In this period, the two differing accounts of chalk formations (organic remains or a chemical precipitate?) had not yet been resolved. The nature of chalk was in turn the pivot for debates about evidence of subsistence and elevation of sea-beds, and gradualist or catastrophic accounts of geological change.
14. *Dr. Fitton*: William Henry Fitton (1780–1861) was a physician and geologist; he was a prominent member of the Royal Geological Society and an authority on chalk stratigraphy in south of England.
15. *of the strait*: the reference is to the Spanish naval officer Antonio de Cordorba (1740–1811); his work was translated into English under the title *A Voyage of Discovery to the Strait of Magellan* (London, 1820).
16. *hills without wood*: from John Hawkesworth (ed.), *An Account of the Voyages...for Making Discoveries in the Southern Hemisphere*, 3 vols (London, 1773) vol. 1, p. 153. Samuel Wallis (1728–95) circumnavigated the globe in HMS *Dolphin*, 1766 to 1768, in a failed search for a southern continent; one of his lasting contributions was a description of Tahiti in terms that fired the erotic imagination of Europe. Philip Carteret (1733–96), commanding HMS *Swallow*, left England in company with Wallis, but their ships were separated on the south-west coast of South America; Carteret returned to England in 1769, having also sailed too far north to sight Australia.
17. *lessen my anxiety*: Louis Antoine de Bougainville (1729–1811), a French naval officer, led a colonizing expedition to the Malouines (Falkands) in 1763 and led the first and very successful French circumnavigation in 1766–9. His account, *Voyage Autour du Monde* (Paris, 1771), was very influential, establishing the description of the noble savage of the Pacific islands and especially, like Wallis, the utopic vision of Tahiti.
18. *M.A. Duclos Guyot*: Nicolas-Pierre Duclos-Guyot was second-in-command to Bougainville in the French circumnavigation of 1766–9.
19. *Admiral Burney and Mr. Dalrymple:* James Burney (1751–1821) was a naval officer and writer who went as a young sailor with James Cook on his second and fatal third voyage

to the Pacific. His sister was the novelist Fanny Burney and he authored a lasting summary of the European voyages of discovery in the Pacific, *A Chronological History* 2 vols (London, 1803); Alexander Dalrymple (1737–1808) was a prominent British hydrographer. Nominated by the Royal Society to lead an expedition to observe the Transit of Venus and search for a southern continent, he was rejected by the Admiralty in favour of James Cook.

Chapter II

1. *The Sailing Directions*: Starting under Sir William Parry (1790–1855), Hydrographer to the Royal Navy from 1823 to 1829, the Admiralty began to publish detailed descriptions of local conditions and coastlines to accompany its charts. These were known as 'sailing directions' or, later, 'pilot guides' or 'pilots'.
2. *irrelevant here*: variously called seaweed or 'fixed kelp' by navigators before and after this time, it was described so often as a navigational marker. Its appearance could be a guide to shoals or rocks; yet, because the seaweed grew from such depths, it was not as dangerous an indication as it might be in European waters.
3. Patella trichi-formis, *Lin*: James Sowerby (1757–1822), a botanical authority, a mineralogist and a renowned illustrator, who was particularly known for his multi-volume *English Botany* (1790–1813). 'Lin.' refers to a classification following the Linnean binomial system.
4. *the sounding-lead:* a critical navigational tool, the sounding lead was a weighted cylinder on a rope marked in lengths; the weight could be fitted with a lump of sticky substance, usually tallow, that would scrape the bottom and bring up evidence of the sand, mud or rock – or, as here, shell specimens.
5. *Wearing every ten minutes*: wearing referred to changing the direction of the ship without putting the sails directly into the wind, as in tacking; it would be a maneouvre to protect the ship in strong winds.
6. *race ... Second Narrow*: 'race' refers to the strong current in a narrow channel, here the tidal current; the whole passage here is typical of the way the narrative was related to the logbook, describing a sustained period of considerable discomfort and even danger in a factual tone.
7. *of his services*: Cutler has not been definitely identified, but there was a Benjamin S. Cutler of Stonington, Connecticut, known to have been whaling and sealing in the region several years earlier, in 1820–2, as recorded by inscriptions found in the Antarctic in 1957.
8. *The tallest of the Indians*: a first reference to the question of the size of natives and the myth of giants which had circulated since the time of Magellan; see General Introduction.
9. *name was Maria:* Maria became an important figure, frequently mentioned in King's account. See the detailed encounter, pp. 78–84.
10. *Guachos of Buenos Ayres:* this independent, often mixed-race people were known as expert horsemen and skilled labour for the cattle trade as well as fierce fighters in the various conflicts of the early nineteenth century; their way of life was in retreat as landownership and farming practices changed around Buenos Ayres.
11. *the Pampas Indians*: this is probably King's phrasing as Fitzroy was always careful to identify native tribes more specifically and would have called these the Telhuelhet. The guachos and bolas also attracted Darwin's attention in 1832; he provides a similar description in his own account of the voyage, and a vivid anecdote:

One day, as I was amusing myself by galloping and whirling the balls round my head, by accident the free one struck a bush, and its revolving motion being thus destroyed, it immediately fell to the ground, and, like magic, caught one hind leg of my horse; the other ball was then jerked out of my hand, and the horse fairly secured...The Gauchos roared with laughter; they cried out that they had seen every sort of animal caught, but had never before seen a man caught by himself (Darwin, *Voyage*, p. 76).

12. *of Coichi's head*: a rare specific indication of the kind of scrutiny Europeans gave to a native body. Although these measurements seem like the physical anthropologists' investigations of racial difference that become standard later in the century, the inspiration here was different. It was partly the reputation of the natives as giants, but also the study of physiognomy, or reading character from facial appearances.
13. *Bankes and Solander*: Daniel Solander (1733–82) was a Swedish naturalist, a protégé of Linneas, who joined Joseph Banks (misspelt here as 'Bankes') on the *Endeavour* voyage of 1768–71; he settled in Banks's London household thereafter as secretary, and as keeper of natural history at the British Museum.
14. *very old ... to the fire*: this description is of the old woman, carefully transported, is interesting as a contrast to later tales of old women being massacred in times of dearth, which shocked and fascinated the British. See, e.g., Volume 3, pp. 152–3.
15. *tons in burthen:* this was the usual size of sealing ships; the *Adventure* was over three times as large. A remark in the conclusion to the *Origin of Species* (1859) significantly echoed this passage. Darwin noted how, once we understand natural selection, we then 'no longer look at an organic being as a savage looks at a ship, as at something wholly beyond his comprehension' (*Origin*, p. 456).

Chapter III

1. *Sarmiento:* Pedro Sarmiento de Gamboa (1532–92), Spanish explorer and cartographer of the Strait of Magellan, who surveyed the area in 1578–9. Both King and Fitzroy always paid tribute to his navigational abilities and descriptions of the Strait.
2. *que era artillero*: King quotes the passage that definitively gives the source of the mountain's name, a respectful gesture to the original Spanish exploration of the strait. Antonio de Herrerra y Tordesillas (1559–1625) was a Spanish historian and secretary of state to Philip IV; he published an influential four-volume general history of the colonies of the New World. King here refers to his *Description*, originally vol. 1 of the longer work.
3. *Our decked boat ... for the latitude*: this passage indicates the range of surveying instruments used in the work of the voyages: the theodolite, the transit circle, the altitude circle, the barometer, the sextant and the hygrometer. The sextant, theodolite, barometer and hygrometer were portable enough to carry on excursions, while the transit and altitude circles were larger and more delicate, and would be used on shore, near the ship.
4. *send an expedition*: the failure to capture Drake was a famous tale of rivalry in the New World, which would appeal to King's audience. The vice-roy of Peru, based at Lima, was Don Francisco de Toledo (1515–82).
5. *Duke of Alva:* Fernando Alvarez de Toledo y Pimento (1507–82), third Duke of Alva or Alba, was a soldier and a chief advisor to Philip II; he was notorious for his brutality and ambition and, for Protestant readers, familiar as the oppressor of the Protestant Netherlands under Philip II.

6. *cruelly neglected colony:* in this history of Port Famine, King rehearses a partisan account of Spanish disaster and betrayal, known to English readers through Burney. The long cast of minor characters are more fully identified in a later English edition of Sarmiento's account by Clements R. Markham (1895), president of the Royal Geographical Society. They are: Fray Antonio, the Fransciscan; Andres de Viedma, captain of artillery; Tome Hernandez, a soldier saved by Cavendish; Diego Flores de Valdes, Diego de Ribera, both captains accompanying Sarmiento; Juan Suarez, a nephew of Sarmiento, and finally the British naval officer, Thomas Cavendish (1560–92), who circumnavigated the globe in 1586–8, capturing several Spanish ships and their gold along the way. At the end of the nineteenth century, Markham, drawing on an investigation of Spanish sources, was less flattering to the British side of the story than King and Burney; he condemned Cavendish in particular for abandoning survivors at Port Famine.
7. *by Andrew Mericke:* Capt. Andrew Merick, in the merchant vessel *Delight of Bristol,* found a Spanish survivor at Port Famine.
8. *Ercilla Araucana:* quoting the Spanish epic poet, Ercilla (1533–94). The passage clearly identifies Magellan as the man who first found the way and who gave his name to the passage. The intent is to show that even Spanish tradition rejected the effort to diminish Magellan. The poem, *La Araucana* (1569–1589), an account of the Spanish conquests against the valorous but doomed Araucanian people, is discussed by Fitzroy, Volume 3, pp. 315–16, 359–61.
9. Micropterus brachypterus: In his usual style King enlists a range of continental and British authorities: Dr Fleming is possibly John Fleming (1747–1829), a doctor and botanist who had served in the Indian Medical Service, but more likely King meant the naturalist by the same name, John Fleming (1785–1847), not a doctor, who authored important zoological and geological works; Coenrad Jacob Temminck (1777–87), a Dutch naturalist, was director of the museum of natural history at Leyden; M. Louis Claude de Freycinet (1779–1842), a French naval officer whose work as an early cartographer of the Australian coast King knew and admired; M. Jean René Constant Quoy (1790–1869) and M. Joseph Paul Gaimard (1796–1858) were the two naturalists on the French expedition to the Pacific in *L'Astrolade* under Jean Dumont D'Urville, in 1826–9.
10. *fizgig:* or fishgig, a barbed pole used in angling to spear hooked fish.
11. *T.R. Forster:* The translator was John (Johan) Reinhold Forster (1729–98), probably with the assistance of his son Georg (145–94). Forster was a Lutheran pastor, naturalist and cartographer who had settled in London in 1766, looking for scientific work, after travels in Russia.
12. *collected in the Strait*: George Louis Le Clerc, Comte de Buffon (1707–88), the French naturalist whose encyclopedic *Histoire naturelle* was a standard reference for many decades.
13. *Daniell's hygrometer*: John Frederic Daniell (1790–1845) was a London-based natural philosopher and entrepreneur. A founding member of the Society for the Diffusion of Useful Knowledge and the Chemical Society, he had a strong interest in meteorology and Alpine geology and had designed a dew-point thermometer in 1819.
14. *Jones' portable hygrometers*: William (*c.* 1762–1831) and Samuel Jones (1770–1859) were prominent navigational and optical instrument-makers in London in the first half of the nineteenth century. Hygrometers, as glass instruments, were fragile and King would have carried several on an important excursion.
15. *red caps and medals:* Fitzroy later recorded that red was a colour of peace and friendship for the Fuegians; see Volume 3, p. 148.

Chapter IV

1. *of which we once heard so much*: the destructive rollers of the South Atlantic were notorious among sailors, and said to precede a gale.
2. *reliquiae deluvianae*: i.e., remains of the flood, referring to the geological theories that the fragments and boulders on mountain peaks were deposited as floods retreated; it was usually considered as a periodic event in the earth's history rather than simply a reference to the Biblical deluge as a singular event.
3. Cordova al Magalleanes, p.65: the passage quoted here described Cordova's speculation that the sudden thud was not thunder as they first imagined but a volcanic explosion, a conclusion they drew from the rocks they found nearby, which reminded them of slag from iron foundries.
4. *shrubs of Port Famine*: All significant plants to the British mariners: fuchsia indicated a sheltered location; beriberis (elsewhere identified as *beriberis ilicifolia*) was considered an antiscorbutic; winter's bark tree (*Drimys winteri*) was used as a stimulant and an antiscorbutic. The latter was named for Captain John Winter, of the *Elizabeth*, who navigated through the Strait with Sir Francis Drake in 1578.
5. *Horrentibus umbris*: Virgil, *Aeneid*, book 1, ll. 310–12: the lines occur as the shipwrecked Aeneas determines to explore the wild and mountainous harbour into which the gods' storm has driven his fleet.
6. *'Ho-say, ho-say'*: King and his officers used this phrase to mean 'come, come' but it is unclear whether he thought it part of the local language, or simply an expression the natives understood that had come from another language, i.e. Spanish; it is not part of FitzRoy's list of Fuegian vocabulary. See Volume 4, p. 335.
7. *the Indians of Admiralty Sound*: a small tribe to the north that hunted guanaco and a species of deer known as huemul; the comment is typical of the efforts of the British to make sense of the various peoples of the regions, who were very different in appearance and culture. See Volume 3, pp. 109–23.
8. *polar voyages under Sir Edward Parry*: Parry (1790–1855) made four voyages to the far north: in 1818–20, 1821–3 and 1824–5 and 1827. This description of the deaths of Ainsworth and Corkhill was widely reprinted in the British press.
9. *sepulchers our dead*: From the well-known opening stanza of Lord Byron, *The Corsair* (1814), a tribute to the manly life of the sailor, contrasting his vigorous experience, however short, to the weakening luxuries of life ashore.

Chapter V

1. *on a sealing voyage*: James Weddell (1787–1834) was a former British naval officer (1810–16) and a sealing captain; he made several voyages to the region, reaching a record latitude of 74 15' degrees south in a voyage of 1822–4. On his return, he published *A Voyage Towards the South Pole* (1825). Matthew Brisbane (*c.* 1787–1833) – clearly an equally remarkable seaman and adventurer – had accompanied Weddell in a 65-ton cutter, the *Beaufoy*. Later, in 1826–7, Brisbane was master of *Prince of Saxe-Coburg*, a larger vessel, and wrecked as King and Stokes describe. He had a further dramatic history of shipwrecks before settling in the Falklands, where he was killed. See Volume 3.
2. *Voyage autour du monde.' 1767*: despite the original title (with an error in the date – it was first published in Paris in 1771) this is probably from the English translation of

Johan Reingold Foster (1772), p. 168, or from Bougainville via Burney's *Chronological History* – a typical example of the layered descriptions and authorities in the *Narrative*.

3. *part of the former*: the Bougainville text, dated 1766, was indeed fragmentary. It referred to a hazardous journey uncertain harbours, and closed with what seemed to be a list of officers and scientific staff, possibly those who made the ascent and left the memorial: the astronomer Pierre-Antoine Veron; the physician and naturalist Philibert Commerçon; the hydrographer and engineer Charles Routier de Romainville; the surgeon Francois Vives; and Landais, Lavery-de-Roi and Fontaine, who were officers of the *Etoile*, one of the two ships on the voyage. Bougainville was detained for weeks in January 1768 in this bay, but does not mention the memorial.
4. *Anno MDCCLXXXIX*: A translation of this more complete memorial is:

   To the kind traveller[seafarer] [I bid] health. In the year of our Lord 1788 a most renowned man, DD Antonius de Cordova Laso de Vega, made a second journey in two ships *(whose names were the SS Casilda and the Eulalia)* for the purpose of pursuing the exploration of the Strait of Magellan and at the same time its shores, harbors, and other noteworthy ... from Cadiz the fleet departed on the fifth of October [and] on the twelfth of the same month saw the New ... From the north southward ... of the wretched ... [we] departed on the second of November. On the nineteenth of November, after surveying the shores of Patagonia, [the fleet] arrived at the mouth of the Strait. Finally, after very considerable mortal dangers and terrors, on the [open] sea as much as in the Strait, were overcome with courage and steadfastness, and after all the natural harbors and soundings on both shores had been noted very carefully, [we] arrived at this natural harbor, [called] Port Joseph or Gallant, on the seventh day of January where, as an everlasting memorial of the fact, [we] left this monument of the most Holy Cross on the hill. These deeds were accomplished at the king's own command in the reign of the noble and great Charles III. This was placed on the 24 of January, in the year 1789.

   King's text should be compared to Cordova's own description of his crossing from Cadiz to Tierra del Fuego (pp. 4–12), as not all dates match; he described unfavourable winds in early November, perhaps the 'wretched' of his text above, which decided him to sail south rather than to Brazil; he also briefly mentioned leaving the memorial at Mountain de la Cruz (p. 48). Cordova and his men were of course unaware that Charles III of Spain had died in December 1788, and had been succeeded by Charles IV.
5. *gripe and false keel*: the purpose of the false keel – timber attached in several pieces to the underneath of the keel – was to protect it against damage, just as described here; the gripe piece connected the keel and the hull.
6. *arming the lead*: i.e., the sounding lead in which the tallow picked up traces of the ocean bottom.
7. *calls them 'arbours'*: e.g. Narbrough, p. 74.
8. *with the utmost precision*: a remark on this ability to mimic was a standard aspect of the European encounter with the native populations of Tierra del Fuego. Europeans found the power of mimicry alternatively impressive and amusing but also contemptible, as a sign of child-like or animal intelligence. See Further Reading, Taussig (1997).
9. *old navigator above-mentioned said ... land to behold*: 'From Cape Quod into the South Sea, I called this land South-Desolation, it being so desolate land to behold.' Narbrough, p. 78.

10. *large bell tent:* the marquee was a typical officer's tent in the British army of the late eighteenth century, roughly rectangular; the bell tent, as its name suggests, was circular and smaller with a central pole.

Chapter VI

1. *account of this place*: Byron said he had nowhere seen trees as tall, 'the dimension of some of these trees being equal to a main-mast of a first rate man of war' (Byron, *Narrative*, pp. 73–4).
2. *Three men and 'Maria'*: Maria was a Tehuelche leader in the region, the daughter of the cacique Vicente as well as the sister of Bysante, and well known to European sailors. These descriptions of encounters with Maria are a chief source of her history. She has become a folk hero of Argentine, and a fictionalized biography has been published: Arnoldo Canclini, *Maria Reina Tehuelche* (Castellano, 2006). Her reign is tentatively dated from the early 1820s, when Louis Vernet, a French merchant appointed governor of the Falkland Islands (Malvinas) by the government at Buenos Ayres, met her. She died in 1840 or 1843. See also Fitzroy's account of a later visit, Volume 3, pp. 143–4, 261.
3. *muchos cueros*: 'Cacique' was a Spanish and Portuguese term for leader, based originally on a word in the Taino language that had been encountered by the Spanish in the Caribbean.
4. *dam's god Setebos*: Shakespeare, *Tempest* I.ii.372. Caliban fears Prospero as even more powerful than Setebos, the god of his mother. Pigafetta listed the natives' 'greater deity' as 'Setebos' in a vocabulary based on the 1520 encounters in 1520 (Burney, vol. 1, p. 37).
5. *Lewis ... American sealing-vessel:* the sealer Lewis is unidentified.
6. *signifying 'people'*: Falkner, p. 110.
7. *Horse Indians, and Canoe Indians*: King and Fitzroy both made efforts to distinguish among the different groups they encountered, and previous travellers had noticed many linguistic and physical differences as well. Fitzroy discusses the history, range and occupations of the different indigenous peoples of this region at length in Volume 3 pp. 110–65.
8. *'aqua ardiente,' or spirits:* Spanish for spirits; literally, burning water.
9. *most trifling offence:* Byron tells of a father who violently knocked over his son, about three years' old, after the boy had dropped a basket of sea-eggs in the surf (Byron, *Narrative*, 148–9); it became even more widely known after Darwin repeated the story in his account, as a comment on the impossibility of 'domestic affections' or intelligence emerging in such a brutal land and people (Darwin, *Voyage*, 178). The reader should contrast this with Stokes's observations in the previous chapter, p. 71.
10. *for my son:* Philip Gidley King (1817–1904) was a midshipman on the *Adventure*; he later served with Fitzroy on the *Beagle* for the second voyage, leaving her to rejoin his family in Australia in January 1836, where he farmed estate and managed a mining company.
11. *some former navigators*: King's compilation of witnesses in the following pages addressed a controversy surrounding the size of the Southern American natives, a controversy which by the 1820s was somewhat tired; see General Introduction.
12. *seven feet six inches:* King's account, including the quotation from Pigafetta, here is lifted from Burney's comprehensive summary of the descriptions of the 'giants' (Burney, vol. 1, pp. 33–6). Burney identifies some obscure sources, repeated here by King: P. Martyr de Angers, or Pietro Matire, author of a summary of Magellan's voyage collected in the

1520s for the emperor Charles V; and Maximus Transylvanus, an Italian secretary to Charles V who compiled a summary around the same time. But Herrara and Pigafetta's accounts of the voyage of Magellan were Burney's chief sources.

13. *on the subject:* again, straight from Burney. The voyage of Spanish navigator Garcia Jofre De Loyasa (1490–1526) is described in Burney, vol. 1, 126–43.
14. *author of the 'World Encompassed'*: i.e. the narrative of Francis Drake's voyage, said to be based largely on journals of Francis Fletcher (Drake's chaplain).
15. *being unusually tall*: King now turned to Sarmiento's account, based on encounters in 1579–80.
16. *not considered credible:* Samuel Purchas (bapt. 1577–1626) was a clergyman in Essex whose major work was a compilation of written and oral travel accounts, *Hakluytus Posthumous, or Purchas Pilgrimes*, 4 vols (London, 1624–5). Knyvet was a traveller of the Elizabethan period. Burney considered Knyvet a sensational and unreliable source, referring to his poor 'character for veracity' (Burney, vol. 2, p. 106).
17. *by Oliver Van Noort:* in this episode in 1599, which Burney censored as an 'extermination', the Utrecht navigator Van Noort (1558–1627) on the first circumnavigation, trapped and massacred a large group of natives, including many women and children, in retaliation for an attack on his own men some weeks before. He also took several children away on board with him. See Burney, vol. 2, pp. 213–14.
18. *Spilbergen (1615)*: Joris Spilbergen of the Dutch East India Company, travelled to the East via the Strait of Magellan in 1614. The voyage is described in Burney, vol. 2, pp. 330–53.
19. *Le Maire and Schouten*: again from Burney, vol. 2, pp. 355–453. Jacob Le Maire (*c.* 1585–1616) and Willem Cornelius Schouten (*c.* 1567–1625) travelled in 1615–1617 through the Straits to the Pacific in search of the southern continent as representatives of a group of Dutch merchants, modelled after the East India Company. They were the first to find the route between Tierra del Fuego and Staten Island, anemd Le Maire Strait.
20. *Mr. Wood*: Mr John Wood, an officer with Narbrough; the encounter is described in Narbrough, pp. 49–50.
21. *They are prodigious stout ... in my life*: Charles Clarke, "An Account of the Very Tall Men, Seen Near the Streights of Magellan..." *Philosophical Transactions of the Royal Society of London*, 57 (1767), p. 78.
22. *of Burgos*: a foot measure used in Spain from the mid-eighteenth century, known as the Burgos and Castile foot. The voyage of the *Santa Maria* was that of Cordova.

Chapter VII

1. *inhabitants of the argonauta:* naturalists were divided on whether the octopus found in the argonauta shell was a parasite or its original inhabitant. The latter position was confirmed by mid-century. For details of the debate, see Dumeril and De Blainville, 'Report of a Notice By M. Rang, respecting the Inhabitant of the Argonaut', *The Magazine of Natural History*, vol. 1 n.s. (August 1837), pp. 393–403.
2. *found on the charts:* Fitzroy also discussed vigia as a general problem, obviously of great import to those who built up navigation charts; see Volume 3, pp. 42, 100.
3. *companions of Malespina:* Don Felipe Bauzá (1794–1834), was a hydrographer in the Spanish navy and served with Malespina on the circumnavigation of 1789–94; he moved to Gibraltar and then London in 1823, a political refugee from the rule of Ferdinand VII. Armed with many Spanish charts, and already well-known to the British hydrographic

and scientific community, he became close to many of them, including King. He advised Beaufort regarding the instructions for Fitzroy's second voyage to Tierra del Fuego.

4. *the southern Patagonians*: Falkner, *Description*, pp. 118–120. The unfleshed skeleton was then enshrined above ground under a low tent; King and his officers saw and visited some of these Tehuelhet tombs.
5. *Mr. Low ... master of the Uxbridge sealer*: King and Mr Low, an American sealer probably based in Stonington, Connecticut, offered each other assistance on several occasions. There were two Lows, brothers, in the region sealing, with the ships *Adeona, Uxbridge* and *Mercy*. One brother (master of the *Adeona*) died in June 1828 at Port Famine while King was waiting for the return of Stokes and the *Beagle*, and King performed the burial service.
6. *Syn. Of Chemistry, iv. 576*: King here refers to Thomas Thomson, *A System of Chemistry* 4 vols (Edinburgh, 1802) which treated the medicinal value of bezoars as a superstition; and quotes from the Latin treatise by George Marcgraff (Marcgrave) (1610–44), *Tractatus topographicus et meteorologicus Brasiliae, cum eclipsi solari* (1658) p. 292. Translated, the extract reads:

   > These animals produce bezoar stones in a certain fold of their stomach: they are greatly valued against poisons and ill-natured fevers, for strengthening and rekindling the heart, and for other effects. They are created from herbs of outstanding virtue, which they eat from a natural instinct in order to preserve their health and overcome diseases and poisons. These stones are found in the more mature among these animals, and they are sometimes so large that I bore one back to Italy which weighed more than thirty-two ounces.

   Marcgraff was a Dutch physician and naturalist, who travelled to Brazil in 1638 and died there. His extensive collections were sent back to Holland (supplying the famous Leyden museum of natural history, among others) and his writings appeared in Johannes De Laet (ed.), *Historia Rerum Naturalium Brasiliae* (1648).

Chapter VIII

1. *along the beach*: Weddell decided that the river here must have been blocked and diverted by sandbanks since the time of Malaspina's description and chart of 1790 (Weddell, pp. 196–7).
2. *my warmest thanks:* King's remarks here were in sharp contrast to his regrets regarding Fitzroy's appointment by Admiral Otway, which superceded King's appointment of Lieutenant Skyring. See below, pp. 161–2.
3. *he made prisoner*: In the skirmish near San Juan river in 1580, with inhabitants Sarmiento described as giants, Sarmiento's purser was injured; Sarmiento captured one man and took him to Spain, where he was presented at the court of Philip II. Sarmiento, pp. 244–5; Markham (ed.), *Narratives of Sarmineto*, pp. 135–6.
4. *Vigors ... Lesson*: Nicholas Vigors (1785–1840) was a London naturalist; Réné Primevère Lesson (1794–1849) was a French surgeon and naturalist who had travelled in *La Coquille* on a circumnavigation led by Louis Duperry in 1822–5; he became a specialist in the birds of the southern hemisphere. His *Manuel d'Ornothologie* was published in 1828.
5. *late Astronomer Royal:* John Pond (1767–1836) was Astronomer-Royal from 1811 to 1835.

6. *Sarmiento:* King quotes to establish priority for the name. 'This mount is the one known in the old descriptions as the Bell of Roldan [Magellan's artillery officer].' Sarmiento, p. 213.
7. *striking and picturesque:* King may here be thinking not of Cordova but Bougainville, who had commented dramatically on the landscape, noting its sheer rugged cliffs, and mountains covered with snow that seemed as 'old as the world': 'Les côtes dans tout cet espace sont boisées et escarpées; celles de la Terre-de-Feu paraissent hachées par plusieurs détroits. Leur aspect est horrible; les montagnes y sont couvertes d'une neige bleue aussi ancienne que le monde.' Bougainville, p. 82.

Chapter IX

1. *in his second voyage:* Parry's second voyage was 1821 to 1823; his account was published as *Journal of a Second Voyage for the Discovery of a North-West Passage* (London: John Murray, 1824) It held several graphic descriptions of eating blubber, including the eagerness of children for the treat, which seemed 'like sugar-candy' to an English child (p. 178).
2. *Cape Remarquable*: Bougainville noted the whole cape seemed composed of strata of fossilized shells, and called this a testimony to the great transformations of the globe. "Ce cap, élevé de plus de cent cinquante pieds au-dessus du niveau de la mer, est tout entier composé de couches horizontaies de coquilles pétrifiées. J'ai sondé en canot au pied de ce monument qui atteste les grands changements arrivés à notre globe." Bougainville, p. 83.
3. *disease ... been fully expected*: these remarks represent well the thinking about scurvy in the early nineteenth century. The British naval doctor James Lind had carried out experiments showing that citrus fruit was an effective cure in cases of scurvy in 1747, and the consumption of citrus fruit or juice slowly won acceptance as a preventative also in some of the long voyages of the eighteenth century (notably that of Malespina in 1789–94). But, given the practical difficulties of supplying ships with citrus fruit on routes like that of the *Adventure* and *Beagle*, other dietary treatments were equally or more common. The consumption of fresh meat was considered especially important. Since the time of Cook's voyage, authorities connected scurvy to shipboard hygiene, and also to psychological conditions, so that King's treatment of the outbreak included both cleaning and careful efforts to improve morale. See Further Reading, Lawrence (1996).
4. *the prime of life*: Stokes was then still a young man; although his date of birth is obscure. For biographical details, see General Introduction and Table of Ships' Companies.

Chapter X

1. *vapours, clouds, and storms*: These are the opening lines of 'Winter' from James Thomson's *The Seasons*, one of the most widely read poems of the eighteenth and nineteenth centuries; it is said to have gone into more than 400 editions by the 1870s. Thomson (1700–48) composed the poem from 1726 to 1830, with 'Winter' as the original section. The opening welcomed the gloom and storms of winter as the prompt to 'exalted thought/And heavenly musing' rather than to despair and suicide, but King does not comment further on the irony.
2. *whose gravel gold'*: Sir John Denham (*c.* 1614–69) was a Royalist, a courtier and MP and a poet, wrote *Cooper's Hill* in 1642; these lines are from a revised version of 1655 (ll. 165–6). The poem used the subject of the Thames to frame reflections on history and the rule of kings, and the lines here patriotically remind the reader that the virtues and bounty

of the river at the heart of England surpasses the gold found in more exotic rivers. The surveying work of King and Fitzroy, interestingly, fell right between two goldrush eras: there was considerable gold-mining in Brazil in the first half of the eighteenth century, though it was in decline by the time of King's visit, while the discoveries in California and Australia lay two decades ahead. Clearly the officers of *Adventure* and *Beagle* did not dismiss the possibility of discovering mineral riches in South America.

Chapter XI

1. *'the soul of man dies in him'*: James Thomson, *The Seasons: Winter*, l. 61. See above, p. 294 (note 1).
2. *hardly withstand it'*: Bulkeley, *Voyage to the South Seas*, 66. Bulkley's account of the monotonous weather and suffering in 1740 to 1741 after the wreck of the *Wager* was more cheerless reading for Stokes.
3. *Yacana Cunnees*: see Falkner, *Description*, 98ff. Falkner collected a small vocabulary, to which Fitzroy added. See Volume 4, p. 335.
4. *Leverian Museum*: a collection of natural history curiosities on display in London from 1775 to 1806 developed by Sir Ashton Lever (1729–88).
5. *It is with the condor ... diminish in size*: Alexander Humboldt wrote an essay on the condor in *Recueil d'observations de zoologie et d'anatomie comparée* (1811) pp. 26–31. The remark here comparing the condor to the Patagonian mythic size is not found in this essay but the comparison was familiar in English natural history writing. It was attributed, for instance, to French ornithologist Louis Jean Pierre Vieillet (1748–1831) in the article on the condor in the *Penny Cyclopedia* (London, 1837), vol. 7, p. 439.
6. *were at war:* the war between Brazil and Argentina, also known as the Cisplatine War, over the independence of Uruguay (1825–8) ran in the favour of Argentinian forces on land, but the superior resources of the Brazilian meant that war dragged on through a blockade of River Plate.
7. *the Commodore's ... of eighteen guns*: The Commodore referred to here by King seems to be George Mason, an English officer who had joined the Brazilian navy in 1822 and commanded the *Maranhão* in naval action from 1824 to 1827; but he seems not to have been present during this episode. The commander of the Brazilian forces blockading the Plate during this period was Commodore James Norton (1789–1835) a former officer of the Royal Navy, and, like Mason, a recruit to the Brazilian navy. The peace treaty of Montevideo giving Uruguay its independence from both states was negotiated with British mediation at the end of August 1828.
8. *the admiral*: The admiral of Brazilian navy was Rodrigo Pinto Guedes (1762–1845); the Latin tag *magna componere parvis* means 'to compare great things to small'. Captain Henry Dundas had a career similar to Fitzroy – he trained at the Royal Naval College a few years before Fitzroy, went to sea in 1815 and served in the Mediterranean, the Caribbean and in South America. He became captain in 1825 and *Sapphire* was his second command, and he retired on half-pay in 1830.
9. *Senor Juanico ... resident at Monte Video*: Don Francisco Juanico was described by another English traveller of this period as a merchant and magistrate, whose house was the resort of the British elite in the city, including the English and French naval officers stationed there. See Beaumont, *Travels in Buenos Ayres* (London: James Ridgway, 1828) pp. 7–8.
10. *a peace ... imperiously demanded*: peace was concluded with the assistance of John Ponsonby (1770–1855), a diplomat sent to resolve the dispute and protect the increasingly

important British commercial interests in Buenos Aires and Montevideo. King's support for the Buenos Aires side of the question was, loosely speaking, the British liberal position, as Brazil wished to maintain Uruguay as a dependent province within its monarchy (or empire, as Pedro I and II titled themselves emperor), whereas the confederation of Argentine provinces supported Uruguayan independence. At the same time, however, the political relations between Buenos Aires and Montevideo were not friendly and grew much worse over the next decade. King's remarks seem a reaction to naval insult in the River Plate rather than a political assessment.

11. *Captain Henry Foster ... his pendulum voyage:* Henry Foster (1796–1831) was a naval officer with much experience in both South America and the arctic. He had surveyed the River Plate in 1819, then joined Capt Basil Hall's expedition in 1820 to 1823, doing geophysical measurements on the west coast of South America. He then made three voyages to the Arctic, one with Edward Sabine and two with Edward Parry. His work in *Chanticleer* from 1828 to 1831 was geophysical as well as navigational – the pendulum measurements sought to determine very precisely the elliptical figure of the earth, and were connected to the objectives of the earlier arctic voyages. He drowned in the River Chagres in Panama in February 1831. See below, p. 297 (note 11).

Chapter XII

1. *not used to it*: Narbrough, *Account*, p. 29. Narbrough was from Norfolk, but clearly classified himself as 'used to it,' adding 'I could travel as far in an hour as many of my men in two'.
2. *eleven or twelve feet in length:* see King's discussion of 'giants', pp. 86–91.
3. *according to Falkner:* see Falkner, *Description*, 118–20 and Fitzroy's ethnographic description in Volume 3, p. 113.
4. *Lieutenant Kendall:* Edward Nicholas Kendall (1800–1845); he accompanied Sir John Franklin's polar expedition in 1825–7 before joining Foster on *Chanticleer*. He later worked as a land surveyor for the Colonial Office, determining boundaries between Maine and New Brunswick, before ending his career as a superintendant of a steam packet company in Southampton.
5. *Kater's Azimuth:* Henry Kater (1777–1835), an army captain and surveyor, developed several scientific instruments for astronomy and geodesy. His compass was invented in 1811, and featured a mirror that allowed the observer to view the scale while keeping a distant object in the instrument's sighting panel.
6. *siphon barometer ... Bunten of Paris*: Mountain barometers needed to be portable and durable, but they were also large, as the tube holding the mercury had to be 2–3 feet long to register pressure changes due to altitude. The instrument-maker Bunten developed a siphon barometer for the French scientist Joseph Gay-Lussac (1778–1850). It was a U-shaped tube with two main reservoirs to hold mercury separated by a capillary tube of several inches; one of the reservoirs had a small opening for air to enter, and the pressure it exerted on the surface of the mercury could be read off a scale, but the air was prevented from flowing further into the instrument (and hence contaminating the measurement) by the capillary tube. Sir Henry Englefield (1752–1822), a British antiquarian with scientific interests, had developed a mountain barometer for travellers in the early nineteenth century: it was built with a cistern to contain the mercury at the bottom of the instrument which was inverted for measurement; the mercury level in the reservoir could be adjusted for temperature by means of a leather sac tightened by a screw. This was

known as the Fortin principle, after Jean-Nicholas Fortin (1750–1811), another Paris precision instrument-maker. In Englefield's model, the tube, about three feet long, was fixed in a protective wooden block that sealed it against air and the whole was then built into a tube about the size of a 'common walking stick', a merit that emphasized its portability. See *The Philosophical Magazine*, 30 (1808), p. 46 for Englefield's description, and Middleton, *History of the Barometer* (1994).

7. *theodolite, it was yet useful*: the transit theodolite measured vertical and horizontal angles; Fitzroy perhaps means the instrument's vertical supports and scales were damaged, but the horizontal movement was unaffected.
8. *than Kater's Peak*: Fitzroy named the landmark after Henry Kater, who had recently developed a pendulum instrument which measured slight variations in gravitational force or local gravity measurement, when used in combination with a precision clock to measure the period of the pendulum; observers could build up geodesic observations for the shape of the earth. Foster's expedition used this instrument.
9. *Daniell's Meteorological Essay*: John Frederic Daniell (1790–1845). His *Meteorological Essays and Observations* (1823) was the leading British study of the subject for this period.
10. *Foster's pluviameter*: i.e., a rain gauge.
11. *was unfortunately drowned ... in a canoe*: see above, p. 296 (note 11) for Foster's history and death in Panama. The obituary and summary of his work was published in *Memoirs of the Royal Astronomical Society* of *London* 5 (1833) pp. 407–10.
12. *General Pinto, the Director:* Francisco Antonio Pinto Diaz (1775–1858), a liberal, was then president of Chile; he developed the republican Constitution of 1828, but opposition to his presidency from conservative factions in the independence movement led to the civil war of 1829–30.
13. *account of Chile:* John Miers (1789–1879) was a botanist and engineer who had travelled to South America in 1818 to go into mining and instead settled in Buenos Ayres for several years and botanized. He returned to England in 1825 and published *Travels in Chile and La Plata* (London: Baldwin, 1826).
14. *house of Mr. Caldcleugh:* Alexander Caldcleugh (1795–1858) first went to South America in 1819 as diplomatic private secretary, living in Rio. Through 1821, he travelled widely to all the major cities in South America and published an influential description of the region's history, geography and commerce (*Travels in South America*, 2 vols (London: John Murray, 1825). He returned to Chile in 1829 and spent the rest of his life there, developing mines, building a railway and farming an estate.
15. *Don Jose Santiago Aldunate*: José Santiago Aldunate (1796–1864) was by this time a seasoned leader in the Chilean independence movement, involved in action against Spain since 1813. He replaced the defeated Spanish soldier Quintanilla as governor of Chiloé in 1826. In the 1840s, he rose to minister of war in the Chilean government.

Chapter XIII

1. *three Portugese seamen ... a year before*: The seamen were not identified further in the *Narrative*, but their case was not unique. As Fitzroy would have known, they were as likely to be deserters as cast off by their captain (as they claimed). Titus Coan (pp.150–2) mentions several cases of sailors who deserted and lived temporarily with the natives near Gregory Bay.

2. *the lower hills: ignus fatuus* or will o'the wisp was a phosphorescent light, often associated with low-lying swampy ground; see Fitzroy's investigations of phosphorescence at sea (St Elmo's fire), Volume 3, pp. 70–1.
3. *for a craniologist*: Fitzroy shows here his familiarity with the standard interpretations of character drawn from physiognomy and phrenology (craniology) treatises that were popular in Britain. A narrow bridge of the nose was an indication of weakness and indecision, a retreating chin marked effeminacy, while a wide or half-open mouth was a sign of cowardice and deceitfulness. See Further Reading: Cooter (1984) and Pearl (2010) .
4. *hawsers and kedges:* a hawser was a strong rope; kedges were small anchors that could also be used to warp the ship along, as described here.
5. *wall-sided:* a nautical term used to describe a ship with steep, near vertical sides.
6. *week's serious illness:* In June 1670, Narbrough wrote cheerfully that 'a Man hath an excellent stomach here; I can eat Foxes and Kites as savourily as if it were Mutton ... Nothing comes amiss to our stomachs' (p. 49).
7. *Red Lion:* this reference is obscure; possibly Fitzroy refers to the common inn sign and heraldic symbol known as the Red Lion, which stands upright on hind legs, with arms outstretched, like the signalling and jumping that Fitzroy describes.
8. *Donkin Cove*: Bryan Donkin (1768–1855) was an inventor, entrepreneur, instrument maker and civil engineer. His company of Donkin, Hall and Gamble purchased the British patent for Francois Appert's food-preserving method in 1810 and began supplying tinned meat to the Royal Navy in 1813. It is possible Fitzroy had met him through London's scientific circles: Donkin was a member of the Royal Astronomical Society, and had developed instruments and standards for the Royal Greenwich Observatory; he was also a member of the Society of Arts (from 1803), a founding member of the Institute of Civil Engineers (1822) and a fellow of the Royal Society. He was elected to the latter in 1836, just before the *Beagle* returned from her second voyage, fifteen years before Fitzroy's own election.
9. *Captain King:* see p. 104 and Volume 2, p. 5.

Chapter XIV

1. *Castle of Otranto*: an early gothic novel written by Horace Walpole and published in 1764; the setting was supposedly the northern regions of Italy.
2. *Ulloa's memory ... in a peninsula*: i.e., he has recharted the island as a peninsula but kept the name of Ulloa.
3. *Chanticleer ... passed through rapidly*: *Chanticleer* was Captain Foster's vessel (see p. 297 (note 11). Cutler has not been identified, but the *Uxor* of Stonington, Connecticut, a 96-ton brig, is recorded as whaling in the South Atlantic and Patagonia in the early 1830s. It was wrecked and sank in 1841 (Starbuck, *History of the American Whale-Fishery*).
4. *Appendix*: see below, Volume 2, p. 184.
5. *out of their pay*: seamen paid for their own clothing from their wages. On a long voyage like this one, the ship would keep a store of cloth that the seaman could put on account to make clothing; but Fitzroy was making the point that the men could hardly be expected to keep their clothes in good repair under these rough conditions.
6. *a Kater's compass could scarcely be used*: i.e., a handheld instrument, so Fitzroy remarked here that the wind was so strong a man could not brace himself and observe.
7. *best bower ... lost one fluke*: bower anchors, also called sheet anchors, were carried on each side of the bow.

8. *fabled Sphinx:* the sphinx at Giza, in Egypt, part of the British visual imagination since the eighteenth century. The vogue for Egyptian history and architecture was strong in the first decades of the nineteenth century, fostered in part by the massive work of Egyptology commissioned by Napoleon, *Description de l'Egypte,* appeared between 1809 and 1822. That work featured a dramatic engraving of the crumbling sphinx looming over a handful of travellers that was widely reproduced in popular and abridged editions in the 1820s, and may be the image Fitzroy called to mind.

Chapter XV

1. *The Hoxsley schooner arrived from Valparaiso*: the *Hoxsley* schooner is unidentified; a reference in the following chapter calls it a 'national schooner' built at Chiloé. The description suggests it may have been a kind of official mail service, such as those partially supported by the Admiralty, which operated regularly from Falmouth to Brazil from 1808, to Buenos Aires by 1813, and by 1828 to the west coast as well (via the Caribbean and Panama). In terms of communications networks, these voyages of King and Fitzroy were at the end of an era: the existing services were replaced by Royal Mail Steam packet to the Caribbean in 1839 and, for the west coast of South America, the Pacific Steam Navigation company, which floated a prospectus in 1839 and operated from the early 1840s.
2. *W. H. Smyth ... had served*: William Henry Smyth was a noted Royal Navy hydrographer (1788–1865) who surveyed in the Mediterranean for over a decade, 1813 to 1824, in the ship that became King's *Adventure* (see General Introduction). He ended his sea-going career in 1824 and became active in London's scientific community as a member of Royal Astronomical Society, the Royal Society of London (1826) and a founding member of the Royal Geographical Society (1830). He worked in the Mediterranean and trained a number of British surveying officers. Skyring served with him in 1820 and Graves was a midshipman with Smyth in 1824 - he drew up the large chart summarizing the Mediterranean surveys for public use that was published in 1825 (Smyth, *The Mediterranean: a Memoir* (London: J. W. Parker, 1854), p. 373).
3. *to the admiral*: ie. Admiral Burney.
4. *'Ancon sin Salida' ... by Admiral Burney*: Burney, vol. 2, p. 16. The name meant bay, or inlet, with no exit.

Chapter XVI

1. *little known to the world*: the archipelago of Chiloe had been strategically important to European colonizers since the first landing by the Spanish in 1558, as it guarded the route from the Atlantic, and provided a recovery point for crews after the hardships of the Strait. (This was an advantage it offered as well to Darwin in 1835, who wrote cheerfully about the islands in chapter 15 of his volume, *Voyage*, pp. 218–22.) A Jesuit mission was established there in 1608, and the main town was fortified in the mid-eighteenth century. In the wars of independence, Chiloe was royalist, the last Spanish stronghold in Chile until the Spanish governor Antonio de Quintanilla was defeated in 1826.
2. *Don Pedro de Valdivia*: (*c.* 1500–53) the Spanish conquistador, and the first governor of Chile; he died in the wars against the Aurancians.
3. *Salinas, chap. xvii. p. 485*: Pedro González de Agüeros, *Descripcion historial de la provincia y archipielago de Chilóe* was published in Madrid in 1791; an English translation was published as an appendix to Molina, *The Geographical, Natural and Civil History of*

Chile in 1809. Buenaventura Salinas y Cordova (1592?–1653), author of *Memorial de las historias del Nuevo Mundo Pirú* (Lima, 1630) was a Franciscan who lived in Lima.

4. *upon his resignation:* Chile was politically unstable for several years after the final defeat of Spanish; Aldunate left Chiloe in 1828 and fought several campaigns in the civil wars of the period.
5. *the present time:* i.e. about eighty years ago. Byron spent several weeks on Chiloe, under Spanish guard as Spain and England were at war, after the shipwreck in the Strait in 1741 and vividly described the Spanish and indigenous communities of Chiloe. (*Narrative of the Loss of the Wager*, pp. 145–76).
6. *an official report*: this report – the source of much of the statistical information in the following pages – has not been identified.
7. *Mercurio Chileno, No. 14 p. 160*: see below, for the botanist Bertero. *Mercurio Chileno* was a botanical work, published in sixteen numbers in Santiago in 1829.
8. *Clouds, and storms:* the same phrase from James Thomson's poem that was quoted earlier in Stokes's account. See p. 151.
9. *coast of California*: Russian traders established fur-trading settlements in California in the early nineteenth century, initially to hunt otter and later to supply Alaskan colonies with food; the chief one was Fort Ross, north of San Francisco, founded in 1812. The Russians sold the land and left in 1841.
10. *Byron's 'Painted Duck'*: Byron described not a duck but a goose, which he shot in the islands north of the western opening of the Strait of Magellan (called Guaianeco Islands by King and Fitzroy, now called Byron and Wager Islands): 'the painted goose, whose plumage is variegated with the most lively colours', p. 43.
11. *termed* Hylactes: King's nomenclature survives for two species, *Hylactes castaneus* and *Hylactes megapodius*. His lists of new bird specimens were presented to the Zoological Society 14 December 1830 and 25 January 1831 (*Proceedings of the Zoological Society*, 1 (1830–2), pp. 14–16, 29–30).
12. *Febres's Dictionary:* Andres Febres, *Arte de la lengua del Reyno de Chile* (Lima, 1765).
13. *no se come*: i.e., something that is not eaten.
14. *tertian fevers*: a severe malarial fever.
15. *Spotted fever (tabardillo)*: a term for typhus, which had symptoms of high fever followed by a rash.

Chapter XVII

1. *Miller's Memoirs:* William Miller (1795–1861) was a former British artillery officer who joined the Chilean independence movement and became a close associate of Simon Bolivar. He returned to England in 1826. King uses his account of the battle against the Spanish in Chiloe in 1826, where Miller was wounded. Others mentioned in this description are Col. (later General) Ramon Freire or Freyre (1787–1851), who served as Chilean president on several occasions in this turbulent period and led a failed attack on Chiloe in 1824; General Antonio Quintanilla (1789–1863) governor of Chiloe from 1818 to 1826, a soldier sent from Spain in 1813 to resist the Chilean revolution; Col. (later General) José Santiago Aldunate, who fought against the Spanish and replaced Quintanilla as governor of Chiloé; Admiral Manuel Jose Blanco (1790–1876), a leader of the Chilean navy who had served with Lord Cochrane and who became president shortly after the Chiloe battle. Captain Bell sailed in *Lautaro* as part of the Chilean fleet

in the battle for Chiloé. Tupper, Maruri and Asagra are mentioned by Miller, p. 316, but have not been further identified.

2. *Appendix*: see Volume 2, p. 184.
3. *chronometer-makers at Liverpool*: the firm of Robert and John Roskell, later Roskell & Son, made chronometers, wall-clocks and pocket-watches. This reference implies that family members carried on the business abroad.
4. *to Rio de Janeiro:* King does not mention here that he took on board some passengers at Valparaiso, M. Bertero, the botanist mentioned later in the chapter (below, note 9), Alexander Caldcleugh and William Bollaert, a chemist and mining consultant, and a former laboratory assistant at the Royal Institution in London, who had come out from England in 1825 to survey mines in Peru. Caldcleugh and Bertero went as far as Juan Fernandez, but Bollaert returned round the Straits with King, and describes the journey in his *Antiquarian, Ethnological and Other Researches in New Granada, Ecuador, Peru and Chile* (London: Trubner & Co, 1860) pp. 189–92.
5. *Anson's voyage*: Anson landed on the island in May 1741 in an effort to treat his fleet's outbreak of scurvy and used it as a base for several months to attack Spanish ships.
6. *Don Joachim Larrain*: the Larraín family was a prominent Chilean family established in Chile since the seventeenth century. Jose Joaquin Larraín leased the island from the government in 1829 in return for maintaining a certain number of convicts there. The island had a colourful history: It was a notorious refuge for pirates in the sixteenth and seventeenth centuries. The Scottish seaman Alexander Selkirk was a castaway there for over four years until rescued in 1709; he reputedly served Daniel Defoe as the model for Robinson Crusoe (the novel published in 1719). After 1750, the Chilean government settled the island as a prison, although the nascent colony was devastated by the earthquake of 1751. After 1834, it was briefly governed by the Englishman and soldier, Thomas Sutcliffe (1790?–1849) before he was dismissed in 1838.
7. *Woodes Rogers*: Woodes Rogers (*c.* 1679–1732) was the merchant captain who rescued Selkirk in 1709 and publicized that tale and an account of the island of Juan Fernandez in his *Cruising Voyage Round the World* (London, 1712). From 1718 to 1721, he served as the first governor of the Bahamas and returned for a second appointment, from 1728 to his death.
8. *Noticias Secretas*: i.e., Jorges Juan y Santacilia and Antonio Ulloa, *Noticias Secretas de America* (1748); it was republished in London by Taylor in 1826, translated with notes by David Barry. *A Voyage to South America*, also by Juan and Ulloa, was also highly regarded as a source of information about the west coast of South America.
9. *Captain Hall's interesting journal*: Basil Hall (1788–1844) was a naval officer with many scientific interests; as captain of the *Conway*, he served in South America, mostly on the west coast, from 1820 to 1823, and published an influential account of his experiences after leaving the navy in 1823.
10. *of the island:* Carlo Luigi Guiseppe Bertero (1789–1831) was an Italian physician and botanist who began his collecting career as a ship's surgeon in the Caribbean in 1816–21; he travelled through Chile collecting from 1827 to 1830 after this excursion botanizing with Alexander Caldcleugh and Captain King in Juan Fernandez in 1830, he died in shipwreck on the return leg of an expedition to Tahiti in 1831.
11. *exceedingly correct:* ie. Anson's *Voyage Round the World*. The *Centurion* was the single ship of the fleet that completed the circumnavigation.
12. *before her eyes:* Hall spells the name of the chief Peneléo. He had learned of the savage murder of prisoners and of this widow-hostage while trying to learn the fate of a captured

American captain called Moison; he abandoned her cause when he found the young widow was (in his view) already reconciled to her new lot with Peneléo (Hall, vol. 1, pp. 314–23).

13. *Rouse, our consul:* Henry William Rouse was appointed vice-consul in Chile in 1823; he became consul at Concepcion in 1827; moved to Valparaiso in 1837 and then became Acting Consul-General in 1847–50; see his journey with Fitzroy, Volume 3, p. 367.
14. *fasten their mantles:* Andres Febres, *Arte de la lengua del Reyno de Chile* (Lima, 1765), p. 632.
15. *a patent log:* an instrument dragged from the ship; its revolutions could be registered and used to estimate speed and distance.
16. *schooner Industry, New Bedford*: this may be a whaler by this name that was in the area, but which listed Westport, Connecticut as its home port. It visited the Atlantic and Cape de Verde whaling grounds repeatedly through the 1820s and 30s (Starbuck, *History of the American Whale Fishery*).
17. *to be discerned:* this comet was observed in March and April 1830 by several observers at sea and on land; King wrote to the Royal Astronomical Society to attribute its sighting by Wickham on 18 March 1830 (*Monthly Notices of the Royal Astronomical Society*, 1 (December 1830) pp. 195–6); the astronomer Thomas Henderson, at the observatory at the Cape of Good Hope (1830–4) summarized the findings in 1834. *Monthly Notices*, 3 (June 1834), p. 59.
18. *Dom Pernetty's history*: Antoine-Joseph Pernety (1716–1801) was with Bougainville in 1763–64 expedition to the Falklands. He wrote *Histoire d'un voyage aux isles Malouines, fait en 1763 & 1764; avec des observations sur le detroit de Magellan, et sur les Patagons*, 2 vols (Paris, 1770).
19. *called the Dryad:* Hudson Bay Company records show this ship was on the Pacific northwest from 1831 to 1836 engaged in the fur trade; its masters are listed as Alexander Duncan and Charles Kipling.

Chapter XVIII

1. *of the Wager*: Byron (*Narrative*, pp. 76–7) recorded the discovery by the assistant surgeon of the *Wager*, Mr Robert Elliot, of a chamber underground where bodies of several men were laid out on biers; he speculated that it was the sacred burial place of their caciques, Cf. p.158.
2. *left behind:* the *Wager* shipwreck ended in mutiny, with most of the men following two officers back through the Strait of Magellan in small boats in an effort to reach Rio de Janeiro. When one of the smaller boats was capsized and lost, four of the weakest marines were left behind. Byron described them as indifferent to their fate, 'so exceedingly disheartened and worn out were they (*Narrative*, p. 73). Eventually, three groups of survivors from the Wager, including Byron and captain Cheap, made it back to England, with dramatic tales of hardship from each party. A court-martial was held in 1746, but the mutineers escaped hanging.
3. *coast of Chiloe*: Captain Cheap and a small group loyal to him made their way to Chiloe and then Santiago and Valpariso. There were only four survivors, including John Byron, then a midshipman.
4. *the gallant admiral*: James Saumarez (1757–1836) was a hero of the Napoleonic wars and admiral at Plymouth Command, one of the top offices in the Admiralty in 1824 to 1827, just before King departed on the *Adventure*.

5. *Sir Edward Owen*: Owen (1771–1849) was commander-in-chief of the West Indies squadron of the Royal Navy from 1822 to 1826. Mitchell had served under him during this period as a midshipman on the *Gloucester*, and was promoted from the *Gloucester* after acts of bravery in rescuing men gone overboard.